Cellulose

Cellulose
Structure, Modification and Hydrolysis

Edited by
RAYMOND A. YOUNG
and
ROGER M. ROWELL

University of Wisconsin
Madison, Wisconsin
and
USDA Forest Products Laboratory
Madison, Wisconsin

A WILEY-INTERSCIENCE PUBLICATION
JOHN WILEY & SONS
New York • Chichester • Brisbane • Toronto • Singapore

Library of Congress Cataloging in Publication Data:

Main Entry under title:
Cellulose: structure, modification, and hydrolysis.

"A Wiley-Interscience publication."
Includes bibliographies and index.
1. Cellulose.   I. Young, Raymond Allen, 1945-
II. Rowell, Roger M.
QD321.C395 1986   547.7'82   85-26578
ISBN 0-471-82761-4
Printed in the United States of America

10 9 8 7 6 5 4 3 2 1

Contributors

JOHN BLACKWELL Department of Macromolecular Science, Case Western Reserve University, Cleveland, Ohio, 44106

H. CHANZY Centre de Recherches sur les Macromolécules Végétales, CNRS Affiliated with the Scientific and Medical University of Grenoble, Saint-Martin-d'Heres, France

ANTHONY H. CONNER USDA, Forest Service, Forest Products Laboratory, Madison, Wisconsin, 53705

KEVIN W. DOWNEY Department of Chemical Engineering, Michigan State University, East Lansing, Michigan, 48824

VIDAR EKLUND Technology Research Centre, Neste Oy, Kulloo, Finland

KURT EKMAN Technology Research Centre, Neste Oy, Kulloo, Finland

JAN FORS Technology Research Centre, Neste Oy, Kulloo, Finland

J. L. GARNETT School of Chemistry, University of New South Wales, Kensington, New South Wales, Australia

JOHN F. HARRIS USDA, Forest Service, Forest Products Laboratory, Madison, Wisconsin, 53705

J. M. HAUDIN Centre de Mise en Forme des Matériaux, Ecole Nationale Supérieure des Mines de Paris, Sophia Antipolis, Valbonne, France

MARTIN C. HAWLEY Department of Chemical Engineering, Michigan State University, East Lansing, Michigan, 48824

TAKAHISA HAYASHI Ajinoinoto Central Research Laboratories, Kawasaki, Japan

B. HENRISSAT Centre de Recherches sur les Macromolécules Végétales, CNRS, Affiliated with the Scientific and Medical University of Grenoble Saint-Martin-d'Heres, France

CHARLES G. HILL, Jr. Department of Chemical Engineering, University of Wisconsin, Madison, Wisconsin, 53706

JOUKO I. HUTTUNEN Technology Research Centre, Neste Oy, Kulloo, Finland

ERDOĞAN KIRAN Department of Chemical Engineering, University of Maine, Orono, Maine, 04469

W. H. KLAUSMEIER Sylvatex Corporation, Burke, Virginia, 22015

DAVID KURZ Department of Macromolecular Science, Case Western Reserve University, Cleveland, Ohio, 44106

DEREK T. A. LAMPORT MSU-DOE Plant Research Laboratory, Michigan State University, East Lansing, Michigan, 48824

DAVID M. LEE Department of Macromolecular Science, Case Western Reserve University, Cleveland, Ohio, 44106

GORDON MACLACHLAN Biology Department, McGill University, 1205 Avenue Docteur Penfield, Montreal, Quebec, Canada, H3A 1B1

S. MORIYAMA Research Center, Toyo Engineering Corporation, Mobara, Chiba, Japan

RAMANI NARAYAN Laboratory of Renewable Resources Engineering, Purdue University, West Lafayette, Indiana, 47907

P. NAVARD Centre de Mise en Forme des Matériaux, Ecole Nationale Supérieure des Mines de Paris, Sophia Antipolis, Valbonne, France

R. D. PRESTON Astbury Department of Biophysics, University of Leeds, Leeds, England

JACQUES REUBEN Research Center, Hercules Incorporated, Wilmington, Delaware, 19899

T. SAIDA Research Center, Toyo Engineering Corporation, Mobara, Chiba, Japan

ANATOLE SARKO Department of Chemistry and the Cellulose Research Institute, State University of New York, College of Environmental Science and Forestry, Syracuse, New York, 13210

JOHAN-FREDRIK SELIN Technology Research Centre, Neste Oy, Kulloo, Finland

SUSAN M. SELKE Department of Chemical Engineering, Michigan State University, East Lansing, Michigan, 48824

P. SIXOU Laboratoire de Physique de la Matière Condensée, CNRS, Université de Nice, Nice, France

MAO-YAO SU Department of Macromolecular Science, Case Western Reserve University, Cleveland, Ohio, 44106

A. TEN BOSCH Laboratoire de Physique de la Matière Condensée, CNRS, Université de Nice, Nice, France

George T. TSAO Laboratory of Renewable Resources Engineering, Purdue University, West Lafayette, Indiana, 47907

OLLI T. TURUNEN Technology Research Centre, Neste Oy, Kulloo, Finland

MORRIS WAYMAN Department of Chemical Engineering and Applied Chemistry, University of Toronto, Toronto, Ontario, Canada, M5S 1A4

BARRY F. WOOD Department of Chemical Engineering, University of Wisconsin, Madison, Wisconsin, 53706

RAYMOND A. YOUNG Department of Forestry, School of Natural Resources and School of Family Resources and Consumer Sciences, University of Wisconsin, Madison, Wisconsin, 53706

P. ZUGENMAIER Institute of Physical Chemistry, Technical University of Clausthal, Clausthal-Zellerfeld, Federal Republic of Germany

This volume is dedicated to the 1983 Anselme Payen Award Winner, Reginald D. Preston. Dr. Preston has received international recognition for his contributions to the physics and chemistry of cellulose and particularly for his investigations on the structure of cellulose and the plant cell wall. Our knowledge of the plant cell wall emanates directly from his pioneering research. Dr. Preston is currently Professor Emeritus and former Head of the Astbury Department of Biophysics, University of Leeds, England. He kindly agreed to prepare the first chapter, "Natural Celluloses," for this book.

The Anselme Payen Award is named for the distinguished 19th-century French scientist and industrialist who discovered and characterized cellulose as the main fibrous component found universally in plant cell walls. The award is presented annually by the Cellulose, Paper, and Textile Division of the American Chemical Society to honor and encourage outstanding professional contributions to the science and technology of cellulose and its allied products. The award is recognized internationally as the most significant honor in the field.

Preface

Polymer chemistry had its beginnings with the characterization of cellulose, but interest in cellulose waned with the advent of cheap synthetic polymers based on petrochemicals. However, several recent developments have indicated that there is currently strong worldwide interest in cellulose chemistry and technology. The excellent attendance at the Cellulose Research Conference in Syracuse in 1982, the Symposium on Cellulose and Cellulose Derivatives which we organized for the American Chemical Society (ACS) Meeting in Washington, D.C., in 1983, and the Cellucon '84 Conference in Wales in 1984 attests to the importance of cellulose science in many laboratories.

There have been significant developments in cellulose chemistry and technology in the last few years and a number of the advances are presented in this book. Many of the contributors also participated in the Symposium on Cellulose and Cellulose Derivatives and the Anselme Payen Award Symposium honoring Professor R. D. Preston at the ACS Meeting in Washington, D.C., in 1982. The contributors are world leaders in their respective fields and were requested to emphasize work from their own laboratories in their respective chapters. Appropriate references to related work are also documented in each chapter. We decided to emphasize specific important developments in cellulose science, since it would be impossible to cover adequately in one volume all aspects of cellulose chemistry and technology. Many industrialized countries are represented in the book including Australia, Canada, England, Finland, France, Germany, Japan, and the United States.

The book is divided into four parts. Part 1, Cellulose Structure and Biosynthesis, covers patterns of cellulose biosynthesis in the cell and structural characterization of crystalline cellulose. In Chapter 1, Dr. Preston describes the fine structure of plant cell walls. During his scientific lifetime the cell wall has passed from a relatively insignificant, inert outer shell of plant cells, of interest to just a few enthusiasts, to the basis of themes central to plant science and demanding investigation by many both pure and applied scientists. Modern methods of fiber diffraction analysis and model selection are described in Chapter 2, and the methods utilized for the recent discovery of the parallel arrangement of cellulose molecules in native cellulose are documented in Chapters 1–3. Further developments on x-ray analysis of cellulose–solvent complexes with ethylene-diamine and hydrazine are covered in detail in Chapter 3. Some recent advances in cellulose biosynthesis, such as hormonal control of metabolism and the role of extensin in the cell wall, are described in Chapters 4 and 5, respectively. In Chapter 6, the structure, swelling, and bonding of cellulose fibers are tied together in a unique fashion through a review of the current literature. Wettability, hornification, and dry forming of cellulose fibers are discussed in this treatment.

In Part 2, Cellulose Modification, reactions of specialized interest are described in five chapters. The formation of cellulose carbamate is outlined in Chapter 7. The degree of substitution required for a soluble product is low, and applications of this derivative to formation of regenerated cellulose fibers, films, and sponges are described in detail. A mathematical description of the reactions of two cellulose ethers, carboxymethylcellulose and hydroxyethylcellulose, is given in Chapter 8. The reactivity of the three hydroxyl groups at positions 2, 3, and 6 on the glucosyl unit of cellulose offers a variety of possibilities for making useful derivatives. Chapters 9 and 10 discuss recent developments in grafting of cellulose by ultraviolet and anionic methods, respectively. The effects of sensitizers, solvents, acids, and salt additives on photografting of cellulose with vinyl monomers and the mechanism of photografting are discussed in Chapter 9. New approaches to anionic grafting of cellulose based on "living synthetic polymers" are described in Chapter 10. With this method, effective molecular-weight control and a narrow molecular-weight distribution of the graft can be achieved, and homopolymer-ization can be eliminated. The last chapter in this section describes a new method, based on organosolv pulping, for production of cellulose acetate. The Eastman process for production of acetic anhydride and wood gasification are discussed as components of the process.

Part 3 is devoted to Cellulose Liquid Crystals. Stiff-chain synthetic macro-molecules in concentrated solution yield mesophases, and it has been found that anisotropic solutions can also be obtained with semirigid polymers such as cellulose derivatives. Considerable interest in cellulose liquid crystals has de-veloped because fibers of improved strength and modulus can be obtained by the spinning of these mesophases. The ultimate goal would be to produce a regenerated fiber with the strength and crystallinity of cotton but without the many disadvantages of the viscose process. The structure and characteristics of liquid crystals from cellulose derivatives are described in Chapters 13 and 14. The theoretical predictions of phase diagrams are reviewed, and the experimental behavior of optical and rheological properties are described in these chapters. In Chapter 15, the rheology of cellulose–N-methylmorpholine–N-oxide mono-hydrate in isotropic phases and the spinning of this hydrate are described. Good fibers are also produced by this method without special care.

In Part 4, Cellulose Hydrolysis and Degradation, aspects of acid, enzyme, and thermal degradation of cellulose are given in-depth treatment. A comparison of various methods of acid degradation of cellulose is given in Chapter 15. Details of SO_2 hydrolysis of cellulosics and a description of associated equipment are given in this chapter. A two-stage dilute sulfuric acid cellulose hydrolysis process is described in Chapter 16 and a hydrogen fluoride process in Chapter 17. A complete listing of the BASIC computer program used to simulate hydrolysis of cellulose with two-stage, dilute sulfuric acid (Chapter 16) is given in Appendix I. In Chapter 18, the various methods for pretreatment of lignocellulosics are reviewed, and the operation of a continuous process for pretreatment and enzymatic saccharification of cellulosics is described. Patterns of enzymatic breakdown of cellulose crystals are given in Chapter 19, and thermal degradation of cellulosics is treated in Chapter 20.

The purpose of the book is to cover recent developments in cellulose chemistry and technology through experts in the field. We feel this goal has been duly met.

Raymond A. Young
Roger M. Rowell

Madison, Wisconsin
January 1986

Acknowledgments

The contributors and editors have received many constructive comments and assistance throughout the preparation of this book. We especially thank Professor Ronald L. Giese, Chairman of the Department of Forestry, University of Wisconsin–Madison, for his continued support of our program. Appreciation is expressed to Pauline Miller for retyping many parts of the manuscript.

Acknowledgments for assistance and support for specific chapters are as follows: Chapter 1, Mr. W. D. Brain and Mr. L. Childs, Astbury Department of Biophysics, University of Leeds, for reproduction of illustrations; Chapter 2, National Science Foundation grants GP27997, CHE7501560, CHE7727749, CHE8107534, and PCM8320548; Chapter 3, National Science Foundation grant DMR81-07130 (Polymer Program); Chapter 4, Dr. E. T. Reese (United Army Laboratories) for the gift of cellulase preparation from *Streptomyces griseus* and Dr. K. Matsuda (Tohoku University, Japan) for the gift of *Aspergillus oryzae* enzyme preparation; financial support by grants from the Natural Sciences and Engineering Research Council of Canada and from the Programme des Formation de Chercheurs et d'Action Concertée du Quebec; Chapter 5, Extensin Club members Jim Smith, E. Patrick Muldoon, Jim Willard, Marcia Kieliszewski, and Jerome Quets; support by DOE contract DE-ACO2-76ERO-1338, NSF grant PCM83-15901, and USDA grant CRCR-1-1424; Chapter 6, John Klungness, Dan Caulfield, and Ralph Scott of the USDA Forest Products Laboratory, and my graduate students David Barkalow and Richard Helm for proofreading the manuscript, Professor John Berg, Von Byrd, Alfred French, S. Haig Zeronian, Noelie Bertoniere, A. M. Scallan, Tom Lindström, and Ludwig Rebenfeld for supplying current literature, and Norah Cashin for assistance throughout the process; Chapter 7, the management of Neste Oy for providing excellent working conditions and for permission to publish this work, Mr. Leo Mandell, who was one of the first to work on this project, and the skillful and dedicated assistance of the laboratory technical staff; Chapter 9, the Australian Institute of Nuclear Science and Engineering, the Australian Research Grants Committee, and the

Australian Atomic Energy Commission for financial assistance; Chapter 12, J. M. Gilli, F. Fried, P. Maissa, J. F. Pinton, and M. J. Seurin, collaborators; Chapter 13, support from Deutsche Forschungsgemeinschaft, and Dipl.-Chem. U. Vogt for the use of some of the results of her dissertation prior to publication; Chapter 15, a grant from the Natural Sciences and Engineering Research Council of Canada; Chapter 16, Virgil Schwandt and Marilyn Effland for sugar analyses, and Linda Lorenz for technical assistance; Chapter 18, the Research Association for Petroleum Alternatives Development of the Japanese Government; Chapter 19, Dr. M. Schulein (Novo Company, Denmark) for his generous gifts of enzymes, Mr. D. Grison for his excellent participation in this work, and Dr. W. Winter for his help in the preparation of this manuscript.

R.A.Y.
R.M.R.

Contents

PART 1

Cellulose Structure and Biosynthesis

1

Natural Celluloses

R. D. PRESTON
Astbury Department of Biophysics, University of Leeds, Leeds, England

It is now almost 150 years since Anselme Payen discovered and isolated from green plants the substance subsequently given the name cellulose and this material has been under continuous investigation ever since. Cellulose differs from most other polysaccharides found in plants (a) in consisting of molecular chains that are very long, (b) in containing only one repeating hexose residue, and (c), except for a (1–3)-linked xylan and a (1–4)-linked mannan which replace it in some algae (15), in occurring naturally in a crystalline state. Of these differences, crystallinity was the first to be established, by Carl von Nägeli in 1858 (1). This constituted the first serious use of the polarizing microscope, and the relevant interpretations were contested at the time, particulary by Hofmeister, and final proof had to wait some 80 years until the discovery of the diffraction of x-rays by crystals.

The molecular chain concept of cellulose structure took somewhat longer to develop (2) largely because the inability of chemists to detect reducing end groups led them to think in terms of cyclic dimers, considered to be associated into colloidal particles in order to account for the high molecular weight. The final turning point came when Sponsler and Dore (3), working from first principles since neither of them was a crystallographer, showed that the x-ray diagram was consistent with the existence of long parallel chains spaced regularly the same distance apart in the crystal. They deduced a single-chain unit cell with dimensions $a = 6.1$ Å, b (fiber axis) $= 10.3$ Å, $c = 5.4$ Å, $\beta = 88°$. That they assumed an alternation of (1–1) and (4–4) links along the chain does not detract from the importance of their findings. A crystallographically more acceptable unit cell was deduced 2 years later by Meyer and Mark (4), a two-chain cell with $a = 8.35$ Å, b(fiber axis) $= 10.3$ Å, $c = 7.9$ Å, $\beta = 84°$. In this, as in the later improved model of Meyer and Misch (5) (Fig. 1.1), the linkage between successive residues in the chain is uniformly (1–4) in harmony

3

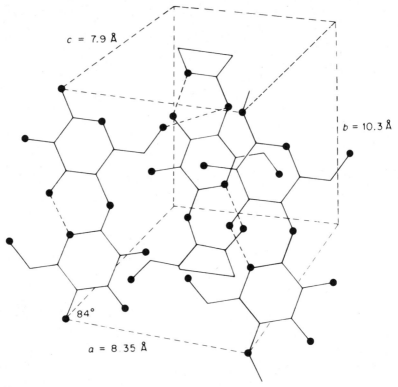

c = 7.9 Å

b = 10.3 Å

84°

a = 8.35 Å

Figure 1.1 The unit cell of cellulose according to Meyer and Misch; solid circles represent oxygen, hydrogen omitted; dashed lines represent hydrogen bonds.

with the chemical evidence published by Haworth at about the same time. A difficulty with the two earlier models—that the fiber axis parameter of 10.3 Å is shorter than the fully extended length of a cellobiose residue—is removed in the Meyer and Misch cell by kinking the residue through the formation of a hydrogen bond between C3 of one glucose residue and the ring oxygen of the next (incorporated in Fig. 1.1). Evidence that this is correct has accumulated since that time. In both the Meyer and Mark and the Meyer and Misch models the central chain lies upside down with reference to the corner chains, which means that one-half of the chains point one way and one-half the other. There was at that time no solid evidence for this, and this so-called antiparallel arrangement has remained a matter of controversy.

THE UNIT CELL

To follow first the question of the crystalline structure of cellulose, the next development was the claim by Honjo and Watanabe (6) that the indexing of reflections found in the electron diffraction diagram of *Valonia*, which, following

Sponsler, I had introduced into wall studies, demanded a unit cell four times larger than the Meyer and Misch cell, with $a = 16.7$ Å, $b = 10.3$ Å, $c = 1.58$ Å, $\beta = 84°$. This eight-chain unit cell soon received support both for *Valonia* cellulose (7) and *Chaetomorpha* cellulose (8). Now since eight chains rather than two must in some way differ from each other, this introduced a much higher order of complexity. More recently still, two groups of workers have re-examined the problem using both x-ray diffraction techniques and the more modern methods of conformational analysis. Nieduszynski and Atkins (9) first demonstrated that eight different unit cells are possible (four of which are illustrated in Figure 1.2), all with antiparallel chains lying in sheets. Although none of these unit cells fully explained all the reflections in the x-ray diagram, it seemed for a time that the problem of the crystalline structure of cellulose was nearing a solution. However, 14 years later Gardner and Blackwell (10) and Sarko and Muggli (11) reopened the question. Assuming a two-chain cell (which must be regarded as at least a slight weakness) and working a least-squares refinement of structure they have shown that, to a high degree of probability, the chains lie parallel, not antiparallel. The same conclusion has since also been reached for cellulose triacetate (11a). This produces something of a paradox, since regenerated cellulose (cellulose II), which can be produced from natural cellulose (cellulose I) without dissolution, is demonstrably antiparallel. There are, however, possible ways out, one of which will be mentioned below.

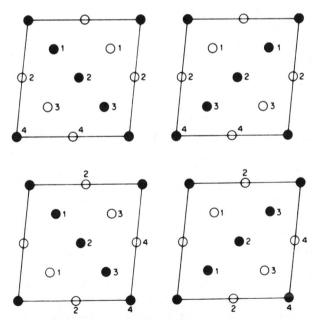

Figure 1.2 The basal planes of four possible eight-chain unit cells of cellulose. Solid and open circles represent chains in antiparallel arrangement. The numbers represent the multiples of $b/4$ by which the chain is translated upward with respect to the corner chains. There are four other such cells with negative displacements. (After Nieduszynski and Atkins (9).)

For some purposes in the study of cellulose it is irrelevant whether the chains are parallel or antiparallel; for others it is of the essence. In considering the biosynthesis of cellulose, for instance, or its degradation by destructive organisms, it is essential to know in this amount of detail the nature of the material to be synthesized or destroyed.

THE MICROFIBRIL

Whether the chains within the crystallite are parallel or antiparallel, in the whole wall they are antiparallel. To see this, it is as well to look first at the supramolecular structure of cellulose. Ever since the time of Nehemia Grew (1683), account has been taken of the fibrous nature of plant cell walls, and the earlier literature pays considerable attention to the nature of the fibrils seen under the polarizing, or the ordinary, light microscope when the wall has been mechanically disrupted, or implied in the untreated wall by the frequent appearance of striations. Almost immediately after the first appearance of the x-ray diagram, this fibrillar concept was applied at the molecular level, since the diagram showed that the crystallites are long and thin (12).

With the advent of electron microscopy it became evident that, at the intermediate range of size then revealed, the wall contains long, thin threads, named microfibrils (13, 14). It later transpired that these occur broadly in two sizes, each of them very long. In higher plants these bodies are normally 10 nm wide by 5 nm thick, whereas in some algae, notably *Valonia* (8) and the Cladophorales (8, 15), they are about twice this size in each dimension. The electron diffraction diagram soon confirmed that these are cellulosic (16). They clearly consist in the main of cellulose chains lying parallel to microfibril length in a bundle some 5 nm wide in higher plants (12, 17) and 17 by 11.4 nm in the algae mentioned above (17–19). The concept that the broader microfibrils are constructed by the aggregation of narrower microfibrils does not therefore carry much weight (20). The very recent demonstration (45) of lattice lines spaced 5.4 Å apart distributed uninterruptedly over the broader face of a *Valonia* microfibril appears to confirm the concept.

When microfibrils are isolated from cell walls by chemical extraction to remove the polysaccharides in which they are imbedded, hydrolysis of the undisrupted microfibrils mostly yields sugars other than glucose. They are not negligible in amount except in the group of algae mentioned above, in which no sugars other than glucose have been detected. In all other plants the yield of nonglucose sugars (normally xylose or mannose) ranges from 15% in higher plants to 50% in some seaweeds (21). There must therefore lie on the surface of the microfibril strongly adsorbed chains of sugars other than glucose, and these may be xylan, mannan (25), xyloglucan (22), or glucomannan. These, presumably with some intermixed glucan chains (since they are often not present in quantity sufficient to cover the whole surface of the microfibrils), must make up the width of the microfibril to 10 nm in a region that may be called the cortex. Further, when cellulose is treated with strong sulfuric acid and titrated back to pH 3, the microfibrils fall into short rods which still yield the x-ray diagram of cellulose (23, 24) and on hydrolysis give no sugar other than glucose (25). This observation carries with it the implication that the central crystalline core is interrupted at points along the microfibril, possibly by an intermixture of glucan and nonglucan chains. Support for this concept is found in the staining of microfibrils with silver ions, which stain hemicelluloses but not cellulose. A possible structure for the microfibril deduced in this way is presented in Figure 1.3.

As found in nature, these microfibrils are always imbedded in a matrix of other polysaccharides and, in some tissues, of nonsugar compounds such as lignin in wood and in phloem sclerenchyma, and cutin and suberin in the outer wall of the epidermis. The chemical nature of the polysaccharides varies from tissue to tissue and from species to species, but physically they fall into four classes: (1) unbranched straight chains such as (1–4)-linked xylan or mannan; (2) branched chains such as xyloglucan or galactoglucomannan also (1–4)-linked; (3) chains of sugar residues that are (1–3)-linked and are therefore helical; and (4) chains of pectin, the methyl ester of pectic acid containing galacturonic acid residues interrupted by occasional rhamnose residues and therefore nowadays referred to as polyrhamnogalacturonic acid. All except the branched chains can be extracted from the walls and crystallized so that their structure is known (15, 26, 27). They all consist of chains that are rod-shaped. Branched chains have been extracted from some bacteria, and it turns out that the side chains, instead of standing out from the main chain, are folded back to hydrogen bond to the main chain (28, 29) so that these too are rodlike. Conceptually, therefore, the rodlike cellulose microfibrils are imbedded in a mixture of chains that are also rod-shaped. By using polarized infrared spectroscopy (30) it has been shown that the chains of xylan lie parallel to the microfibrils, and using x-ray diffraction analysis it has been shown that the chains of some other matrix polysaccharides

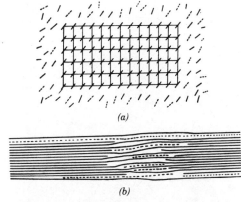

(a)

(b)

Figure 1.3 Diagrammatic representation of the structure of a cellulose microfibril. (a) transverse view; (b) longitudinal view. Solid lines, glucan chains; broken lines, chains of other sugars. The rectangles in the crystalline center of (a) are the basal planes of the Sponsler and Dore unit cells.

lie similarly parallel (31, 32). The mutual distribution of microfibrils and matrix polysaccharides (and lignin where it occurs) and the nature of the bonding between them are, however, still unknown. The general consensus seems to be that extensive hydrogen bonding is involved, and wall models have been proposed carrying hydrogen bonds and primary valence bonds in various degrees, though none are wholly satisfactory. It has been suggested that nuclear magnetic resonance spectroscopy might be helpful.

WALL ORGANIZATION

While a plant cell is growing it is surrounded by a thin wall which accommodates its dimensions to the increasing size of the cell. This is the primary wall. As the cell ceases to grow, the wall becomes thickened by apposition upon the primary wall of a sequence of lamellae that are collectively called the secondary wall. The thin primary wall and the thick secondary wall must be taken separately, and we will begin with the secondary wall. In at least some of the algae this separation into two distinct wall types is not found, because the wall becomes thick as the cell grows. These walls will be included with the secondary walls.

Within each lamella of the wall the microfibrils usually lie more or less parallel to each other, though here and there twisted around each other. The common direction is such that the microfibrils form a helix round the cell, the axis of the helix in elongated cells coinciding with the length of the cell. The pitch, and sometimes the sign, of the helix varies between lamellae, giving the well-known crossed microfibrillar structure (Fig. 1.4). The epitome of this, and the organism

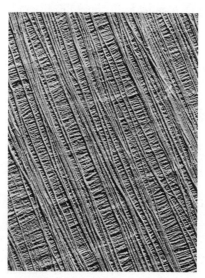

Figure 1.4 Electron micrograph of lamellae from the wall of *Chaetomorpha melagonium*; magnification 24,000×, shadowed with Pd/Au.

in which it was first conclusively proved, is *Valonia*, which takes the form of vesicles that can be as large as a hen's egg. Crossed fibrillar structure was first demonstrated by x-ray diffraction (33) and has since been confirmed by electron microscopy (13) (Fig. 1.4). In *Valonia*, as more recently also shown for *Cladophora* and *Chaetomorpha* (8) and for the unicells *Glaucocystis* and *Oocystis* (34), the pitch changes repeatedly through the wall from slow to fast to slow and so forth, with in some species a third orientation which roughly bisects the angle between the other two (Fig. 1.5). In any area of wall the two major directions lie roughly at right angles, though there is some variation through the wall thickness since the cell is growing. Though the signs and pitches vary from species to species, they are constant within species so that both must be under genetic control. The run of the microfibrils in *Valonia* (35) is presented in Figure 1.6. In the cylindrical cells of *Cladophora* (8, 36) and *Chaetomorpha* (8), the microfibrils run similarly in two or three helices, but the poles are obscured by the flat end walls. Similar overall structures are found in *Glaucocystis* and *Oocystis* (34) which are microscopic in size. The presence of two or three coaxial helices in cells ranging in size from a hen's egg down to a few microns must be expected to foreshadow similar arrangements in the secondary walls of those cells of higher plants that present them.

This expectation is fulfilled for tracheids and fibers in both softwoods and hardwoods and for phloem fibers such as hemp, ramie, jute, sisal, and so forth (37). The prototype is the tracheid of conifer wood. This was shown long ago by Bailey to possess in general a three-layered secondary wall labeled by him (Fig. 1.7) S1 (outermost), S2, and S3. These terms have become standard. S3 is missing in some species such as spruce, and variations occur in tension and compression wood. It was early recognized (see refs. in 38) that the mean cellulose chain direction, as determined by polarization microscopy, ran in a helix around

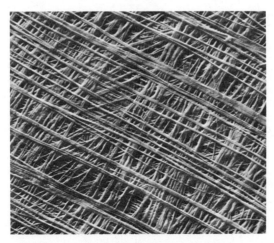

Figure 1.5   As in Figure 1.4 but using *Cladophora prolifera*, magnification ∼ 12,000×. Structures shown in Figures 1.4 and 1.5 are typical also of *Valonia*.

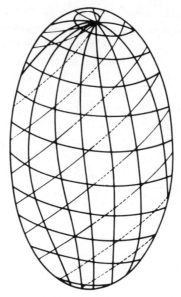

Figure 1.6 Schematic diagram of the run of the three microfibril directions in a vesicle of *Valonia*.

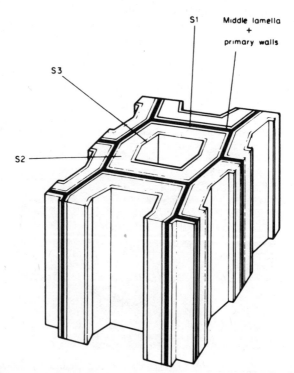

Figure 1.7 Diagrammatic representation of a segment of a conifer tracheid surrounded, as in the xylem, by six others.

the cell. The angle the helical vector makes with the cell length (now called the microfibril angle) varies with length according to the relation $L = a + b\cot\theta$ (39). This relation has been found to hold for other plant fibers—sisal (40), bamboo (41), various textile fibers (42). It therefore seems to be widespread among fibrous cells. Bailey and Vestal (43) had pointed out, however, that the wall was more complicated than this simple picture allowed, and they believed, on evidence that had other possible explanations, that three helices occurred in the wall: a flat helix in S1 and S3 with a steep helix in S2, though they referred to these as transverse and longitudinal orientations, respectively. After some controversy it became apparent on the basis of rigorous application of polarization microscopy (44) that Bailey was correct, and electron microscopy subsequently confirmed it (Fig. 1.8) (15). The central layer of the wall, S2, is usually thicker than either S1 or S3, and, accordingly, determination of chain direction by polarization microscopy gives values close to the run of microfibrils in S2. X-ray diffraction shows that the angular separation of the (weak) arcs from S1 and S3 makes them readily distinguishable from those due to S2 so that the run of the microfibrils in S2 can again be determined. Both methods are used in the investigation of structure-related properties. All three helices individually follow the L/θ relationship though with different values of the constants a and b. A schematic view of the whole structure is presented in Figure 1.9.

The question remains whether the microfibrils all run in the same direction. Since the unit cell is monoclinic, there is in the structure an up-and-down sense even if the chains lie antiparallel. When an x-ray beam is passed through a wall parallel to a microfibril direction, there are in the diagram two arcs, one on each side, corresponding to spacings of 6.1 Å and two arcs, in a line at right

Figure 1.8 Electron micrograph of a replica of a split tracheid wall of *Picea sitchensis* showing the S2 and (upper left) S1 layers. The microfibrils are about one half the size of those in Figures 1.4 and 1.5.

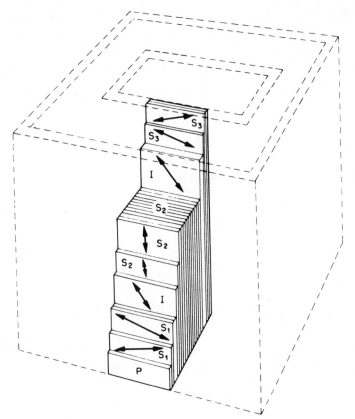

Figure 1.9 Schematic view of a sector of the wall in a segment of a conifer tracheid. Double-ended arrows show the run of the microfibrils.

angles, corresponding to 5.4 Å. There should therefore be along an oblique line two arcs representing a spacing of 3.9 Å. Instead there are four, along two oblique lines crossing each other (15). Hence there must occur, among the relevant set of microfibrils, two unit cells, one upside down relative to the other. Hence there must be two sets of microfibrils lying antiparallel. This could be of special importance if the chains in each individual microfibril point the same way. It might, for instance, lead to an explanation of the paradox mentioned above, that conversion of cellulose I (assumed parallel) leads to a cellulose II which is antiparallel (also see Chap. 2).

It should also be noted that planes of spacing 6.1 Å tend to lie parallel to the wall surface in what is called uniplanar orientation, as originally noted by Sponsler and Dore (3) and confirmed by Preston and Astbury (33). Because in surface view of the wall the microfibrils present their broader faces, these planes must lie within the microfibril parallel to the broader face. This has been elegantly confirmed in a very recent brief statement (45) recording lattice planes spaced

5.4 Å apart in a high-resolution electron micrograph of a *Valonia* microfibril. The orientation of the chains within the unit cell is therefore completely specified.

STRUCTURE AND MECHANICAL PROPERTIES

The helical nature of these cells is important in determining their mechanical properties, and the variability of the angle offers the opportunity to examine the factors involved. This was first recognized by Brown et al. (46) with reference to Young's Modulus of cotton hairs and later confirmed (47). The importance of the fibril angle for Young's Modulus was made more emphatic when it was shown (48) that in sisal fibers (which allow a much wider choice of angle than does cotton) Young's Modulus for $\theta = 10°$ is some 30 times greater than for $\theta = 50°$. The first attempt to correlate structure quantitatively with the Modulus was made, however, by Hearle (49) for a variety of vegetable fibers. His model of the cell assumed the cell to be solid (i.e., with no lumen) and that extension involved lengthening of the helical winding, extension of the helix like a spiral spring and a compression of the matrix surrounding the microfibrils (assumed to be isotropic). In spite of the limitations of the model, the fit of his relation between E and θ must be regarded as good except for values of θ less than about 10° (15).

More recently, attention has turned to conifer wood as experimental material. This offers advantages since, in passing from the pith to the bark across any radial line in a tree trunk, the length of the tracheids increases, and so does therefore the steepness of the microfibrillar helices. It is therefore possible to choose a series of specimens with different average θ but with approximately the same chemical composition. Both Cowdrey and Preston (50) and Cave (51) have made use of this by measuring E across the growth rings and have derived structural models upon which the variations found may be based; these models differ from each other, but the two forecast variation of E with θ in much the same way.

The models have it in common that the cells are considered to be endless cylinders, that the only helix to be considered is that in S2, and that the bordered pits always found on tracheid walls (see 15) are ignored. This last point is not a serious matter, since it is known that these pits, far from providing points of weakness, are stronger than the rest of the wall. Cowdrey and Preston considered two models. The first resembled that of Hearle except that the compression factor was ignored since the wall is thin and the lumen relatively wide. The relation between the compliance, J, and θ was deduced to be

$$\frac{J \ln(r_o/r_i)}{r_o^2} = \frac{v(q \tan^2\theta + 2n)}{a^2 qn} \tag{1}$$

where r_o is the outside radius and r_i the inside radius of the S2 layer, q is the microfibril Young's Modulus, n its torsional rigidity, and a the microfibril radius (considered of circular cross section). v is a space factor to take into account that the wall does not consist wholly of microfibrils.

This relation fits the experimental points (Fig. 1.10) except for values of θ above about 45°. However, reasonable values of v, a, r_o, and r_i give values of q six orders of magnitude too high. If a helical spring model is to be acceptable, the radius of the fibrils forming the helix must be about 5 μm so that many microfibrils must be considered aggregated into a single mechanical unit.

Accordingly, Cowdrey and Preston explored a different model assuming the wall to be homogeneous with anisotropic elastic properties with orthorhombic symmetry. The axes of symmetry are defined by the angle θ, and the elastic properties combine those of the microfibrils and the matrix. This led to the relation

$$J = a_o + a_1 \sin^2\theta + a_2 \sin^4\theta \qquad (2)$$

where the a's contain the elastic constants of the wall material.

A curve of this kind fitted to the experimental points is shown in Figure 1.11. This is almost linear with $\sin^2\theta$, since the term in $\sin^4\theta$ is small, and the fit must be considered good. From this fitted curve, values for Young's Modulus can be calculated both parallel and perpendicular to the microfibrils. Remembering that these refer to the wall substance, not the microfibrils alone, they are found to

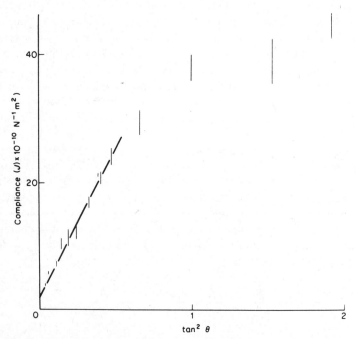

Figure 1.10 The relation between the compliance and $\tan^2\theta$ for strips of *Picea* wood. Vertical lines give the spread of the experimental results.

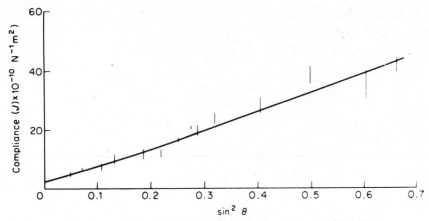

Figure 1.11 As in Figure 1.10, except that the relation is with $\sin^2\theta$.

lie well inside the acceptable range. The weakness of this approach, however, is that nothing is said either about the microfibrils or the matrix, or of the mutual displacements that must occur.

An entirely different approach was adopted by Cave (51). Though he used basically a model similar to the second model described above, he introduced a device that allowed for the inhibition in whole wood of the tendency of individual tracheids to twist on extension. Further, rather than evaluate Young's Moduli by curve fitting, he assumed values given by Mark (52) and calculated the Young's Moduli of the cells to compare with his observed values. Some complication was introduced by his choice of specimen cross section very much larger than that used by Cowdrey and Preston so that he extended some 200 times more tracheids than they did. This means that there will be within each specimen a range of θ through some tens of degrees. He therefore adopted a mathematical trick whereby all values of θ, assumed to follow a Gaussian distribution, lie in one wall. To prevent twisting, Cave further introduced into the one wall a second helix of opposite sign to the first since this resembled the bonded wall pairs of adjacent cells. The validity of this concept is clearly doubtful. Nevertheless, Cave deduced again a quartic function of $\sin\theta$. His curve relating Young's Modulus to θ is presented in Figure 1.12, and the fit is good except for values of θ below about 15°. In evaluating Cave's findings in relation to those derived from the second model of Cowdrey and Preston, it is difficult to decide whether it is preferable to ignore the potential twisting of the extending tracheids, which is very small, or to allow for it by a device that does not comply with the reality of the situation.

Neither model is, however, completely satisfactory, and the relation between Young's Modulus and structure is still not fully understood. The next step may be removal from the models of the condition that the matrix is isotropic, but this must await further definition of conditions in the matrix.

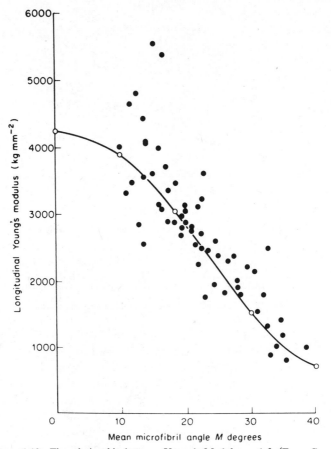

Figure 1.12   The relationship between Young's Modulus and θ. (From Cave (51).)

The mechanical properties of the primary wall, which prove to be relevant for its circumstances, are of a different kind, associated with a different structure. The detailed structure will be briefly described later. In cells that are increasing, or due to increase, in length the general run of the microfibrils as judged either by polarization microscopy or by x-ray diffraction analysis lies transversely to cell length or in a slow helix (15, 53). The mechanical properties usually cannot be examined, since the walls are delicate and lie mostly in a tissue from which they cannot readily be extracted without serious disturbance. Recourse has therefore to be made to cells that grow in isolation, and attention has been focused on the internodal cells of a green alga, *Nitella*, which grow rather rapidly from about 0.5 μm in length to several centimeters while remaining cylindrical. These cells are just large enough that strips of wall can be removed and extended either parallel to cell length or perpendicular to it. The initial Young's Modulus is then found to be about five times as great transversely as longitudinally, in harmony with the structure (54). This continues as long as the cell continues to grow; on cessation of growth the ratio falls to 2.

The longitudinal Young's Modulus, but not the transverse, varies inversely with the rate of growth, and at first sight this seems to indicate that the elastic modulus is the factor controlling the rate of growth. This, however, can hardly be. During growth the wall is extended by stresses set up within it by the hydrostatic pressure within the cell consequent upon the intake of water induced by the osmotic pressure of the cell contents. This hydrostatic pressure remains constant, or even falls, during growth so that the wall cannot be increasingly elastically deformed. On the contrary, the wall is clearly expanding under constant load, and this makes it likely that the mechanical property involved is *creep*. This has been fully confirmed (54). Wall strips cut longitudinally show creep rates consistent with the rate of growth, but transverse strips do not creep even under loads close to the breaking point. The evidence is clear that the viscoelastic properties concerned are those of the matrix. Moreover, this is not the whole story however, convincing as it is, that control of growth is by creep. The rate of creep of an extracted wall strip is much greater than the rate of growth of the cell from which it was taken. The metabolic machinery of the cell must be able to control wall creep, and this is generally agreed. This line of investigation is being followed more recently by Taiz and his colleagues (55).

The inverse problem, the effect of extension on the wall structure, is complex. While a cell is extending in length, the net orientation of what we now know to be microfibrils remains constant even for a hundredfold extension, whereas mechanically induced extension of only 10% changes the net orientation from transverse to longitudinal (56). This remained a paradox until the advent of electron microscopy, when Roelofsen (57) noticed that the microfibrils of the innermost face of the wall tend to lie transversely (although with considerable angular dispersion), whereas those of the outermost zone are oriented more nearly at random or even axially. Since the wall remains the same thickness throughout, in spite of the fact that extension must make it thinner, new wall material must be laid down continuously as the cell expands. This new material, deposited by the cytoplasm, in effect moves outward as the cell extends, as further new material is progressively laid down on the inner face. Since this is a continuous process, there must be a gradient of strain across the wall (from inside to outside) and with it a gradient of reorientation of the microfibrils. This is the Multinet Growth Hypothesis of Roelofson, so named because he conceived the microfibrils of the inner wall face as forming a network. It has two parts: the constantly increasing strain of a wall zone as, in effect, it passes through the wall giving the strain gradient; and the constant net transverse orientation of the microfibrils laid down on the inner wall face at any time, giving the orientation gradient. These two do not necessarily always go together, and if the orientation at the inner face is not constant there is no discernible reorientation through the wall, though reorientation must have occurred.

The strain at any point within the wall can easily be calculated (58) whence, given the starting conditions at the inner face, the gradient of orientation may be deduced. Figure 1.13 gives this rate of strain, and the reorientation if the microfibrils at the inner face lie in a helix making an angle of either 5° or 10° to the transverse. This figure explains the lack of average reorientation as the wall extends since at any time during growth the strain at any point, and therefore the microfibrillar orientation, must be the same.

The basis of the Multinet Growth Hypothesis has recently been questioned by Roland and Vian (59), but their evidence has been called into question by Wardrop et al. (60) and by Preston (58).

As already mentioned, Taiz and his colleagues have taken up the problem of wall extension and cell growth and have published some interesting findings. Among other things, they have demonstrated that only an inner part of a primary wall is capable of sustaining a load since, as Figure 1.13 shows, the strain on a wall zone as in effect it passes out through the wall soon reaches a point at which the zone must fail. Richmond (61) puts the dimension of this inner part

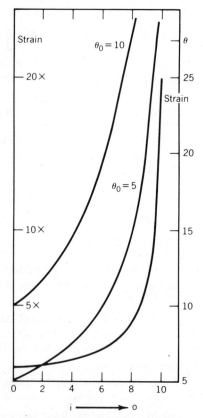

Figure 1.13 The gradient of strain across growing cell wall from the inner face (i) to the outer (o); together with the reorientation of the microfibrils if their preferred orientation lies at either 5° or 10° to the transverse.

at only one-quarter of the wall thickness. This is important, because growth hormones are known to affect the extension process, and it is essential to know how much of the wall it is necessary and sufficient to modify. This finding enables the load-bearing part of the wall to be even more narrowly limited. By making reasonable assumptions it can be shown that with the walls of internodal cells of *Nitella*, to which all these statements refer, an innermost zone about 0.5 μm thick (in a wall which is 8 μm thick) bears by far the greater part of the total stress on the wall (Preston, unpublished). If this can be translated to the primary walls of higher plants (and there is as yet no certainty that it can), then only the innermost part of the wall a few microfibrils thick bears most of the load induced by the hydrostatic pressure of the cell. Nature therefore seems to have exploited to the full the effectiveness of fiber-reinforced materials.

CELLULOSE BIOSYNTHESIS

It is a somewhat odd circumstance that, at a time when so many of the complex materials produced by living organisms have successfully been synthesized outside the cell, the extracellular synthesis of cellulose has not yet been reported (though see below), and even the synthesis inside the cell is not understood. A number of investigators have claimed a synthesis, but these claims have always proved false. Leading from the original finding of Glaser in 1956 of synthesis, from particulate extracts of *Acetobacter xylinum* treated with uridine–diphosphate–glucose (UDPG), of an alkali-insoluble polysaccharide, a number of other workers have claimed the synthesis of cellulose using extracts from a variety of plants. For example, Batra and Hassid (62), using extracts from germinating mung beans, have claimed to show that the guanosine derivative (GDPG), not UDPG is the precursor from which cellulose is formed. Using their materials and methods, however, Robinson and Preston (63) have concluded that the alkali-insoluble product from either GDPG or UDPG is not cellulose but possibly a mixture of oligosaccharides. A major deficiency in all attempts to synthesize cellulose outside a cell has been that, whatever the nature of the product, the yield was always very much less than what would have been obtained from the particulate fraction had that still been part of a living cell. Electron micrographs of the fraction (63) revealed that it is composed largely of membranous material, and it was therefore possible to conclude that the synthetase was held on membranes that carried also other, degradative enzymes and/or that disruption of the membranes during extraction destroyed mutual associations necessary for synthesis. Most recently, however, Aloni et al. (64, 65) have apparently succeeded in solubilizing the enzyme and have claimed the synthesis from UDPG (now generally accepted as the precursor of choice) of an alkali-insoluble polysaccharide of high molecular weight and which Benzamin (private communication) considers to be cellulose. This claim has now, however, been withdrawn.

The site of synthesis has been a matter of speculation and remains so. Apart from claims for one species, *Pleurochrysis* (66), no evidence has been produced for synthesis in Golgi vesicles (which appear to be concerned with other wall polysaccharides). The structure and organization of cellulose speak against such a mechanism.

Several structural features indicate that microfibrils are constructed as a unit, not as separate chains. Apart from a very few species, plants synthesize cellulose as cellulose I. If synthesis involved construction of individual chains and their subsequent aggregation, the crystalline form would be cellulose II. Again, in most plant walls, the microfibrils, even if lying parallel, are occasionally twisted round each other. It is difficult to see how this could be achieved through the initial synthesis of separate chains. It seems more likely that microfibrils are synthesized by end synthesis—that is, by the agency of synthetase particles attached to the end of microfibrils. Structural evidence for this was foreshadowed by the observation (16, 67, 68) of granules on the inner face of the wall of a variety of cells, sometimes associated in groups from which microfibrils appeared to spread. These were later demonstrated more clearly (8) with *Chaetomorpha* (Fig. 1.14) and led directly to the formulation of what has come to be called the ordered granule hypothesis. This hypothesis is based not only on these observations but also on a general overview of the architecture of cell walls, the relevant features of which are enumerated below. They are discussed more fully elsewhere (15).

1.   For the reasons given above, end synthesis of microfibrils is indicated.

2.   The microfibrils, though in some cells lying almost at random in the plane of the wall, lie mostly in some specific orientation. The alignment may be very perfect and sometimes related to cell dimensions. It is not easy to see how this could come about unless synthetase granules, attached to the ends of micro-

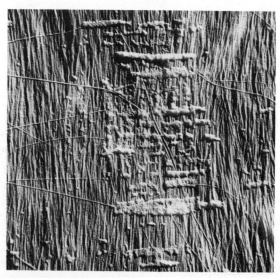

Figure 1.14   Electron micrograph of innermost wall lamella of a plasmolysed cell of *Ch. melagonium*; magnification 20,400×, shadowed Pd/Au.

fibrils, are situated in, on, or near the plasmalemma and unless neither the granules nor the microfibrils move during synthesis.

3. The microfibrils of a single wall often lie in two different directions, occasionally with a third and, in some higher plants, a fourth. In the algae showing these different orientations the orientation changes from one lamella to the next in a regular sequence and without mistakes. The synthetase system must therefore be such as to recognize three, sometimes four, directions and no other.

4. Though the different orientations are usually found in different lamellae, this is not always the case. Microfibrils of one lamella may be interwoven with those of the next (with a different orientation). A microfibril of one lamella may turn through 90° and join those of the next lamella or through 180° and rejoin those of the same lamella. It seems likely that the different lamella are not laid down independently in a regular time sequence but are deposited together. The synthetase system must therefore have the capacity to lay down microfibrils simultaneously in the directions it recognizes.

5. When cells of *Cladophora* or *Chaetomorpha* are plasmolyzed (the cytoplasm pulled away from the wall by osmotic withdrawal of water), the synthetic mechanism still lays down microfibrils over the now spherical surface, but these are now oriented at random. Similarly, just before the cytoplasm breaks up ready for sporulation, the microfibrils are again deposited at random. Hence the shape and integrity of the cytoplasmic surface are significant for the orienting mechanism.

A model that seemed best to meet these requirements was presented in 1963 (69). This required the presence in or on the plasmalemma of synthetase granules arranged in a cubic lattice. They were taken to be such that when a microfibril end came into contact with a granule, synthesis would occur within the granule; the microfibril would then grow through to the next granule so that there were four possible microfibril directions (2 parallel to the sides of a square and 2 diagonals). This suggestion was made at a time when it was thought that the plasmalemma was a mere 7.5 nm thick, far too thin to accommodate such a granular array. In this context, it becomes important to know whether the molecular chains constituting the microfibrils lie parallel or antiparallel. If antiparallel, then the granules must contain two enzymes, whereas parallel chains would allow a simpler system.

Confirmation that granules are present in or on the plasmalemma was immediately forthcoming through the technique of freeze-fracturing in the hands of Moore and Mühlethaler (71), and since that time there have been repeated observations of plasmalemma-associated granules thought to be synthetase granules. The latest recording has been made by Fujita (70) for the fusiform cells of conifer cambium. Sometimes the granules are associated with the ends of short filaments which might be incipient microfibrils (34) and even of long filaments which are almost certainly so (79). It remains to be proved that these are synthetases actively producing cellulose. Ordered granules have been seen, for instance, on the plasmalemma of yeast cells (which do not produce cellulose, however) and, over limited areas, of *Oocystis* (79). There is now a good deal of evidence in favor of the granular hypothesis, but proof is still awaited.

The only recently considered alternative to this hypothesis involves microtubules. These are "hollow" tubes some 23–27 nm wide with a wall about 7 nm thick, composed of spherical particles of the protein tubulin, and of indefinite length. They are ubiquitous in living cells. Although they can occur anywhere within a cell, sometimes associated with the nucleus, they often lie near the plasmalemma. Ever since their discovery by Hepler and Newcombe, and following the conclusion of Marx-Figini and Schultze in 1963 regarding cellulose synthesis in cotton hairs, a possible connection between them and the synthesis of cellulose has been mooted. The microtubules lie some distance from the plasmalemma, and a direct connection with synthesis itself is no longer supported. It has often been reported, however, that microtubules lie parallel to the microfibrils on the other side of the plasmalemma, and this has come to be regarded as causative. Further evidence for such a view has been claimed from the observation that treatment of *Oocystis* with colchicine destroys the microtubules while at the same time maintaining the orientation of the microfibrils but preventing a switch to an orientation at right angles, which would otherwise occur in this crossed microfibrillar wall (80). It can be argued that parallelism between two filamentous bodies does not necessarily imply a direct causal connection, and it must not be overlooked that colchicine reduces the mobility of the plasmalemma and therefore presumably the situation of the granules associated with it, and that these in turn may be the directors of orientation.

If it is assumed that the basic features of an orienting mechanism are uniform over the plant kingdom, then any proposed mechanism must be compatible with all known wall structures. A cursory review even of the brief and incomplete details given above makes it difficult to support any of the microtubule-associated mechanisms so far proposed (reviewed by Lloyd, 81). The complexity of wall architecture seems to demand an entirely different scenario. The principles on which such a scenario must be based can readily be defined (Preston, unpublished) in such a way as to understand all wall structures—including the relation between tracheid length and microfibrils orientation—but this involves structures and associated vectors still to be searched for.

CELLULOSE DEGRADATION

The inverse problem, the degradation of cellulose by microorganisms, is equally complicated not only biochemically but also structurally, since microfibrils are always covered by molecular chains of noncellulosic polysaccharides. The condition in wood cell walls is especially complex since, as evidenced by the fact that untreated wood does not react positively to cellulose stains, the microfibrils, or groups of them, must be isolated by a covering of lignin. With white-rot fungi, which attack lignin preferentially, and brown-rot fungi, which take out cellulose, there must be induced some modification of the whole cell wall complex. In both these cases the fungal hyphae, though penetrating the wall, travel along the cell in the lumen and attack the wall by erosion of the inner face. This does not develop any marked structural features in the wall. There is, however, a group of fungi—the soft-rot fungi—whose activity is manifested by a remarkable "structural" development in the wall.

The difference with these fungi, mostly Ascomycetes and Fungi Imperfecti, is that the hyphae come to lie within the S2 layer and grow through it parallel to the microfibrils. As first noticed by Schacht, they produce cavities there around the hypha, larger than the hypha in a form that is diamond-shaped in surface view (72). Developments since that time have been reviewed by Levy (72), but so far no satisfactory explanation has been given as to how it comes about that a hypha that is not visually crystalline, attacking a wall that is not visually crystalline either, produces a cavity that has a clear crystal shape.

The cavity mostly becomes elongated in the direction of the microfibrils, and the problem resolves itself into the question of how the smoothly "conical" ends are produced. Explanation has been offered along two lines. Each involves the release of enzymes from the hyphae, and this undoubtedly occurs. They are known to remove from wood not only cellulose but also xylan, mannan, and galactan, and in some way to modify lignin (73, 74), and many of these fungi have been shown to produce cellulase, mannase, and xylanase (75). They therefore have the battery of enzymes needed to reach and attack cellulose.

Both proposed explanations of cavity formation throw weight on the wall rather than on the fungus. The first, proposed by Bailey and Vestal (43), imagines that there are planes in the lattice of cellulose along which hydrolysis takes place preferentially. This has received strong support from Frey-Wyssling (76) and has been accepted by a number of workers. Regrettably, this elegant explanation has two weaknesses. First, the acute apical angle of the cavity is not constant but varies fairly widely (77) from, for example, 47° for oak, pine, and larch to 72° for beech. This is hardly consistent with a strictly crystalline interpretation. Second, the hypothesis seems to demand a mutual register between the microfibrils of a wall for which there is no evidence.

The second explanation is a looser one. As first suggested by Roelofsen, the enzymes released by the fungus could travel at different rates along and perpendicular to the microfibril array, and these rates could be so regulated that they produce a diamond-shaped hydrolysis cavity. This has received support (73) in a paper that makes the point that transfer of enzyme attack from one microfibril to the next must occur at a paracrystalline region, which it has been shown must occur periodically along the microfibril (Fig. 1.15). This hypothesis has been re-examined recently (78) in an attempt to give it a quantitative basis.

Let us imagine a fungal hypha FF (Fig. 1.15) lying in contact with an array of microfibrils spaced d apart, and suppose in the first instance that enzymes are released from a single point, P. We consider the effect of the consequences of a random walk of the enzyme through the array, of which a number of possible paths are illustrated in Figure 1.15. By the time an enzyme particle has reached point B along the microfibrils, distance $n(l + x)$ from P (where n is the number of crystalline regions passed and l is the average length of such a region), it will also have passed across the array to a distance $n'(b + d)$ where n' is the number

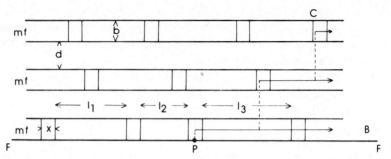

Figure 1.15 Representation of three microfibrils (*mf*) lying *b* units wide and *d* units aparts over the surface *FF* of a soft rot fungus; *x* is the length of the paracrystalline regions, assumed constant.

of microfibrils traversed in this direction. If we draw a line joining these points (Fig. 1.16), then it can be shown that an enzyme particle passing along *AB* to *a* and then traversing the microfibrils will reach point *b* on the line *BC* at the same time that other enzymes reach *B* and *C*. The erosion will at that time have created a cavity limited by line *BC*. Since erosion occurs in both directions along the microfibrils, and since enzymes will also be released from the lower surface

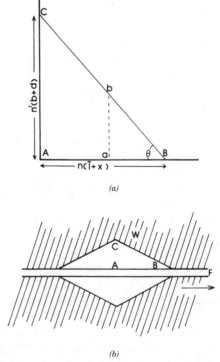

(a)

(b)

Figure 1.16 The generation of a diamond-shaped cavity from Figure 1.15. For explanation, see text.

of the hypha (and, indeed, from all round the hypha in an annulus), the cavity will be diamond-shaped. Since, in practice, enzymes are presumably released not from a single linear annulus, but from an extended zone, the cavity will in fact have flat lower and upper sides and pointed ends, as actually observed.

It should be noted that if the cellulose chains within a microfibril are antiparallel, then two enzymes will again be needed, and they will be able to digest cellulose in both directions along the microfibril. If, on the contrary, they are parallel, only one enzyme will be necessary and the biochemical problem is simpler. The enzyme can work in both directions because half the microfibrils are "upside down" with reference to the others, as shown above. The details of the hypothesis set clear limits to the possible values of the apical angle of the cavity while allowing the variation actually observed. The hypothesis itself is clearly an oversimplification, since cavities can widen without lengthening, but the consequent difficulties seem not insuperable. This is, however, yet another problem concerning cellulose which is still not completely solved.

Final solutions will be approached only as the finer details of the architecture of cellulose and of the whole cell wall are progressively revealed. It is evident that all these details are important from the moment of synthesis, through the properties of the materials produced (both in their functions in the plant and their usefulness for man), down to the last moments of the cell wall as a coherent entity.

REFERENCES

1. A. Frey, *Die Micellartheorie von Carl Nägeli*, Leipzig, 1928.
2. H. I. Bolker, *Natural and Synthetic Polymers*, Dekker, New York, 1974.
3. O. L. Sponsler, and W. H. Dore, *Coll. Symp. Monogr.*, **41**, 174 (1926).
4. K. H. Meyer and H. Mark, *Ber. Dtsch. Chem. Ges.*, **61B**, 593 (1928).
5. K. H. Meyer and H. Misch, *Helv. Chim. Acta*, **20**, 232 (1936).
6. J. Honjo and M. Watanabe, *Nature (Lond.)*, **181**, 326 (1958).
7. D. G. Fischer and J. Mann, *J. Polym. Sci.*, **C12**, 189 (1960).
8. E. Frei and R. D. Preston, *Proc. R. Soc. Lond.*, **B154**, 70; **B155**, 55 (1961).
9. I. A. Nieduszynski and E. D. T. Atkins, *Biochim. Biophys. Acta* **222**, 109 (1970).
10. K. H. Gardner and J. Blackwell, *Biopolymers*, **13** (1975); *Biochim. Biophys. Acta*, **343**, 232 (1974).
11. A. Sarko and R. Muggli, *Macromolecules*, **7**, 486 (1974).
11a. A. J. Stepanovic and A. Sarko, *Polymer*, **19**, 3 (1978).
12. J. Hengstenberg and H. Mark, *Kristallogr.*, **69**, 271 (1928).
13. R. D. Preston, E. Nicolai, R. Reed, and A. Millard, *Nature (Lond.)*, **162**, 665 (1948).
14. A. Frey-Wyssling, K. Mühlethaler, and R. W. G. Wyckoff, *Experientia*, **6**, 12, 475 (1948).
15. R. D. Preston, *Physical Biology of Plant Cell Walls*, Chapman and Hall, London, 1974.
16. R. D. Preston and G. W. Ripley, *Nature (Lond.)*, **174**, 76 (1954).
17. I. A. Nieduszynski and R. D. Preston, *Nature (Lond.)* **225**, 273 (1970).
18. D. F. Caulfield, *Textile Res. J.*, **41**, 267 (1971).

19. A. Bourret, H. Chanzy, and R. Lazaro, *Biopolymers,* **11**, 803 (1972).

20. R. D. Preston, *J. Microsc.,* **93**, 7 (1971).

21. J. Cronhaw, A. Myers, and R. D. Preston, *Biochim. Biophys. Acta,* **27**, 89 (1958).

22. K. Keegstra, K. W. Talmadges, W. D. Bauer, and P. Albersheim, *Plant Physiol.,* **51**, 188 (1973).

23. B. G. Ranby and G. Ribi, *Experientia,* **6**, 2 (1950).

24. G. Ribi, *Arkiv. Kemi,* **2** No. 40,551 (1950).

25. D. T. Dennis and R. D. Preston, *Nature (Lond.)* **191**, 667 (1961).

26. R. D. Preston, *Planta,* **30**, 55 (1979).

27. D. A. Rees, *Polysaccharide Shapes: Outline Studies in Biology*, Chapman and Hall, London, 1977.

28. R. Moorhouse, W. T. Winter, and S. Arnott, *J. Mol. Biol.,* **109**, 373 (1977).

29. E. R. Morris, D. A. Rees, G. Young, M. D. Wilkinshaw, and A. Darke, *J. Mol. Biol.,* **110**, 1 (1977).

30. R. H. Marchessault and C. Y. Liang, *J. Polym. Sci.,* **59**, 357 (1962).

31. P. A. Roelofsen and D. R. Kreger, *J. Exp. Bot.,* **2**, 332 (1951).

32. E. Frei, and R. D. Preston, *Nature (Lond.),* **196**, 130 (1962).

33. R. D. Preston and W. T. Astbury, *Proc. R. Soc. Lond.,* **B122**, 76 (1937).

34. D. G. Robinson and R. D. Preston, *J. Exp. Bot.,* **22**, 635 (1971); *Planta,* **104**, 234 (1972).

35. J. Cronshaw, and R. D. Preston, *Proc. R. Soc. Lond.,* **B146**, 37 (1958).

36. W. T. Astbury and R. D. Preston, *Proc. R. Soc. Lond,* **B129**, 54 (1954).

37. R. D. Preston, *The Molecular Architecture of Plant Cell Walls*, Chapman and Hall, London, 1952.

38. A. Frey-Wyssling, *Submicroscopic Morphology of Protoplasm and its Derivatives*, Elsevier, Amsterdam, 1948.

39. R. D. Preston, *Philos. Trans R. Soc.,* **B224** 131 (1934).

40. R. D. Preston and M. Middlebrook, *J. Text. Inst.,* **40**, T175 (1949).

41. R. D. Preston and K. Singh, *J. Exp. Bot.,* **1**, 214 (1950); **3**, 162 (1952).

42. F. Stern and H. P. Stout, *J. Text. Inst.,* **45**, T1896 (1954).

43. I. W. Bailey and M. R. Vestal, *J. Arnold Arbor.,* **18**, 185, 196 (1937).

44. A. B. Wardrop and R. D. Preston, *Nature (Lond.),* **160**, 911 (1947); *J. Exp. Bot.,* **2**, 20 (1951).

45. J. Sugyama, H. Harada, Y. Fujiyashi, and N. Uyeda, *Makuzai Gakkaishi,* **30**, 98 (1984).

46. K. C. Brown, J. C. Mann, and F. T. Pierce, *Shirley Inst. Mem.,* **9**, 1 (1930).

47. R. Meredith, *J. Text. Inst.,* **27**, T205 (1946).

48. L. C. Spark, G. Darnborough, and R. D. Preston, *J. Text. Inst.,* **49**, T309 (1958).

49. J. W. Hearle, *J. Appl. Polym. Sci.,* **7**, 1207 (1963).

50. D. R. Cowdrey and R. D. Preston, *Proc. R. Soc. Lond.,* **B166**, 245 (1966).

51. J. D. Cave, *Wood Sci. Technol.,* **2**, 628 (1968); **3**, 40 (1969).

52. R. E. Mark, *Cell Wall Mechanics of Tracheids*, Yale University Press, New Haven, CT, 1967.

53. P. A. Roelofsen, *The Plant Cell Wall*, Borntraeger, Berlin 1959.

54. M. C. Probine and R. D. Preston, *J. Exp. Bot.,* **12**, 261 (1961); **13**, 111 (1962).

55. L. Taiz, J.-P. Métraux, and P. Richmond, *Cell Biol. Monogr.,* **8**, 231 (1981).

56. J. Bonner, *Jb. Wiss. Bot.,* **82**, 377 (1935).

57. P. A. Roelofsen, *Biochim. Biophys. Acta,* **7**, 43 (1951); P. A. Roelofsen and A. L. Houwink, *Acta Bot. Neerl.,* **2**, 218 (1952).

58. R. D. Preston, *Planta,* **155**, 356 (1982).

59. J. C. Roland and B. Vian, *Int. Rev. Cytol.,* **61**, 129 (1979).

60. A. B. Wardrop, M. Walters-Arts, and M. M. A. Sassen, *Acta Bot. Nóerl.,* **28**, 313 (1979).

61. P. Richmond, Ph.D. thesis, University of Pennsylvania, Philadelphia, 1977.

62. K. W. Batra and W. Z. Hassid, *Plant Physiol.,* **44**, 755 (1969).

63. D. G. Robinson and R. D. Preston, *Biochim. Biophys. Acta,* **273**, 336 (1972).

64. Y. Aloni, D. P. Delmer, and M. Benzamin, *Proc. Natl. Acad. Sci. USA,* **79**, 6448 (1982).

65. Y. Aloni, R. Cohen, M. Benzamin, and D. P. Delmer, *J. Biol. Chem.,* **258**, 4419 (1983).

66. R. M. Brown, W. W. Franke, H. Kleinig, H. Falke, and P. Sette, *J. Cell Biol.,* **45**, 246 (1970).

67. R. D. Preston, E. Nicolai, and B. Kuyper, *J. Exp. Bot.,* **4**, 40 (1953).

68. W. T. Williams, R. D. Preston, and G. W., Ripley, *J. Exp. Bot.,* **6**, 451 (1955).

69. R. D. Preston, in *The Formation of Wood in Forest Trees*, M. Zimmerman, ed., Academic, New York, 1964.

70. M. Fujita, Thesis, Kyoto University, Japan, 1981.

71. H. Moor and K. Mühlethaler, *J. Cell Biol.* **17**, 609 (1963).

72. J. F. Levy, in *Advances in Botanical Research*, R. D. Preston, ed., Academic, New York, Vol. 2, p. 323.

73. M. P. Levi and R. D. Preston, *Holzforschung,* **19**, 185 (1965).

74. K. Seifert. *Holz als Roh-u-Werkstoff,* **14**, 208 (1966).

75. T. Nilsson, *Bull. R. Coll. For., Stockholm, Sweden*, No. 117 (1974).

76. A. Frey-Wyssling, *Papierfabrikant,* **36**, 212 (1938).

77. H. Courtois, *Holzforsch. Holzwerwert,* **15**, 88 (1963).

78. R. D. Preston, *Wood Sci. Technol.,* **13**, 155 (1979).

79. R. M. Brown, Jr., submitted (1984).

80. H. Quader and D. G. Robinson, *Berl. Dtsch. Bot. Ges.,* **94**, 75 (1981).

81. C. W. Lloyd, *Int. Rev. Cytol.,* **86**, 1 (1984).

2

Recent X-Ray Crystallographic Studies of Celluloses

ANATOLE SARKO
Department of Chemistry and the Cellulose Research Institute,
State University of New York, College of Environmental Science and Forestry,
Syracuse, New York

The structure of cellulose has been the subject of sustained study for more than a century. A high level of interest has been maintained largely because of the nature and utility of this substance—a unique polysaccharide among the most diverse, widespread, and complex class of natural compounds. It is a simple molecule, yet its biosynthesis is more complicated than we know. It is naturally crystalline, exhibits crystalline polymorphism, and possesses a highly ordered, fibrous morphology. Of the many widely utilized natural substances, cellulose is one of the most important commercial raw materials, the parent substance for a large variety of chemical derivatives.

Earlier, traditional chemical studies of cellulose structure contributed greatly to our understanding of the macromolecular state. Later, as instrumental techniques for the study of structure were developed, cellulose was one of the first to be analyzed with the new techniques. A prominent example is x-ray diffraction—no sooner was it discovered by Laue than diffraction diagrams of cellulose were reported. More recently, when the technique of solid-state nuclear magnetic resonance (NMR) was perfected, spectra of cellulose were again among the first to be recorded.

Impetus for the study of cellulose by these techniques arises from the knowledge that a thorough understanding of the physical structural features of a polymer—chain conformation, supermolecular morphology, molecular interactions in solid and solution states—is crucial for the understanding of its properties. It is also recognized that a detailed characterization of a relatively

29

simple polysaccharide such as cellulose can provide useful information on the principles that govern the structure and organization of polysaccharides in general. This has, in fact, greatly aided in the study of other, chemically more complex polysaccharides.

The x-ray diffraction techniques applied to fibrous polymers have become much more sophisticated in recent years. The major component in their successful use with celluloses and other polysaccharides has been the harnessing of the computer as an exhaustive model evaluation tool. Sound stereochemical principles have been incorporated into the analysis, and elegant refinement software has been written. Before examining the results obtained with celluloses using these techniques, a brief description of the methodology would be instructive.

MODERN METHODS OF FIBER DIFFRACTION ANALYSIS

Although the majority of structure analyses of cellulose and other polysaccharides have used x-ray diffraction techniques, other diffraction methods—notably electron diffraction—are assuming an increasing importance. With some exceptions, the principles and applications of the various methods to polysaccharides do not differ materially. As a group, however, these methods differ considerably from those used in the classical, single-crystal diffraction analysis. The main reason for this divergence stems from the nature and morphology of the polysaccharide materials. In most instances the latter are polycrystalline rather than single crystalline, as is the case with small molecules or some polymers such as globular proteins. The polycrystallinity, coupled with small crystallite size and a morphology that is only partly crystalline, limits not only the quantity of the diffraction data available from such materials but also their resolution. As a result, direct methods or methods in which the crystal structure is determined solely from the diffraction data cannot be applied to them. Instead, modeling methods must be used in which an exact correspondence of the observed and calculated diffraction data is sought through the refinement of the most probable model of the structure. Three important elements forming the basis of this method are (a) selection of the initial model(s), (b) preliminary, stereochemical refinement of the structure, and (c) final refinement with diffraction data.

Model Selection

In this initial step, a model of the molecule is established that simultaneously satisfies the requirements of the crystal lattice (i.e., the unit cell) and the stereochemistry of the monomer residue. The former is usually easily determined from the x-ray fiber pattern, and the latter is commonly determined on the basis of data accumulated from single-crystal carbohydrate structure analyses. A growing body of such data is reinforcing the assumption that the conformation of the sugar residues in the polysaccharide molecule is a minimum-energy one with predictable bond lengths, bond angles, and torsion angles (1, 2). The latter parameters fall into relatively narrow ranges, so a reasonable model based on average parameters can be established even in the absence of specific crystallographic data for a particular sugar (1). For common hexopyranoses it is now known that the chair conformation of the pyranose ring is the minimum-energy one—for example, $^4C_1$ for glucopyranose and its derivatives.

It has also been shown in a growing number of analyses that the main features of the chain conformation of the crystalline polysaccharide are controlled by rotations about the bonds linking the sugar residues. For all but the $1 \rightarrow 6$ linkages, two such rotations exist: the ϕ angle describing the rotation about the $C(1)-O(1)$ bond, and the ψ angle for the rotation about the $O(1)-C(x)$ bond. (For the $1 \rightarrow 6$ linkages there are three rotations—ϕ: $C(1)-O(1)$, ψ: $O(1)-C(6)'$, and ω: $C(6)'-C(5)'$.) Variations in bond lengths, bond angles, and other torsion angles serve only to change the detailed features of the conformation. It is thus a relatively simple matter to find all allowed chain conformations using the ϕ, ψ search procedure and molecular mechanics methods. For example, in such a procedure, an arbitrary initial chain model is established based on "average" sugar residues linked together via the desired linkages, as shown in Figure 2.1a. The entire conformational space available to this chain model is then searched by systematically varying the ϕ, ψ angles (e.g., in $10°$ steps), calculating the conformational potential energy for each combination of the angles. This can be conveniently accomplished by summing over all pairwise atomic interactions, using the equation:

$$E = \Sigma E_{ij}$$
$$E_{ij} = E_{nb} + E_{hb} + E_{el} + \ldots \tag{1}$$

where E is the total conformational potential energy, E_{ij} is the pairwise interactions energy between atoms i and j, and E_{nb}, E_{hb}, E_{el} are, respectively, the nonbonded, hydrogen-bonding, and electrostatic energy terms. (Other energy terms may be included in this equation, as desired (3).) In practice, it has been found that the nonbonded and hydrogen-bond energies are the most important energy components. The nonbonded energy could be calculated, for example, using the Lennard-Jones equation:

$$E_{nb} = Ar_{ij}^{-12} + Br_{ij}^{-6} \tag{2}$$

where r_{ij} is the distance between atoms i and j, and A and B are specific constants for a given atom pair. Similarly, the hydrogen-bond energy could be calculated using a simple, nondirectional equation (4):

$$E_{nb} = Cr_{ij}^{-3} \tag{3}$$

where C is a constant specific for the ij pair. A number of other equations for the nonbonded and hydrogen-bond energies, as well as for the electrostatic energy terms, are available (3).

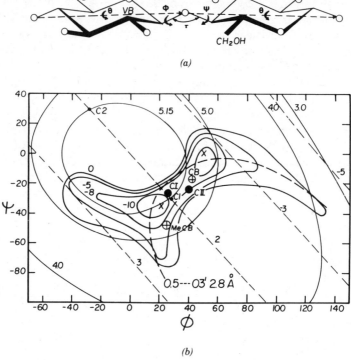

(a)

(b)

Figure 2.1 (*a*) Model of cellulose chain used in the initial conformational analysis. (θ,ψ = rotations about bonds to the glycosidic oxygen; τ = bond angle at the glycosidic oxygen; VB = *virtual bond* or the vector connecting consecutive glycosidic oxygens; θ = rotation of the entire residue about VB). (*b*) Conformational potential energy contour map for cellulose as a function of θ,ψ rotations. Energy contours in kcal/mole; other solid contours designate *h* (Å), and dashed line contours designate *n* (negative for left-handed helices). Crosses designate energy minima; filled circles CI and CII designate the conformations found in celluloses I and II, respectively. The points C1 and C2 designate *n* = 2, *h* = 5.15 Å conformations (cf. Fig. 2.2); circled crosses CB and MeCB designate the conformations of cellobiose and β-methyl cellobioside in the respective crystal structures.

A typical resulting ϕ,ψ energy map ("Ramachandran plot") for cellulose is shown in Figure 2.1*b*. The allowed region—that is, the region containing conformations free from short, nonbonded contacts and those with reasonable hydrogen bonds—is seen to be relatively small. The superimposed *n* and *h* contours (where *n* is the number of residues per helix turn and *h* is the rise per monomer residue along the helix axis) indicate the allowed region to be roughly bounded by -3_1 and 3_1 helices (i.e., left- and righthanded threefold helices), centered on the 2_1 helix, with *h* values of the order of 5 Å. The region of the energy minimum coincides with the presence of the interresidue O(5)–O(3)′ hydrogen bond and, most importantly, contains the conformations present in cellobiose crystal strutures (5, 6). (As discussed later, all crystalline cellulose conformations fall into the minimum energy region.)

An alternative procedure, based on the rotation about the *virtual bond* (VB) linking successive glycosidic oxygens (see Fig. 2.1*a*), is frequently more rapid and more convenient to use, especially when the values of n and h are known from the diffraction data (7, 8). In this procedure, the residues are rotated about the VB (e.g., in 10° steps), monitoring both the glycosidic bond angle τ and the conformational energy E (of Eq. (1).) The main advantages of this procedure are two—a quick identification of the most probable chain conformation, and the absence of restrictions necessitating searching the conformational space with a fixed τ angle, as in the ϕ,ψ procedure. A typical τ versus θ plot (where θ is the rotation angle about the VB) for cellulose is shown in Figure 2.2. Only two regions of probable conformations are evident in this plot—centered about θ values of 120° and 220°, and exhibiting τ angles ranging from 110° to 120°. The conformation marked C1 corresponds to the actual $n = 2$ and $h = 5.15$ Å conformation seen in celluloses, whereas the conformation marked C2 is a disallowed one, characterized by a very high conformational potential energy (see the ϕ,ψ map, Fig. 2.1*b*).

Both procedures work well in practice, particularly for two-bond glycosidic linkage structures such as cellulose. For three-bond (i.e., $1 \rightarrow 6$) linkages and for structures exhibiting more than one set of ϕ,ψ rotations (e.g., multiresidue heteropolysaccharides), correct results are also obtained although at the expense of considerably more searching. In the latter structures, one generally varies only two rotations at a time, keeping the others constant. This procedure is then cycled through all variables, altering two at a time, until all of the conformational space has been searched.

It should be understood that the key to the effectiveness of the described procedures lies in the very nature of the chemical structure of the polysaccharides.

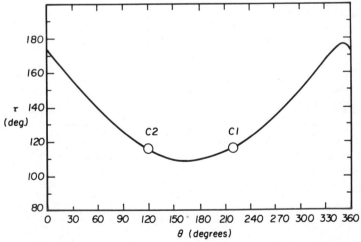

Figure 2.2 Variation of the glycosidic bridge angle τ as a function of the rotation θ of the residue about the *virtual bond* (cf. Fig. 2.1*a*).

Because the minimum energy conformations of carbohydrate residues in polysaccharide molecules are cyclic and not open-chain, the search variables are reduced drastically—in most cases to two parameters—with the bulk of the conformation remaining essentially invariant during the search. In this respect, chemically complex polysaccharide molecules may be easier to handle than some other, chemically more simple polymers.

It should also be understood that these conformation-invariant procedures are used only during the initial phase of structure analysis—the identification of the most probable model(s). As shown below, in the subsequent steps of structure refinement all conformational features should be considered variable. To do otherwise may result in an inability to complete the structure analysis or, at best, serious errors in the final structure.

Stereochemical Refinement

Once the most plausible chain model has been established (or all probable models have been identified), the refinement procedure, which will ultimately lead to the final structure, begins. In principle, such refinement could be immediately done with diffraction data (i.e., against the set of observed structure factor amplitudes); however, in practice it is more expedient to insert a preliminary, stereochemical refinement step. The main reason for doing so is that stereochemical refinement proceeds about 10 times faster than the crystallographic refinement and, in most instances, leads to very nearly the final structure. The models resulting from this step can then be relatively quickly evaluated in the final, crystallographic refinement step.

In stereochemical refinement, the objectives are to find the most probable positions of the chains in the unit cell and to refine their conformations, orientations, and packing to a state of minimum energy—in other words, using crystallographic jargon, to find the most probable "phasing model" suitable for final refinement against the diffraction data. Because crystal structures are generally minimum-energy structures—or at least reside in clearly identifiable local energy minima—stereochemical energy minimization procedures are well suited to finding such *phasing* models. The same molecular mechanics principles and potential energy functions are used as were in the initial step when finding suitable chain models. The number of refinable parameters (variables) can be quite large: all bond lengths, bond angles, and torsion angles, and all chain positioning parameters (rotations and translations relative to as many coordinate axes as necessary). If symmetry information (i.e., the space group) is available from diffraction data, it should be used. The following "energy" equation has been found to be convenient, efficient, and easy to use in practice (8):

$$E = \Sigma \sigma_i^{-2}(a_i - a_{io})^2 + W\Sigma E_{ij}$$

$$E_{ij} = w_{ij}(d_{ij} - d_{ijo})^2 \text{ for } d_{ij} < d_{ijo} \tag{4}$$

$$E_{ij} = 0 \qquad\qquad \text{for } d_{ij} \geqslant d_{ijo}$$

where a_i, a_{io}, and σ_i represent, respectively, the value of a bonded parameter (bond length, bond angle, torsion angle), its most probable value, and the standard deviation of the latter; d_{ij}, d_{ijo}, and w_{ij} represent, respectively, the nonbonded distance between atoms i and j, its most probable value, and weight; W is the weight of the nonbonded term; and the summations are taken over all bonded parameters and pairwise nonbonded distances. The values of the most probable parameters a_{io} and d_{ijo}, as well as their standard deviations and weights, are taken from single-crystal structure analyses of appropriate carbohydrates (8).

In all polysaccharide structures analyzed by us to date using these procedures, it has been found that the minimization of the stereochemical energy proceeded hand-in-hand with the maximization of the number of hydrogen bonds. In this respect, polysaccharide structures resemble crystalline carbohydrates. It is thus very important to monitor the possible development of hydrogen bonds during refinement and to include their energies at all times.

Crystallographic Refinement

In this final refinement step the structure is brought into complete agreement with the diffraction intensity data. In the event there is more than one probable model resulting from the stereochemical refinement, crystallographic refinement will determine the correct model. Thus, all reasonable models must be carried through the final refinement.

The procedure is very similar to that used in stereochemical refinement. The variables remain the same; however, instead of energy minimization, the crystallographic residuals R or R'' are calculated and minimized during the refinement:

$$
R = \frac{\sum \|F_c| - |F_o\|}{\sum |F_o|}
$$

$$
R'' = \left\{ \frac{\sum w (|F_c| - |F_o|)^2}{\sum w |F_o|^2} \right\}^{1/2}
$$

(5)

where F_c and F_o are the calculated and observed structure factor amplitudes, respectively, w are the individual reflection weights, and the summations are taken over all (observed and unobserved) reflections in the recorded diffraction space. It is generally better to use the weighted residual R'', because it incorporates weights that reflect the varying degree of accuracy of measuring diffraction intensities. It should also be understood that the success of the final refinement depends critically on both the number of F_o in the recorded data set and the quality and resolution of the data. For this reason it is very important to maximize the number of recorded reflections and to measure their integrated intensities with the utmost care. Because the x-ray fiber patterns usually exhibit moderate to severe degrees of reflection overlap, some form of intensity resolution must be employed. Fortunately, a number of excellent resolution methods are available (9, 10).

Notwithstanding the care taken in the recording of the data set, the latter never contains enough reflections to permit a complete, detailed structure refinement with such data alone. The usual manifestation of insufficiency of data is the appearance of some unrealistic stereochemical features during the refinement—short nonbonded contacts, bond angles outside a reasonable range, and so on. This is particularly true when the structure contains water or other solvent of crystallization. To guard against the occurrence of such features, the final crystallographic refinement is best carried out under stereochemical constraints. This can be accomplished by minimizing the function (8):

$$\Phi = fR + (1-f)E \qquad (6)$$

where R represents any of the residuals, E is the energy calculated in Eq. (4), and the fraction f is chosen such that the energy term in the function is just large enough to prevent the development of bad stereochemistry.

In most instances, the use of the above procedures has resulted in well-refined structures, as evidenced by the values of the residuals in the 0.1–0.2 range.

STRUCTURE OF CRYSTALLINE POLYMORPHS

The number of polysaccharides whose structures have been examined by crystallographic techniques is growing rapidly. Among them, cellulose has received considerable attention since the mid 1970s (11–19). The main points of interest concerning its structure are in two areas: polymorphy of its crystalline forms, and the supermolecular morphology of its fibrous structure. Another area of widespread interest—the mechanism of mercerization—is intimately associated with structure, spanning the two structural levels. In our investigations, considerable effort has been spent on the elucidation of the structures of the crystalline polymorphs and, more recently, on the interrelated aspects of alkali mercerization and fiber morphology.

Crystalline Polymorphism

The polymorphism of cellulose crystallinity is well known. The native cellulose, or cellulose I, exhibits a diffraction pattern that is recognizably different from those of the three "man-made" cellulose crystal forms known as celluloses II, III, and IV, all of which also exhibit distinctive patterns (20). The main point of interest related to the polymorphism arises from the observation that cellulose I occurs only as a natural product, a result of biosynthesis. Once it has been mercerized or solubilized and regenerated, the transformation (to cellulose II) remains irreversible. Other conversion steps, including some chemical derivatizations, may or may not be reversible, depending on the conditions of the transformations. The current knowledge concerning the reversibility of the conversions of all four polymorphs, as well as such derivatives as cellulose triacetate (CTA), is schematically illustrated in Figure 2.3.

A number of crystallographic studies on cellulose polymorphs have been carried out to obtain information on this interesting phenomenon. As a result, the crystal structures of all cellulose polymorphs (except that of III_{II}) are now reasonably well known (11–19), as are the structures of CTA I and CTA II (21, 22). From these structural studies, a partial understanding of the phenomenon is emerging.

As shown in Figure 2.3, there are essentially two structural families of celluloses—the parallel-chain and the antiparallel-chain structures. Native cellulose I is typical of the parallel-chain structures, and cellulose II is typical of the antiparallel-chain structures. In both, the chain conformations are very nearly identical, corresponding to the minimum-energy conformations predicted by the ϕ,ψ mapping procedure (see Fig. 2.1b). In the structures of both families, the chains pack into sheets in parallel orientation, with stabilization in two directions of the sheet provided by hydrogen bonds. As shown in Figure 2.4, the sheet structures of the two structural families are similar. The main difference between the parallel- and antiparallel-chain structures resides in the three-dimensional packing of the sheets—a parallel packing in the former, and a regularly alternating, up-down arrangement in the latter. The antiparallel packing affords a more extensive, three-dimensional scheme of hydrogen bonding, as can be seen by comparing the structures of cellulose II and cellulose I (see Fig. 2.5). The increased hydrogen bonding results in cellulose II being a more stable, lower-energy structure and partially explains why it cannot be made to revert to the higher-energy, parallel-chain cellulose I structure.

It is possible to retain the parallel-chain structure in some transformations, under certain conditions. For example, cellulose III_I can be easily made from cellulose I by treatment with liquid NH_3 or some amines (14). Similarly, cellulose III_I can be converted to IV_I by heating in glycerol to 260°C (15). These con-

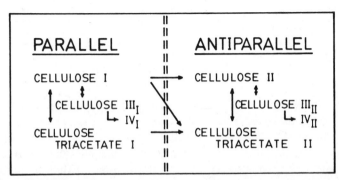

Figure 2.3 Conversions of cellulose polymorphs (including cellulose triacetate) within the parallel and antiparallel structural families. (Courtesy of Gordon & Breach Publishers).

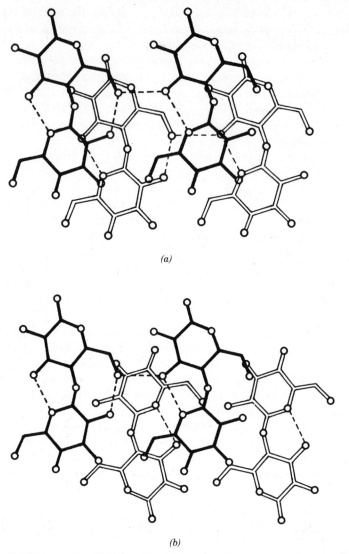

(a)

(b)

Figure 2.4   Parallel (*a*) and antiparallel (*b*) packing of sheets of cellulose chains as seen in celluloses I and II, respectively.

versions are low-swelling in nature and apparently proceed as solid-state transformations. They can be reversed with the return of the structure to that of the
original cellulose I. The same kinds of transformations occur in the antiparallel-
chain family (cellulose II to III$_{II}$ to IV$_{II}$), with the same type of reversibility,
but in this case back to cellulose II. The transformations are thus not determined
by a particular chain packing scheme.

   A comparison of the crystal structures of all of the cellulose polymorphs (Fig.
2.5) affords a reasonably good (but not complete) view of their relative stabilities.

For example, the parallel-chain structure of cellulose III$_I$ appears to possess about the same degree of stability as cellulose I, simply on the basis of the number of hydrogen bonds. This may explain its readily obtained transformation from cellulose I and an easy reversion to the latter. On the other hand, by the same criterion the parallel cellulose IV$_I$ should differ even less from cellulose I in stability, yet the interconversion between these two structures is quite difficult. The same trends are seen in the antiparallel family of structures: Cellulose II to III$_{II}$ transformation is facile and reversible, whereas IV$_{II}$ to II needs much more drastic conditions, in spite of the fact that cellulose II appears to be considerably the more stable of the two structures.

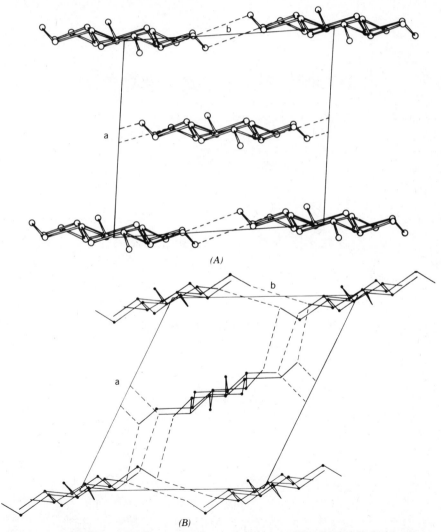

(A)

(B)

Figure 2.5 Projections of cellulose unit cells onto the x,y plane: (*A*) cellulose I (15), (*B*) cellulose II (13), (*C*) cellulose III$_I$ (14), (*D*) cellulose IV$_I$ (17), (*E*) cellulose IV$_I$ (17). (A-C Courtesy of the American Chemical Society, D and E Courtesy of the National Research Council of Canada).

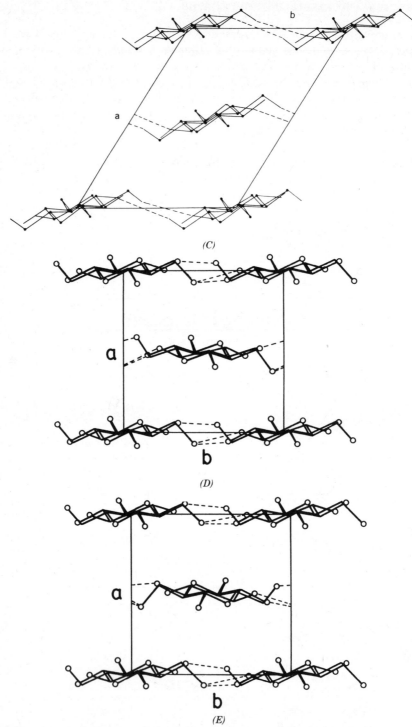

(C)

(D)

(E)

Figure 2.5 *continued.*

Quantitative, theoretical evaluations of the stabilities of the polymorphs (as shown in Fig. 2.6 in terms of potential energy surfaces) confirm in part the structural and experimental observations (23). As can be seen in Figure 2.6, all of the polymorphs are correctly predicted to reside in local energy minima. Cellulose II is predicted, also correctly, to be the most stable of all polymorphs. At the same time, the stability calculations provide litte information on the

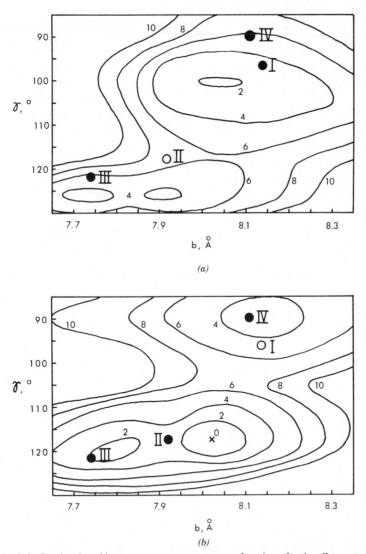

(a)

(b)

Figure 2.6 Predicted packing energy contour maps as a function of unit cell parameters b and γ for (a) parallel celluloses and (b) antiparallel celluloses. Contours are in arbitrary units; filled circles designate positions of known cellulose polymorphs; open circles designate positions of unknown polymorphs.

pathways of the transformations. It is clear that influences other than simple potential energies need to be investigated for reasonable explanations of the observed conversions to be formulated—kinetic factors, chain dynamics, interactions with other molecules (e.g., solvent), and so on.

One fact, however, is clear: The parallel-chain cellulose I structure forms only under very special circumstances, such as are present in biosynthetic mechanisms. The crucial conditions appear to be a simultaneous, unidirectional synthesis of a number of chains, with an opportunity for crystallization to occur either simultaneously with synthesis or very shortly thereafter. This mechanism appears to be followed in the synthesis of cellulose by the bacterium *Acetobacter xylinum*, which produces a cellulose of a high degree of crystallinity (24).

The same types of crystalline polymorphy and structural reversibility are seen in some derivatives of cellulose. For example, under nonswelling conditions, acetylation of cellulose I yields CTA I, whose crystal structure is also based on parallel chains (21). It reverts upon deacetylation to cellulose I. Under analogous conditions cellulose II yields the antiparallel-chain structure of CTA II, which in turn reverts to cellulose II (22). It can also be shown that under high-swelling or solubilization conditions, acetylation of cellulose I yields CTA II directly, which reverts to cellulose II on deacetylation. As shown in Figure 2.7, the crystalline packing of chains in the two structures is considerably different, but we do not know the significance of this difference in terms of stability. A quantitative evaluation of the stability has not been made, and a quick estimation of it is not possible in view of the absence of such strong attractive forces as the hydrogen bond. It is tempting to speculate that because of the presence of only weak, nonbonded interactions, the two CTA structures should not differ significantly in stability. Therefore, the direct transformation of cellulose I to CTA II under high-swelling conditions may simply reflect the acetylation of an already antiparallel structure that has formed during swelling and prior to reaction. (In this context, compare the results of mercerization studies described below).

Nonetheless, the observed transformations of celluloses I and II to CTA I and CTA II clearly support the parallel–antiparallel division of the structures of cellulose polymorphs.

Mercerization and Fibrous Morphology

The existence of two different chain–polarity structures in crystalline celluloses raises some interesting questions. For example, cellulose I can be mercerized with alkali to cellulose II in what appears to be a solid-state phase transformation, proceeding without loss of fibrous morphology. How this process occurs while the chain polarity changes from a parallel to an antiparallel crystal structure is not known. Diffraction studies of several native celluloses have shown that their crystallites are much smaller in diameter than the fibers into which they are aggregated (25, 26). Electron microscopy and diffraction studies have further shown these crystallites to be single crystals and of approximately the same

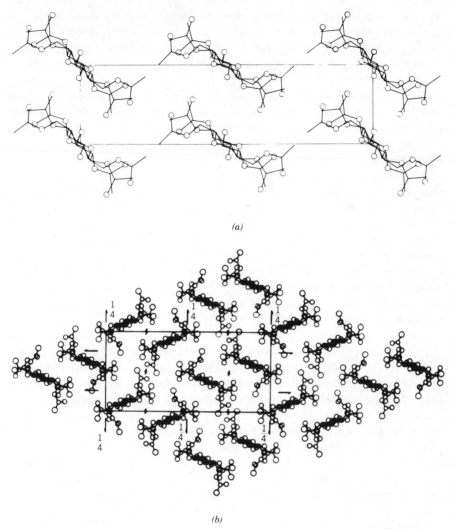

(a)

(b)

Figure 2.7 Projections of the unit cells of (*a*) cellulose triacetate I (21) and (*b*) cellulose triacetate II (22). (*a* - Courtesy of Butterworths Scientific, London, *b* - Courtesy of American Chemical Society).

lateral dimension as the microfibrils, the apparent smallest subdivision of the fiber (27). How these crystals aggregate into fibers and what is the nature of the partly crystalline, partly noncrystalline morphology, are also not completely understood.

Some light has been shed on both questions in a study of ramie mercerization (28, 29). By keeping ramie fibers immobilized in a glass capillary tube, it was found that the process of alkali mercerization could be slowed down sufficiently so that intermediate products could be identified and studied by x-ray diffraction.

The most interesting finding was that mercerization under these conditions proceeds through a number of crystalline alkali-cellulose ("Na-cellulose") complexes, in a sequence of apparently crystal-to-crystal phase transformations. During the latter, decrystallization of the preceding structure does not occur prior to crystallization of the succeeding structure; one crystal form simply disappears with the simultaneous appearance of the other crystal form. As shown in Figure 2.8, five Na-cellulose structures (some of them known already from diffraction patterns recorded in the 1930s) were identified as intermediates in this process. Their degrees of crystalline order are high, as shown by the x-ray diffraction diagrams reproduced in Figure 2.9a.

The crystal-to-crystal phase change is particularly evident in the first step of the transformations shown in Figure 2.8—the change from cellulose I to Na-cellulose I. As shown in the series of X-ray diagrams in Figure 2.9b, the two crystal structures are simultaneously present during the process. Furthermore, if the alkali is quickly washed out from the fibers, a mixture of cellulose I and cellulose II crystal structures is obtained. Clearly, cellulose I in this mixture represents the unconverted part of the original cellulose, and cellulose II has resulted from the portion that had been converted to Na-cellulose I. This observation suggests that Na-cellulose I is already an antiparallel-chain crystal structure.

Another interesting observation concerns the structures of Na-cellulose IIA and IIB. Both form only after all vestiges of cellulose I have disappeared and there is an excess of NaOH present (28, 29). Both are based on threefold helical chain conformations with a c-axis repeat of ~15 Å rather than the familiar, twofold, ~10 Å repeat conformation present in cellulose polymorphs. Preliminary results of the crystal structure determination of Na-cellulose IIB indicate that it contains a large amount of NaOH and water in its unit cell (30). The solvent molecules separate the chains and break all nonbonded contacts between them. The structure is stable, which suggests that the threefold chain conformation may be the most stable alkali complex intermediate occurring during these transformations. In contrast, the Na-cellulose I structure is not very stable, which in turn suggests that its formation may be directed by the presence of as yet unconverted cellulose.

A more detailed study of the cellulose I to Na-cellulose I transformation, in which the crystallite sizes of both structures were monitored as a function of

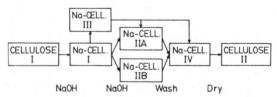

Figure 2.8 Transformation pathways of cellulose I to cellulose II through various crystalline Na-celluloses, as observed during controlled mercerization (28).

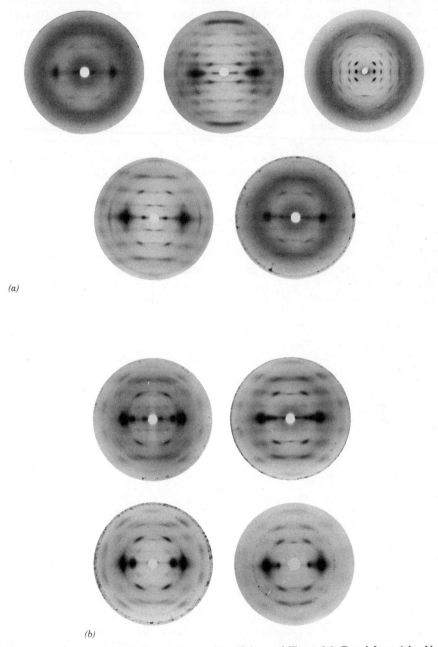

(a)

(b)

Figure 2.9 (*a*) X-ray diffraction patterns of Na-celluloses of Figure 2.8. Top, left to right: Na-celluloses I, IIA, IIb; bottom, left to right: Na-celluloses III, IV (28). (Fiber axis is vertical.) (*b*) Mixed crystal structures observed during the conversions. Top, left to right: cellulose I + Na-cellulose I, cellulose I + cellulose II resulting from the mixture on the left; bottom, left to right: celluloses I and II, shown for comparison (29). (Courtesy of John Wiley & Sons, publisher)

the degree of conversion, has shown that Na-cellulose I begins to form long before the crystallites of cellulose I begin to diminish in size (see Fig. 2.10) (26). This type of transformation is consistent with a mechanism in which the crystal structure of Na-cellulose I develops first in the amorphous or paracrystalline regions of cellulose I. Once present in a more substantial amount, Na-cellulose I crystallites begin to increase in size at the expense of decreasing cellulose I crystallites. This transformation proceeds entirely in the solid state with two crystalline phases coexisting simultaneously (as shown in Fig. 2.9b) and with the outward appearance and dimensions of the fiber not changing significantly. Consequently, mercerization can be pictured to proceed as shown in Figure 2.11. The alkali enters the amorphous regions between the crystallites of cellulose I, converting them to Na-cellulose I. Because a fiber is most likely a statistical arrangement of crystallites (i.e., a random aggregate of parallel-chain microfibrils in approximately equal numbers pointing in both directions along the fiber), the amorphous, interfacial region would contain roughly equal numbers of chains pointing in both "up" and "down" directions. These chains would give rise to an antiparallel Na-cellulose I crystal structure. The Na-cellulose I structure coexists with that of cellulose I until the latter is completely converted, after which Na-cellulose I proceeds to be converted through other structures, principally Na-celluloses IIA and IIB, to the end product, cellulose II. In all of

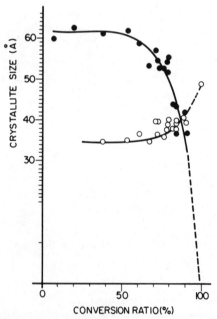

Figure 2.10   Crystallite sizes of ramie cellulose I (solid circles) and Na-cellulose I (open circles) as a function of the conversion ratio observed during the first transformation step in Figure 2.8. (Courtesy of John Wiley & Sons, publisher)

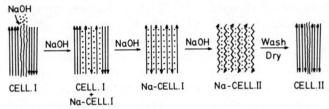

Figure 2.11 Proposed model of structural changes occurring during mercerization (29). (Courtesy of John Wiley & Sons, publisher)

these transformations, the chain motions need not be extensive to effect change from one crystal structure to another. For example, there is no need to fold the chains or otherwise require large-scale movements; the motions can be lateral only. In this the transformations are consistent with the lack of apparent change in the macroscopic dimensions of the fiber.

A model of the morphology of the fiber that is consistent with this mechanism is illustrated in Figure 2.12. The fiber is composed of a large number of microfibrils, substantially single-crystalline throughout and parallel-chain in structure. The aggregation of these microfibrils is random and, therefore, statistically antiparallel. The amorphous regions are represented mostly by the interfacial, surface regions of the microfibrils, containing chains in both orientations. From the crystallite size measurements shown in Figure 2.10, the average diameter of

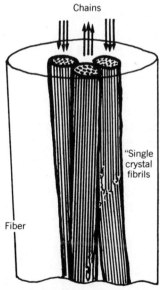

Figure 2.12 Model of a cellulose fiber consistent with structural changes observed during mercerization (29). (Courtesy of John Wiley & Sons, publisher).

the microfibril can be assumed to be 60–70 Å, and the average lateral extent of the intercrystalline region 30–35 Å. Furthermore, since such an amorphous region is mostly composed of chains that lie on the surfaces of the crystallites, the orientation of the amorphous chains must also be nearly as good as that of the crystallites. In other words, the amorphous regions in a cellulose fiber such as ramie are likely to be much more paracrystalline in nature than truly dis-ordered-amorphous. This is supported by the high degree of order evident in the x-ray pattern of Na-cellulose I (see Fig. 2.9a).

In a model such as this, the changes from one crystal structure to another can proceed without major disruptions of the fiber structure and without loss of chain orientation. The change from parallel- to antiparallel-chain structure can easily take place through small lateral shifts of chain segments, loosened from one another by the action of the alkali. In these respects, this model is in good agreement with all of the experimental observations of mercerization of fibers such as ramie. More complex cellulose morphologies, such as that of wood fiber, may be equally well represented by this model, albeit only at the micro-fibrillar level.

SUMMARY

X-ray crystallographic studies of celluloses have been decidedly helpful in in-creasing our understanding of the structure of this fascinating substance. Even if not every study has yielded unequivocal results, all of them taken together are beginning to produce a recognizable solution to the puzzle. An example is the apparent division of celluloses into parallel- and antiparallel-chain structures, a concept that has been controversial for years. The accumulated evidence from diffraction studies is now clearly in favor of this division. In turn, the structural features associated with it are becoming helpful in explaining other experimental observations—for example, those connected with fiber morphology, biosyn-thesis, and mercerization.

Questions still remain regarding these and other structural and mechanistic features. As has been the case in all structural research on polymers, no single technique has been able to provide clear-cut answers to every question. It can only be hoped that as different techniques—among them diffraction—are im-proved and perfected and new techniques are invented, our understanding of the cellulose structure will continue to increase. If, at this point in time, one were to speculate where to look for the next advances, good candidates would be the methods of electron diffraction, solid and solution state NMR, and, perhaps, neutron scattering.

REFERENCES

1. S. Arnott and W. E. Scott, *J. Chem. Soc.* (*Perkin Trans.*), **2**, 324 (1972).

2. G. A. Jeffrey and M. Sundaralingam, *Adv. Carbohydr. Chem. Biochem.*, **30**, 445 (1974); **31**, 347 (1975); **32**, 353 (1976); **34**, 345 (1977); **37**, 373 (1980); **38**, 417 (1981).

3. D. A. Brant, *Annu. Rev. Biophys. Bioeng.,* **1**, 369 (1972).

4. J. Blackwell, A. Sarko, and R. H. Marchessault, *J. Mol. Biol.,* **42** 379 (1969).

5. S. S. C. Chu and G. A. Jeffrey, *Acta Crystallogr.,* **B24**, 830 (1968).

6. J. T. Ham and D. G. Williams, *Acta Crystallogr.,* **B26**, 1373 (1970).

7. A. Sarko and R. H. Marchessault, *J. Am. Chem. Soc.,* **89**, 6454 (1970).

8. P. Zugenmaier and A. Sarko, In *Fiber Diffraction Methods*, A. D. French and K. H. Gardner, eds., ACS Symp. Ser., 141, American Chemical Society, Washington, D.C., 1980, p. 225.

9. A. Sarko, "FIBRXRAY," a FORTRAN program for the resolution of X-ray diffraction intensities.

10. D. P. Miller and R. C. Brannon, In *Fiber Diffraction Methods*, A. D. French and K. H. Gardner, eds., ACS Symp. Ser., 141, American Chemical Society, Washington, D.C. 1980, p. 93.

11. A. Sarko, *Tappi,* **61**, 59 (1978).

12. A. Sarko and R. Muggli, *Macromolecules,* **7**, 486 (1974).

13. A. J. Stipanovic and A. Sarko, *Macromolecules,* **9**, 851 (1976).

14. A. Sarko, J. Southwick, and J. Hayashi, *Macromolecules,* **9**, 857 (1976).

15. C. Woodcock and A. Sarko, *Macromolecules,* **13**, 1183 (1980).

16. E. S. Gardiner and A. Sarko, In *Proceedings of the Ninth Cellulone Conference*, A. Sarko, Ed., *J. Appl. Polym. Sci.: Appl. Polym. Symp.,* **37**, Wiley, New York, 1983, Vol. 1, p. 303.

17. E.S. Gardiner and A. Sarko, *Can. J. Chem.,* **63**, 173 (1985).

18. K. H. Gardner and J. Blackwell, *Biopolymers,* **13**, 1975 (1974).

19. F. J. Kolpak and J. Blackwell, *Macromolecules,* **9**, 273 (1976).

20. R. H. Marchessault and A. Sarko, *Adv. Carbohydr. Chem.,* **22**, 421 (1976).

21. A. J. Stipanovic and A. Sarko, *Polymer,* **19**, 3 (1978).

22. E. Roche, H. Chanzy, M. Boudeulle, R. H. Marchessault, and P. Sundararajan, *Macromolecules,* **11**, 86 (1978).

23. A. Sarko, *Appl. Polym. Symp.,* **28**, 729 (1976).

24. C. H. Haigler, R. M. Brown, Jr., and M. Benziman, *Science,* **210**, 903 (1980).

25. I. Nieduszynski and R. D. Preston, *Nature,* **225**, 273 (1970).

26. H. Nishimura and A. Sarko (to be published).

27. E. Roche and H. Chanzy, *J. Biol. Macromol.,* **3**, 201 (1981).

28. T. Okano and A. Sarko, *J. Appl. Polym. Sci.,* **29**, 4175 (1984).

29. T. Okano and A. Sarko, *J. Appl. Polym. Sci.,* **30**, 325 (1985).

30. T. Okano and A. Sarko (to be published).

3

Structure of Cellulose–Solvent Complexes

JOHN BLACKWELL, DAVID M. LEE, DAVID KURZ, and MAO-YAO SU

Department of Macromolecular Science, Case Western Reserve University, Cleveland, Ohio

There is renewed interest at the present time in organic solvents for cellulose, following the development of several new systems, including dimethylsulfoxide–paraformaldehyde (1), N-methyl-morpholine N-oxide (2), hydrazine at elevated temperature and pressure (3), and lithium chloride-dimethylacetamide (4). Dissolution of cellulose requires disruption of both the intramolecular hydrogen bonding network and the hydrophobic van der Waals forces between the "surfaces" of the ribbonlike cellulose chains. An ideal method to study such cellulose-solvent interactions is to determine the structure of suitable cellulose intercalation complexes by x-ray diffraction. This paper describes our analyses of the ordered cellulose complexes formed by intercalation of ethylene diamine, hydrazine, and water.

The study of the structure of cellulose complexed with small molecules dates from the 1930s (see, e.g., refs. 5–10). Further analyses of these structures have become possible with the determination of the structures of cellulose itself. Detailed molecular models have been refined for both cellulose I (11, 12) and cellulose II (13, 14), and it is useful to review these structures before examination of the complexes.

Figure 3.1 shows the refined structure of cellulose I (11) based on the intensity data for the highly crystalline fibers drawn from the cell walls of the sea alga *Valonia ventricosa*. This material has a monoclinic unit cell with the following dimensions: $a = 16.34$ Å; $b = 15.72$ Å; $c = 10.38$ Å; and $\gamma = 97.0°$, containing disaccharide repeats of eight chains. The reflections with odd h and/or odd k

51

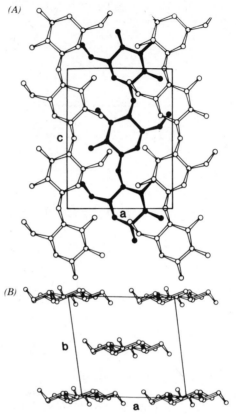

Figure 3.1 Structure of cellulose I: (*A*) *ac* projection; (*B*) *ab* projection. (From ref. 12.)

indices are very weak (most of them are in fact absent), and thus a two-chain unit cell with the *a* and *b* axes halved can serve as an adequate approximation to the full structure. The most important question to be resolved was the polarity of these two chains: Do the two have the same sense (parallel), or do they have alternating sense (antiparallel)? The structure was determined by setting up models with both parallel and antiparallel chain arrangements and refining these using the linked-atom least-squares (LALS) routines (11) to obtain the best agreement between the observed and calculated intensity data. This led to the conclusion that the best model for cellulose I consists of an array of parallel chains, as shown in Figure 3.1. The best antiparallel model could be rejected at a statistical significance of 0.005% (i.e., the parallel model is preferred by 200 to 1). The two residues of the cellobiose repeat are related by a twofold (2_1) screw axis, as indicated by the absence of odd-order 001 reflections, and adjacent chains are staggered by 0.265*c*, consistent with the approximate quarter stagger first proposed by Meyer and Mark (15). Each chain forms two intramolecular hydrogen bonds: 03′–H . . . 05 as proposed by Hermans et al. (16) from a study of space-filling models, and 02–H . . . 06′, as suggested to account for the polarized infrared spectra (17, 18). There is also an intermolecular hydrogen bond, 06–H . . . 03, linking adjacent chains in sheets along the *b* axis. There is no hydrogen bonding along the a axis or along the diagonals of the *ab* face, and the structure must be stabilized in these directions by hydrophobic forces.

The structure refined for cellulose II (13) is shown in Figure 3.2 and was derived from the intensity data for Fortisan rayon. The unit cell is monoclinic

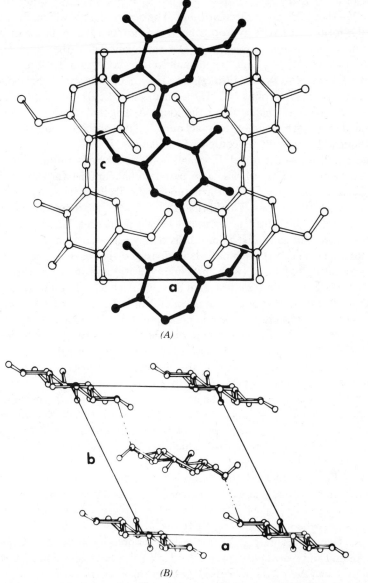

(A)

(B)

Figure 3.2 Structure of cellulose II: (*A*) *ac* projection; (*B*) *ab* projection. (From ref. 14.)

with dimensions $a = 8.01$ Å, $b = 9.04$ Å, $c = 10.36$ Å, and $\gamma = 117.1°$; the space group is approximately $P2_1$, and the cell contains disaccharide repeats of two chains. The best agreement between the observed and calculated intensities was obtained for an antiparallel chain model, and the best contending parallel models were also found to be stereochemically unacceptable. In the axial (ab) projection shown in Figure 3.2B, the corner chains have a conformation similar to that in cellulose I: Each chain forms 03'–H . . . 05 and 02–H . . . 06' intramolecular hydrogen bonds and is linked to its neighbor along the b axis by an 06–H . . . 03 intermolecular bond. The center chains, which have opposite polarity, form a different hydrogen bond pattern as a consequence of having a different conformation for the $-CH_2OH$ side chain (gt rather than tg). These chains have the 03'–H . . . 05 intramolecular bond and are linked to their neighbors along the b axis through 06–H . . . 02 intermolecular bonds. The remaining hydroxyl groups (02–H) form 02–H . . . 02 intermolecular bonds between antiparallel chains along the long ab diagonal. This additional intermolecular hydrogen bonding may account for the observed higher stability of the cellulose II structure as compared to cellulose I.

The parallel I–antiparallel II conclusion has obvious implications for the study of cellulose biosynthesis. The parallel chain model for native cellulose rules out a regularly folded chain structure for the fibrils, which would necessarily lead to antiparallel chains. The fibrils can be viewed as extended-chain polymer single crystals, which lead to optimum tensile properties. The current view of cellulose biosynthesis is that microfibrils are synthesized from rows of enzyme complexes on the cell surface (19). It is envisaged that each of these complexes produces a cellulose chain (with the same polarity) and these are then bundled together as a growing microfibril; that is, there is almost simultaneous synthesis and crystallization. Hence the native microfibrils have the metastable parallel chain cellulose I structure: There is no opportunity for rearrangement to antiparallel cellulose II. Indeed a major difference must exist between cellulose I and II, because otherwise it is difficult to explain why nature does not synthesize the latter. The best explanation is that cellulose I is synthesized because the chains are constrained to be parallel by the biosynthesis mechanism. At this point we note that evidence for chain sense in native cellulose microfibrils has recently been reported by Hieta et al. (20). These workers have shown that silver nitrate labeling of the reducing ends of the cellulose chains results in staining of only one end of each microfibril, which requires a chemical sense to the fibrils (i.e., parallel chains).

A remaining problem is the conversion of cellulose I to cellulose II in the alkali swelling process known as mercerization. There is no difficulty in visualizing a reversal in chain polarity as a result of regeneration, but this does not seem so easy when the geometry of the solid state is partially retained during swelling. Two routes are possible for reversal of polarity during swelling: regular chain folding and rearrangement of extended chains. Extensive folding is regarded as unlikely in view of the increased mechanical strength of mercerized cotton, and electron microscopy shows that mercerized microfibrils have a smooth morphology with no sign of a regular folded chain structure. Hence repacking of extended chains is the likely mechanism for the polymorphic transition. Electron microscopy by Willison and Brown (21) on the cell walls of *Glaucocystis* shows a winding of the microfibrils around the cell, which would lead to a 50/50 mixture of "up" and "down" microfibrils in the bulk cotton fiber. Hence a cotton fiber contains equal numbers of up and down chains: They are simply packaged in small bundles with the same sense. On swelling in alkali of sufficient concentration, these chains are separated and lose track of their previous neighbors. Then on removal of the swelling agent, a particular chain is likely to unite with antiparallel neighbors because of the higher thermodynamic stability of cellulose II. Such a rearrangement mechanism is consistent with the relative difficulty in achieving the conversion to cellulose II (see also Chap. 2). If forms I and II have the same chain polarity, then it is surprising that swelling of cellulose I in ammonia, ethylenediamine, hydrazine, and so on is insufficient to achieve mercerization.

In the work described below we show how the cellulose I and II structures are rearranged when they form complexes with swelling agents. The structures for the complexes were derived by application of the same LALS refinement methods, and it will be seen that in each case the results confirm the parallel I–antiparallel II view of cellulose structure.

COMPLEXES FORMED BY RAMIE AND RAYON FIBERS

Mercerized specimens of delignified native ramie fibers were prepared by repeated soakings in 22% aqueous sodium hydroxide solution. They were separated by thorough washing in water until no more than a trace of residual cellulose I was detected in the x-ray pattern. Fortisan rayon fibers were also used in our experiments.

Complex Formation

Ethylenediamine complexes were prepared by coating the taut ramie and rayon fibers in ethylenediamine for a minimum of 2 hr followed by vacuum drying over P_2O_5 for a minimum of 1 hour. The resultant complexes were stable in air at room humidity. Hydrazine complexes were prepared in the same way but required 36 hr soaking in hydrazine followed by 12 hr vacuum drying. The hydrazine complexes soon reverted to the uncomplexed parent cellulose structure on standing in air at room humidity but were stable under vacuum conditions.

The Fortisan/water complex (cellulose II–hydrate) was prepared by soaking the hydrazine complex in water at room temperature for 1 hr. The complex was unstable but was sufficiently long-lived for an x-ray pattern to be recorded in a helium atmosphere at 93% relative humidity.

Hydrazine contents were estimated by addition of the specimen to 0.1 N HCl followed by potentiometric titration against 0.1 N NaOH solution. Ethylenediamine contents were determined from elemental analyses performed by Galbraith Laboratories, Knoxville, TN.

X-Ray Diffraction

Wide-angle x-ray diffraction patterns were recorded on Kodak No-Screen x-ray film using CuK radiation and a Searle torroidal focusing camera. The d spacings were calibrated with sodium fluoride. The intensities of the x-ray reflections were determined from an x-y grid of optical density obtained using a Photometrics optical densitometer. These data were then corrected for background and for the Lorentz and polarization effects. Intensities with multiple hkl indices were divided in the ratio of the calculated intensities (F^2), and unobserved reflections were assigned a value equal to half the threshold intensity. Atomic coordinates for the cellulose chains were based on those used by Kolpak and Blackwell (13) in their refinement of the structure of cellulose II, with minor adjustments to fit the slightly different fiber repeats. The N-N bond length in hydrazine was set at 1.44 Å as in the crystal structure determined by Collin and Lipscomb (22), and the bond lengths and angles in ethylenediamine were those used by Yokozuki and Kushita (23) for gas diffraction studies.

CELLULOSE–SOLVENT COMPLEXES

Unit Cells

The unit cells determined for hydrazine and ethylenediamine complexes prepared in this laboratory for both cellulose I and cellulose II are listed in Tables 3.1 and 3.2, respectively, where they are compared with those reported previously. Figure 3.3 shows the x-ray fiber diagrams of some of these complexes, selected because of their relevance to the structural work described below.

Examination of Tables 3.1 and 3.2 shows that a relatively large number of structures have been proposed in the past 50 years, especially in the case of the hydrazine complexes. Some of these may result from different indexing of the

TABLE 3.1   Unit Cell Dimensions for Cellulose–Ethylenediamine Complexes

	a (Å)	b (Å)	c (Å)	$(\gamma°)$	Reference
Ramie (I)	12.6	13.1	10.3	135.2	5
Cotton (I)	12.2	12.3	10.3	137.0	10
Ramie (I)	12.37	9.52	10.35	117.8	24
Fortisan (II)	12.81	18.27	10.34	118.0	24
Mercerized ramie (II)	13.42	8.41	10.34	110.1	24

TABLE 3.2   Unit Cells for Cellulose–Hydrazine Complexes

	a (Å)	b (Å)	c (Å)	$(\gamma°)$	Reference
Ramie (I)	10.9	10.9	10.3	127.7	5
Ramie (I)	9.1	8.1	10.38	114.7	7
Cotton (I)	9.68	9.96	10.3	125.2	10
Ramie (I)	9.19	16.39	10.37	97.4	24
Fortisan (II)A	9.37	19.88	10.39	120.0	27
Fortisan (II)B	8.84	23.76	10.38	114.7	28
Mercerized	9.48	15.39	10.37	96.4	24

same d-spacings, especially in cases where x-ray patterns of different quality were obtained. However, the complexes identified in this laboratory give x-ray patterns that are easily distinguishable, and it is clear that they correspond to different structures. Our ramie cellulose I–hydrazine complex has unit cell dimensions that are very different from those reported previously. This is probably explained by the fact that our complex was prepared from hydrazine containing no more than 3% water, as compared to 40% water for the earlier work. Unit cells for cellulose II–hydrazine complexes had not been proposed prior to our work, and so far we have found three such structures: two for Fortisan rayon (designated A and B), and a third for mercerized ramie. These three complexes have all been obtained by what we tried to make identical preparative procedures, and the fact that they are different probably represents some as yet unidentified variable. In our work so far, only complex A has been produced when Fortisan is treated with completely anhydrous hydrazine, and hence water may be an important component of complex B. The different complexes formed by Fortisan and mercerized rayon may reflect the different morphologies seen for these celluloses in the electron microscope (25). However, it seems likely that all three complexes can be formed by any cellulose II, and their existence simply points to the variety of solvent interactions that are possible once the cellulose structures are swollen. Similar conclusions can be drawn with respect to the ethylenediamine complexes, including the existence of different complexes for Fortisan and mercerized ramie. The ethylenediamine complexes are more stable than those of hydrazine, and they are also less susceptible to absorption of atmospheric water.

Ramie Cellulose I–Ethylenediamine Complex

Figure 3.3a shows the x-ray fiber diagram of ramie cellulose I–ethylenediamine complex. The unit cell contains disaccharide units of two chains that have the approximate 2_1 conformation typical of many cellulose structures. Intensities can be estimated for 39 nonmeridional reflections, and there are a further 36 reflections predicted within the 20 range of the observed data that are too weak to be detected. These data are comparable to those used for the refinement of

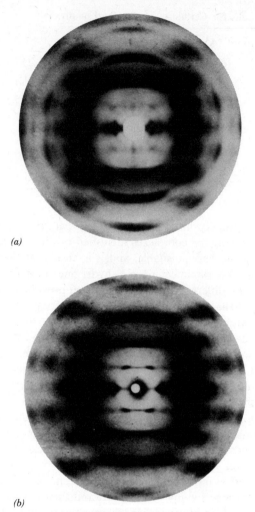

(a)

(b)

Figure 3.3 X-ray diagrams of (*a*) ramie cellulose I–ethylenediamine complex, (*b*) Fortisan cellulose II–hydrazine complex A, (*c*) Fortisan cellulose II–hydrazine complex B, (*d*) Fortisan cellulose II–hydrate.

the structures of uncomplexed cellulose I and II, except that here we also need to define the arrangement of the ethylenediamine molecules. Elemental analysis indicates the presence of one ethylenediamine per anhydroglucose unit. When one examines molecular models, it becomes self-evident that all three hydroxyl groups of the glucose residue must be hydrogen bonded to the complexing molecules, either as a donor or an acceptor. For convenience we attached the ethylenediamine to O3 via a N . . . O hydrogen bond of length 2.8 Å and refined its position, treating it as a flexible side chain. To start with, we considered three basic conformations for the ethylenediamine: *trans, gauche* $^+$, and *gauche* $^-$.

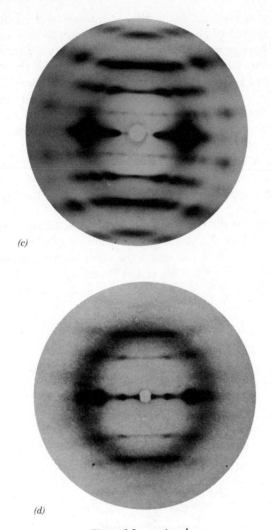

(c)

(d)

Figure 3.3 *continued.*

Two refinable angle parameters are then needed to define the position of the complexing molecule as a pendant side chain.

The first requirement was to determine the positions of the chain axes in the *ab* projection of the unit cell. Gross structural features such as this can be determined without knowledge of the ethylenediamine positions. Good agreement was obtained for stacking of the two chains along the *b* axis, and the other possibilities—stacking along the *a* axis, and a centered structure (i.e., chain axes at 0,0 and 0.5,0.5)—could easily be ruled out. The *b* dimension of 9.52 Å corresponds approximately to twice the thickness of a cellulose chain when in the 2_1 conformation, and hence the complex consists of sheets of chains with ethylenediamine molecules between them.

Our procedure was to set up parallel and antiparallel chain models for all possible combinations of the three ethylenediamine conformations. The antiparallel models have significantly worse agreement than the parallel models and could be ruled out. Given the parallel chain structure for the starting material and that cellulose I is regenerated on removal of the complexing agent, it would be unlikely in any case that complex formation involved a reversal of polarity. Models with different basic conformations for the ethylenediamine molecules could also be ruled out on the basis of both the x-ray agreement and stereochemistry. After all the above refinements, we were left with four models for the complex that were indistinguishable in terms of their agreement with the x-ray data and had no unacceptable bad contacts. However, only one of these could form a hydrogen bond network that included all the available hydroxyl and amine hydrogens.

The final model for the ramie cellulose I–ethylenediamine complex is shown in Figure 3.4 and had a crystallographic residual of $R'' = 0.189$. The two cellulose chains are stacked one above the other, with essentially zero stagger and identical conformations. This is confirmed by the x-ray data: The hkl reflections with odd k indices are mainly very weak or absent, which indicates

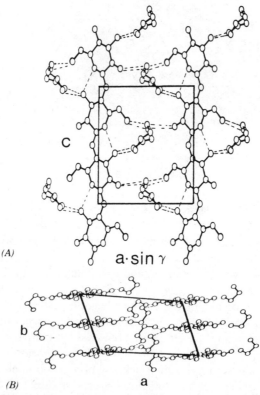

(A)

c

$a \cdot \sin \gamma$

b

(B) a

Figure 3.4 Structure of ramie cellulose I–ethylenediamine complex: (*A*) *ac* sin γ projection; (*B*) *ab* projection. (From ref. 26.)

that the differences between the two halves of the unit cell must be small. Our refinement shows that these differences are in the positions of the two ethylenediamine molecules. The ac projection of the structure shown in Figure 3.4A reveals the difference between these two molecules. Constraint of the ethylenediamines to have the same structure (differing only by translation) increases R'' to 0.233, and the improvement on release of this constraint is significant at the 0.005% level.

Comparison of the structure of this complex with that of native cellulose (Fig. 3.1) shows the extent of the rearrangement that has occurred as a result of the interaction with ethylenediamine. The chains have shifted from a quarter staggered arrangement and now form stacks in register—that is, zero stagger. Thus the interaction with ethylenediamine results in disruption not only of the hydrogen bonding network but also of the hydrophobic forces between the "surfaces" of the ribbonlike chains. The stacks of chains along the b axis are very similar to those that occur in the polymorphic forms of chitin. In the latter structures, chains with the same sense are stacked in register, approximately 4.7 Å apart, and have a similar inclination to the stacking plane. This has the effect of bringing the acetamido groups into register so that a chain of N=H . . . O=C hydrogen bonds are formed along the stacks. Figure 3.3b shows that the register stacking in the cellulose I–ethylenediamine complex brings the bulky CH_2OH side chains directly above one another, creating channels within the structure that are filled by the complexing molecules.

Fortisan Cellulose II–Hydrazine Complex A

The x-ray fiber diagram of Fortisan cellulose II–hydrazine complex A (Fig. 3.3b) is less detailed than that for the ramie cellulose I–ethylenediamine complex treated above. Twenty-six reflections are observed, and these are indexed by a monoclinic unit cell containing disaccharide residues of four independent chains. Essentially this means that approximately twice as much information must be derived from half the data, and it proved impossible to refine the four-chain unit cell by least-squares methods. However, the four-chain unit cell is required for indexing of only four of the observed reflections; the remaining 22 can all be indexed by a two-chain unit cell with the a dimension halved. Thus the differences between the two halves of the unit cell must be very small, and we can approximate the structure to the two-chain unit cell. This use of a subcell to refine the structure was used effectively in the determination of cellulose I, as described above.

Titration showed that the hydrazine content of the complex was very close to one molecule per glucose residue. As before, a study of molecular models showed that each of the three hydroxyl groups of the glucose residue will be hydrogen bonded, either as a donor or an acceptor, to the hydrazine molecules, and these were attached via an O2 . . . N bond, 2.8 Å in length. Parallel and antiparallel models were refined, and the antiparallel options were found preferable throughout, as is the case in uncomplexed cellulose II. Intensity calculations for simple packing models showed that the two-chain unit cell has its chain axes passing through 0,0 and 0,0.5 in the *ab* projection. Models were then set up with chains in the 2_1 helical conformation for all possible combinations of the three staggered positions for the CH_2OH groups (*gg, gt,* and *tg*), and the chain orientations, stagger, and hydrazine positions were refined in each case. The refined models were then compared in terms of their x-ray agreement and stereochemistry and the requirement that all of the $-OH$ and $-NH$ groups should form donor hydrogen bonds. On this basis, all but four of the models could be rejected. The model with the lowest residual ($R'' = 0.207$) is shown in Figure 3.5, but this is not significantly better than the other three models. The latter have essentially the same orientation and stagger of the chains but have different CH_2OH conformations, and this leads to small differences in the hydrazine positions. However, the important features of the structure described below are common to all four refined models. The difficulty in defining the positions of the hydrazines may derive from the differences between the two halves of the four-chain unit cell; disorder of the hydrazines themselves also seems quite likely, and there is the possibility that some water molecules have been incorporated.

As can be seen in Figure 3.5*A*, the antiparallel chains along the *b* axis are arranged with their glycosidic oxygens approximately in register. This has led

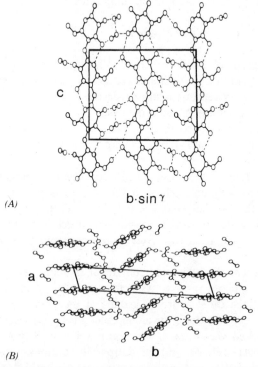

(A)   b·sin γ

(B)   b

Figure 3.5 Structure of Fortisan cellulose II–hydrazine complex A: (*A*) *bc* sin γ projection; (*B*) *ab* projection. (From ref. 27.)

to a situation where the CH_2OH groups point toward one another such that there are channels between the chains, then these channels are filled by the hydrazine molecules. The (parallel) chains are also in register along the a axis. This has been assumed in the approximation to a two-chain unit cell but must be close to being so, based on the weakness of the reflections with odd h indices for the four-chain cell. This means that there are stacks of chains along the a axis, and these also facilitate the formation of channels to be occupied by the complexing molecules. The refinement indicates that there are two types of stacks: The stack of chains passing through the origin in Figure 3.5B are very similar to those in the ramie cellulose I–ethylenediamine complex and also in chitin. In the other stack (the center chains in Fig. 3.5B), the chains are rotated by $\sim30°$. This type of stacking is still stereochemically acceptable, but this is a feature of the structure that needs re-examination if better-quality patterns are ever obtained.

Rearrangement of the cellulose II structure to form the hydrazine complex thus involves disruption of both the hydrogen bonds and the hydrophobic interactions between the chains, as was also the case for cellulose I when complexing with ethylenediamine. In cellulose II, the antiparallel chains are in contact along the short diagonal of the ab face of the unit cell. These chains become separated by the hydrazine molecules, and chains of the same sense rotate about their axes and unit to form the stacks described above. Preliminary information about the structure of complex B (28) points to a rather different structure. The a dimension of the unit cell is approximately the same as the distance across two chains along the short ab diagonal in cellulose II. It appears that complex B is formed by swelling cellulose II while leaving the sheets forming the 110 planes intact. In this respect the structure is similar to the Fortisan cellulose II–hydrate described below (except that complex B is much more crystalline). Work is still in progress on complex B, and the structure will be described at a later date.

Fortisan Cellulose II Hydrate

As has been mentioned above, the ethylenediamine and hydrazine complexes can be restored to the original uncomplexed structures by washing in water. In the case of the cellulose II complexes, an intermediate hydrate structure can be observed. This structure is unstable, and we have not analyzed for residual hydrazine, but the x-ray pattern is very similar to that obtained by Hermans and Weidinger (29) by exposure of regenerated cellulose to steam. They described this structure as a hydrate of cellulose, not be to confused with the old (erroneous) description of cellulose II as "cellulose hydrate." In this regard it is interesting that Hermans and Weidinger were unable to prepare a hydrated cellulose I by the same procedure, and so far we have not detected a hydrate intermediate on reconversion of the cellulose I complexes.

The x-ray pattern of Fortisan cellulose II–hydrate is shown in Figure 3.3c. The specimen used was prepared via the hydrazine complex, but essentially the same results are obtained via the ethylenediamine complex. The pattern is relatively diffuse: Some 12 Bragg reflections can be identified, but the layer lines are streaky and become continuous at d spacing less than 3.5 Å, indicating a relatively disordered structure. The reflections are indexed (30) by a monoclinic unit cell with dimensions: $a = 9.02$ Å, $b = 9.63$ Å, $c = 10.38$ Å, and $\gamma = 116°$; the space group is approximately P2$_1$, and the unit cell contains disaccharide residues of two chains. In view of the poor quality of the data, antiparallel chains were assumed for the structure. Refinement of an unhydrated model (i.e., simply the two rigid cellulose chains) led to two indistinguishable models with approximate chain staggers of $+c/4$ and $-c/4$, with residuals of $r'' \simeq 0.34$. These structures have projections as shown in Figure 3.6. The stacking of antiparallel chains is similar to that along the short ab diagonal in the cellulose II structure, at least for the model with $-c/4$ stagger (as in cellulose II).

There are many possible positions for the water molecules in the two models. Refinement of these positions was not possible with so little intensity data, and the water molecules were inserted by trial and error. In this manner we considered a large number of possible structures, but the only result was an increase in the value of R'' rather than the decrease hoped for. This experience, together with the diffuseness of the x-ray pattern, suggests that the hydrate is not a truly crystalline complex but rather that it is an intercalation structure, with the water molecules in a disordered array between the sheets of chains.

This type of disordered structure can be modeled by the use of "water-weighted" atomic scattering factors, which treats the atoms as though they were surrounded by an atmosphere of disordered water rather than a vacuum. This approach was first used for crystalline proteins and has the effect of reducing the intensities at higher d spacings relative to the remainder. When the water structure and 06 were modeled in this way, the x-ray agreement improved considerably, and R'' values of 0.089 and 0.098 were obtained for the models with $+c/4$ and $-c/4$ stagger, respectively. These values are lower than those typically reported in x-ray studies of polymer structures and simply reflect the low ratio of data to parameters. The two structures are not significantly different

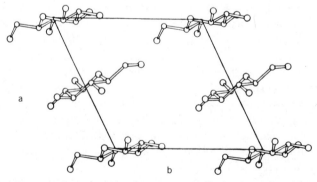

Figure 3.6 Structure of Fortisan cellulose II–hydrate, ab projection showing only one residue per chain. (From ref. 30.)

in terms of their agreement. The model with $+c/4$ stagger seems the more likely structure, since cellulose II has approximately the same stagger, and this is shown in projection in Figure 3.6. The hydrate is obtained from cellulose II by swelling between the 101 planes, but this swelling is irregular, and there is probably a different water structure between successive layers.

SUMMARY

The three cellulose complexes described above give some idea of the effect of solvent molecules on the interactions of cellulose chains. The most crystalline complex is that formed by ramie cellulose I and ethylenediamine. Here the intermolecular hydrogen bonding network is broken down, and the chains shift from a quarter-staggered array to in-register packing. As a result, the ethylenediamine molecules occupy channels between the cellulose chains. This type of packing also occurs in chitin, and it would seem to be a likely structure for many cellulose derivatives. The Fortisan cellulose II–hydrazine complex is less crystalline, and the position of the complexing molecules can be defined with much less certainty. However, it has been shown that the chains have rearranged from a quarter-staggered structure to one where they are packed in register, and they now form stacks analogous to those in the cellulose I–ethylenediamine complex. The results of these refinements are fully consistent with the structures of the parent cellulose polymorphs and add weight to the parallel I–antiparallel II view of cellulose structure. The cellulose II hydrate is the least ordered of the complexes and is seen as a cellulose II intercalation structure, with disordered water molecules between sheets of quarter-staggered antiparallel chains. We are currently investigating the possibility of preparing other complexes via the route used to produce the hydrate — that is, soaking the hydrazine and ethylenediamine complexes in reagents that are not themselves swelling agents but are components of the new solvent systems, such as dimethylsulfoxide and dimethylacetamide.

REFERENCES

1. D. L. Johnson, *U.S. Pat.* 2,371,359 (1970).
2. D. L. Johnson, *Br. Pat.* 1,44,048 (1969).
3. M. H. Litt and G. Kumar, *U.S. Pat.* 4,028, 132 (1977).
4. A. F. Turbak, A. El-Kafrawy, F. W. Snyder, and A. B. Averback, *U. S. Pat.* 4,302,252 (1981).
5. E. Halle, *Kolloid-Zt.*, **69**, 324 (1934).
6. C. Trogus and K. Hess, *Z. Phys. Chem. (Leipzig)*, **B14**, 387 (1931).
7. I. Sakurada and K. Hutino, *Kolloid-Zt.*, **77**, 346 (1936).
8. I. Sakurada and S. Okamura, *Kolloid-Zt.*, **81**, 199 (1937).
9. G. Centola, *Gazz. Chem. Ital.*, **68**, 825 (1938).
10. J. J. Creely, L. Segal, and L. Loeb, *J. Polym. Sci.*, **36**, 205 (1959).

11. K. H. Gardner and J. Blackwell, *Biopolymers,* **13**, 1975 (1974).

12. A. Sarko and R. Muggli, *Macromolecules,* **7**, 486 (1974).

13. F. J. Kolpak and J. Blackwell, *Macromolecules,* **9**, 273 (1976).

14. A. J. Stipanovic and A. Sarko, *Macromolecules,* **9**, 851 (1976).

15. K. H. Meyer and H. Mark, *Ber. Dsch. Chem. Ges.,* **61B**, 593 (1928).

16. P. H. Hermans, J. de Booys, and C. Maan, *Kolloid-Zt.,* **102**, 169 (1943).

17. H. J. Marrinan and J. Mann, *J. Polym. Sci.,* **21**, 301 (1956).

18. D. W. Jones, *J. Polym. Sci.,* **32**, 371 (1958).

19. D. Delmer, *Adv. Carbohydrate Chem. Biochem.,* **41**, 105 (1983).

20. K. Hieta, S. Kuga, and M. Usuda, *Biopolymers,* **23**, 1807 (1984).

21. J. H. M. Willison and R. M. Brown, *J. Cell Biol.,* **77**, 103 (1978).

22. R. N. Collin and W. N. Lipscomb, *Acta Cryst.,* **4**, 10 (1951).

23. A. Yokozuki and K. Kushita, *Bill. Chem. Soc. Jpn.,* **44**, 2926 (1971).

24. D. M. Lee and J. Blackwell, *J. Polym. Sci. Polym. Phys. Eds.,* **19**, 459 (1981).

25. F. J. Kolpak and J. Blackwell, *Text. Res. J.,* **45**, 558 (1975).

26. D. M. Lee, J. Blackwell, and M. H. Litt, *Biopolymers,* **22**, 1383 (1983).

27. D. M. Lee, K. E. Burnfield, and J. Blackwell, *Biopolymers,* **23**, 111 (1983).

28. D. Kurz, M.-Y. Su, and J. Blackwell. (In preparation).

29. P. H. Hermans and A. Weidinger, *J. Colloid Sci.,* **1**, 185 (1946).

30. D. M. Lee and J. Blackwell, *Biopolymers,* **20**, 2165 (1981).

4

Pea Cellulose and Xyloglucan: Biosynthesis and Biodegradation

TAKAHISA HAYASHI[‡] and GORDON MACLACHLAN
Biology Department, McGill University, Montreal, Quebec, Canada

Primary cell walls of young plants contain levels of xyloglucan that almost equal those of cellulose. Xyloglucans all possess a 1,4-β-glucan backbone with 1,6-α-xylosyl residues along the backbone. Strong acid or alkali (e.g., 85% phosphoric acid, 24% KOH) is required to dissolve xyloglucan from cell wall preparations (1–3), where the polysaccharide binds strongly to cellulose microfibrils. It appears to function in young plant cell walls as a cementing matrix material which contributes cross-links and rigidity to the cellulose framework. The integrity of xyloglucan could control the ability of microfibrils to separate and the whole cell to expand during growth.

Plant membranes may contain at least two UDP-glucose 4-β-glucosyltransferases, one of which, together with UDP-xylose and xylosyltransferase, is involved in xyloglucan synthesis, and one that forms cellulose. Although the latter should be active *in vivo* at the plasma membrane (4, 5) and the former in Golgi dictyosomes (6), sucrose gradient analyses of plant membranes have shown that both glucosyltranferase activities can be identified in association with Golgi membranes (7, 8). Recent studies (8, 9) indicate that soybean xyloglucan synthase proceeds by concurrent transfer of glucose and xylose to a nascent xyloglucan acceptor and does not operate by xylosyl transfer to a preformed 1,4-β-glucan. The present study examines properties of xyloglucan synthase in membrane from growing regions of etiolated peas and structure of xyloglucan from that tissue.

Several physiological studies (10–13) with peas have provided evidence for

[†]Current address: Ajinoinoto Central Research Laboratories, 1–1 Suzuki-cho, Kawasaki-ku, Kawasaki 210, Japan

xyloglucan turnover during growth after treatment with auxin. The treatment also induces the development of endo-1,4-β-glucanase activities (EC 3.2.1.4) in peas (14). This raises the question of whether auxin-induced endo-1,4-β-glucanases are responsible for the observed degradation and solubilization of pea xyloglucan in the growing cell wall. Furthermore, after supraoptimal auxin treatment the direction of growth changes from elongation to lateral expansion, which is similar to the phenomenon observed after ethylene treatment (15). Burg and Burg (16) proposed that the auxin effect is due to auxin-induced ethylene production which might be accompanied by changes in orientation of microtubules and cellulose microfibrils (17, 18). Thus the metabolism (biosynthesis and bio-degradation) of cellulose and xyloglucan was studied together with the influence of auxin and ethylene treatments under conditions where both plant regulators promote swelling of growing regions.

BIOSYNTHESIS OF 1,4-β-GLUCANS BY PEA MEMBRANES

When UDP-[^{14}C]glucose is supplied to pea membranes, β-glucans are formed at a rate that is proportional to the amount of membrane protein, incubation time, and substrate concentration. The activities of β-glucan synthases produce at least two kinds of linkages (19): One with low K_m forms 1,4-β-linkages mostly recovered in the alkali-insoluble fraction, and the other with high K_m produces 1,3-β-linkages. The main question addressed here is whether the 1,4-β-glucosyltransferase activity is responsible for the synthesis of pea cellulose and also the 1,4-linked glucan backbone of xyloglucan as the major constituents of the primary cell wall.

The requirement for Mg^{2+} for Golgi-located 4-β-glucosyltransferase is well known in higher plants (20), but there is evidence that, in the soybean system xyloglucan synthase activity requires Mn^{2+} and not Mg^{2+} (9). Pea membranes were incubated with these ions and UDP-[^{14}C]glucose to examine effects on glucosyltransferases. The results are shown in Table 4.1. The activity of xyloglucan glucosyltransferase was stimulated (12-fold) by Mn^{2+}: but Mg^{2+} was less effective in this system. In contrast, the biosynthesis of 1,4-β-glucan (alkali-insoluble) was stimulated mainly (10-fold) by Mg^{2+} and much less by Mn^{2+}. This suggests that there are indeed two separate synthases for 1,4-β-glucan and xyloglucan in peas, as in soybean (9). Biosynthesis of both products occurs at maximum rate when the two ions are provided together, but one does not enhance the effect of the other. This implies that xylosyltransfer does not occur to preformed 1,4-β-glucan during xyloglucan synthesis, but rather that concurrent transfer of both glucose and xylose are obligatory.

Enzymatically synthesized xyloglucans from UDP-[^{14}C]glucose or UDP-[^{14}C]xylose were exhaustively hydrolyzed with crude cellulase preparation from *Streptomyces griceus*, and the hydrolyzates were subjected to Bio-Gel P-2 gel filtration. A hepta- and a pentasaccharide were the only products, and both

TABLE 4.1 Effect of Mg²⁺ or Mn²⁺ on Glucosyltransferases[a]

Metallic Ions (5mM)	Total Glucan	Incorporation into	
		Alkali-Insoluble Glucan	Xyloglucan
	pmol glucose/10 min·mg protein		
None	26	3.6	1.2
Mg²⁺	183	40	2.1
Mn²⁺	120	10	15.4
Mg²⁺ plus Mn²⁺	210	41	15.0

[a] Pea membranes were obtained from the growing regions of pea epicotyls and incubated with 10 μM UDP-[¹⁴C]glucose (total glucan and alkali-insoluble glucan) or 10 μM UDP-[¹⁴C]glucose plus 10 μM UDP-xylose (xyloglucan) in 20 mM HEPES/KOH, pH 7.0, containing metallic ions at 5 mM. After 10 min at 25°C, the mixture was washed with 70% ethanol and chloroform/methanol (2:1), and ¹⁴C was estimated in the residue as the total glucan. Alkali-insoluble glucan was obtained from the residue washed with 24% KOH, and xyloglucan was estimated as the ¹⁴C of recovered isoprimeverose after hydrolysis of the residue with *Aspergillus oryzae* enzymes (21).

were labeled with [¹⁴C]xylose and [¹⁴C]glucose. Further hydrolysis with a crude glycosidase preparation from *Aspergillus oryzae* yielded all of the xylosyl residues of the [¹⁴C]xylose-labeled oligosaccharide as isoprimeverose (6-O-α-D-xylopyranosyl-D-glucopyranose), whereas the [¹⁴C]glucose-labeled oligosaccharide gave isoprimeverose and glucose in proportions of 3:1 and 2:1, respectively, in the two oligosaccharides. Therefore, the structures of the hepta- and pentasaccharides (Fig. 4.1) appears to be the same as those of soybean fragment oligosaccharides (8).

BIODEGRADATION OF 1,4-β-GLUCANS

Two auxin-induced endo-1,4-β-glucanases (buffer-soluble and -insoluble) were purified from pea epicotyls and used to degrade naturally occurring xyloglucans.

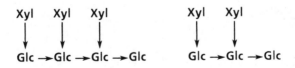

Heptasaccharide   Pentasaccharide

Figure 4.1 Proposed structure of the oligosaccharides obtained from endoglucanase digest of *in vitro* xyloglucan.

Both enzymes reduced the viscosity of solutions of amyloid xyloglucan (*Tamarindus indica*) as well as the staining of pea xyloglucan–iodine complex, with a linear and relatively minor concurrent increase in the number of reducing groups generated, indicating endohydrolysis (22). The digests of pea xyloglucan were fractionated on Bio-Gel P-2 columns, and a nonasaccharide (glucose/xylose/galactose/fucose, 4:3:1:1) and a heptasaccharide (glucose/xylose, 4:3) were obtained from each digest.

The K_m values of the two pea endo-1,4-β-glucanases acting on pea xyloglucan, amyloid xyloglucan, cellohexaose, and CM-cellulose were all remarkably similar (22). However, the V_{max} values with the two xyloglucans were lower for the buffer-insoluble glucanase (10 times) and for the buffer-soluble glucanase (30–50 times). The glucanases possess a binding site that recognizes at least six consecutive 1,4-β-linked glucose units (23), and random limited substitution at C-6 (as in carboxymethylcellulose) does not interfere with binding or the capacity to hydrolyze the substrate (V_{max}). However, the substitution pattern of pea xyloglucan is such that the 1,4-linked glucose backbone is only hydrolyzable by the endo-1,4-β-glucanases at every fourth glucose unit (Fig. 4.2). Such structural constraints probably account for a V_{max} that is lower than that for cellohexaose.

The potential of pea endo-1,4-β-glucanases to degrade the xyloglucan:cellulose complex found in 4% KOH alkali-insoluble material (cell wall ghosts) was measured by incubation with the glucanase for up to 48 hr (Table 4.2). The release of buffer-soluble saccharide was accomplished by a decrease of cellulose-bound xyloglucan content. The absorbance maximum of the cell wall ghosts was also changed from 640 to 550 nm (Fig. 4.3). There was also a marked increase in number of free reducing end groups (determined by reduction with NaBT$_4$) in the xyloglucan molecule that remained bound to cellulose, and a small but detectable increase in the reducing ends of the cellulose. Very little of the cellulosic framework was degraded in these tests, indicating that the xyloglucan component was much more accessible and susceptible to enzymic hydrolysis.

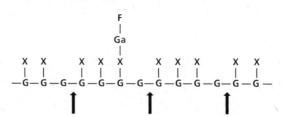

Figure 4.2 Site of hydrolysis in pea xyloglucan molecule by pea endo-1,4-β-glucanases. The site is shown as arrows in the molecule. G, glucose; X, xylose; Ga, galactose; F, fucose. Pea xyloglucan is composed of equal amounts of two subunits, a nonasaccharide (glucose/xylose/galactose/fucose,4:3:1:1) and a heptasaccharide (glucose/xylose,4:3), which appear to be distributed at random, but primarily in alternating sequence (22).

TABLE 4.2   Degradation of Xyloglucan–Cellulose Complex by Pea Endo-1,4-β-Glucanase[a]

Component Assayed	Zero Time	48 hr
Total weight (mg)	6.8	5.2
Ratio of xyloglucan to cellulose	0.7	0.44
Average molecular weight		
Xyloglucan	330 K	1.4 K
Cellulose	930 K	190 K

[a] Ghosts (6.8 mg) were incubated with 10,000 units of purified pea endo-1,4-β-glucanase. The treated ghosts were washed and extracted with 2 mL of 24% KOH containing 2.5 mCi of NaBT$_4$. Average molecular weight of the polysaccharides was calculated as reducing equivalents (tritiated accessible reducing end groups).

HORMONAL CONTROL OF THE METABOLISM OF CELLULOSE AND XYLOGLUCAN

Several enzymic activities involved in the biosynthesis and biodegradation of cellulose and xyloglucan were determined after treatment with plant hormones (Table 4.3). The activities of buffer-soluble and -insoluble endo-1,4-β-glucanases increased markedly only in auxin-treated tissue. Alkali-insoluble glucan synthase activity increased in the tissue only after auxin treatment but decreased after

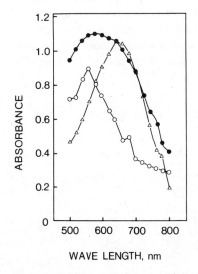

Figure 4.3   Absorption spectra of cell wall ghosts in iodine. The ghosts were stained by the iodine–sodium sulfate method (24). ●—●, Xyloglucan:cellulose complex; △—△, xyloglucan alone in solution; ○—○, *Streptomyces* endoglucanase-treated complex to remove xyloglucan.

TABLE 4.3   Activity Levels of Endo-1,4-β-Glucanases, β-Glucan Synthases, and Xyloglucan Synthase After Treatment with Auxin or Ethylene[a]

Enzyme Assayed	Control (untreated)	+ Auxin	+ Ethylene
	(activity at 48 hr/activity at zero time)		
Endo-1,4-β-glucanases (vs. CM-cellulose)	3.0	18	3.0
β-Glucan synthase (10 μM UDP-[^{14}C]glucose)	0.4	1.3	0.04
β-Glucan synthase (1 mM UDP-[^{14}C]glucose)	2.0	3.8	0.3
Xyloglucan synthase (2 μM UDP-[^{14}C]xylose plus 2 mM UDP-glucose)	0.4	1.0	0.04

[a] β-Glucan synthase activity was estimated as the ^{14}C of 24% KOH-insoluble glucan (5, 9), and xyloglucan synthase activity was estimated as the ^{14}C of recovered isoprimeverose after enzymic hydrolysis (21).

remained at the same level as that in growing tissues but markedly decreased after treatment with ethylene. The results suggest that auxin evokes the synthesis treatment with ethylene. Xyloglucan synthase activity after auxin treatment of cellulose and xyloglucan with concurrent partial xyloglucan degradation, but ethylene markedly inhibits synthetic reactions without evoking hydrolysis.

The changes in xyloglucan and cellulose levels in the tissues following hormone treatments were also examined in order to compare effects *in vivo* with those predictable from assayed changes in enzymic activities (Table 4.3) after hormone treatments. Table 4.4 shows the response and level of cellulose and xyloglucan of the elongating regions (5 mm) of pea epicotyls after auxin or ethylene treatment. Auxin treatment markedly increased the deposition of cellulose and xyloglucan in pea cell walls. It also led to solubilization of part of the normally

TABLE 4.4   Responses in the Elongating Regions of Pea Epicotyls to Auxin and Ethylene

Component	Control	+ Auxin	+ Ethylene
	(value at 48 hr/value at zero time)		
Cellulose	13.0	15.4	5.2
Soluble xyloglucan	0.4	2.6	0.2
Bound xyloglucan	2.1	4.9	1.6
Molecular weight of bound xyloglucan	0.7	0.1	0.7
Swelling (X-section area)	0.9	2.5	2.4

insoluble xyloglucan and to a decrease in the average molecular weight of the polysaccharide. Ethylene treatment, however, inhibited the deposition of cellulose and xyloglucan. Levels of soluble xyloglucan in ethylene-treated tissue were very low. Furthermore, the molecular weight of the bound xyloglucan in the tissue, as assayed by gel filtration, was the same as that in controls and much higher than that in auxin-treated tissue. Nevertheless, the swelling phenomenon that is induced by auxin or ethylene is apparently identical in spite of the very different metabolism.

OVERVIEW—BIOSYNTHESIS AND BIODEGRADATION

Despite 20 years of investigation for cellulose in biochemical studies *in vitro*, the mechanism of biosynthesis of cellulose in higher plants remains uncertain. 1,4-β-Glucan has never been synthesized *in vitro* in any substantial amount or without contaminating 1,3-linkages by the enzyme system of plant plasma membranes where cellulose synthase is believed to be associated. No lipid-linked intermediate for 1,4-β-glucan has been observed in higher plant systems. The formation of a glycoprotein intermediate also has not been confirmed in the system. UDP-glucose may contribute as glucosyl donor for 1,4-β-glucan synthesis as in *Acetobacter xylinum* systems (25), because plant tissues contain the sugar nucleotide at high concentrations (26, 27). Nevertheless, most pea membrane preparations synthesize mainly 1,3-β-glucan from UDP-glucose (19). It has been suggested (7) that 1,4-β-glucan synthase in pea membranes is probably responsible for the synthesis of xyloglucan, because hydrolysis of β-glucan with *Streptomyces griceus* cellulase might have yielded xylosyl-glucose, which has the same chromatographic behavior as laminaribiose. However, the hydrolysis of xyloglucan yielded high-molecular-weight oligosaccharides, a hepta- and a pentasaccharide (Fig. 4.1). These oligosaccharides did not undergo further hydrolysis into isoprimeverose by this particular cellulase preparation, even when higher enzyme concentrations were used or when the reaction time was prolonged. It was also shown that xyloglucan 4-β-glucosylstransferase activity requires Mn^{2+} and not Mg^{2+} in pea system. With respect to the role of the 1,4-β-glucan 4-β-glucosyltransferase that appears to exist in Golgi membranes in higher plants (5, 19), there are two possibilities: that the enzyme serves as a proenzyme of cellulose synthase *en route* to its active site in the plasmamembrane, or that the glucosyltransferase is a free form of xyloglucan glucosyltransferase from the system.

Fragmentation analysis of pea xyloglucan by *Streptomyces* endoglucanase digestion showed that xyloglucan synthesized by pea membranes was mainly built of two kinds of oligosaccharide units (Fig. 4.1)—a pentasaccharide (glucose/xylose,3:2) and a heptasaccharide (glucose/xylose, 4:3)—whereas pea xyloglucan *in vivo* contained a heptasaccharide (glucose/xylose,4:3) and a nonasaccharide (glucose/xylose/galactose/fucose,4:3:1:1) unit (Fig. 4.2). The pentasaccharide unit is probably derived from the heptasaccharide unit (28). The nonasaccharide unit could be formed by additional galactosyl and fucosyl donors (e.g., UDP-galactose and GDP-fucose).

Cellulose microfibrils in the primary walls of growing pea epicotyl cells have a similar appearance whether or not matrix materials are removed by extraction with concentrated alkali (24% KOH). The microfibrils are oriented primarily in a transverse direction on the inner surfaces of walls, and some change direction to travel in longitudinally oriented "ribs" on the outer surface (29). Pea xylogucan:cellulose complex, referred to here as "cell wall ghost," was prepared by extraction of cell wall preparation with 4% KOH containing 0.1% NaBH$_4$(3). Xyloglucan in cell wall ghosts was visualized by light microscopy using iodine staining, by radioautography after labeling with [^{3}H]fucose, by fluorescence microscopy using a fluorescein-lectin (*Ulex europeus* fucose-binding lectin) as probe, or by using Calcofluor and by electron microscopy after shadowing (Fig. 4.4). The polysaccharide clearly occurs both on and between the cellulose microfibrils. It is presumably the interfibrillar component of xyloglucan that was free to bind to iodine and particularly accessible to hydrolysis by pea endo-1,4-β-glucanases.

Auxin-induced pea 1,4-β-glucanase preferentially hydrolyzes and solubilizes xyloglucan over cellulose in pea cell wall ghosts (Table 4.2, Fig. 4.3), thus mimicking the situation *in vivo* when 1,4-β-glucanase activity is induced by auxin. There is a particularly marked decrease in xyloglucan molecular weight, which eventually leads to solubilization. Since these events probably occur *in vivo* mainly on the inner surface of the wall, where microfibrils are transversely oriented, endohydrolysis of xyloglucan at that locus may weaken those cross-linkages that constrain fibril slippage in a lateral direction. However, auxin not only induces endo-1,4-β-glucanase activities but also promotes the deposition of xyloglucan and cellulose (Table 4.4). It may be that auxin-induced glucanase also acts during growth to promote cellular deposition, by introducing new chain ends in pre-existing cellulose where new chain lengthening can occur (30), and to reduce wall rigidity by degrading xyloglucan chains that cross-link cellulose microfibrils (22).

The swelling of pea epicotyls following auxin treatment was accompanied by greatly increased net synthesis of wall fractions, including xyloglucan and cellulose, which were deposited at rates that more than kept pace with growth in volume. In contrast, the swelling following ethylene treatment was attained by relatively little wall deposition. Swelling after auxin treatment may have been facilitated by a weakening of xyloglucan cross-linking between microfibrils by hydrolysis with endo-1,4-β-glucanases, but swelling after ethylene treatment was more readily attributable to weakening of the wall by inhibition of cell wall polysaccharide synthesis rather than by the hydrolysis. Both auxin and ethylene treatments evoke lateral expansion of pea cells owing in part to increased turgor pressure and in part to effects on xyloglucan metabolism—that is, degradation, albeit enhanced deposition in the one instance and inhibited synthesis in the other.

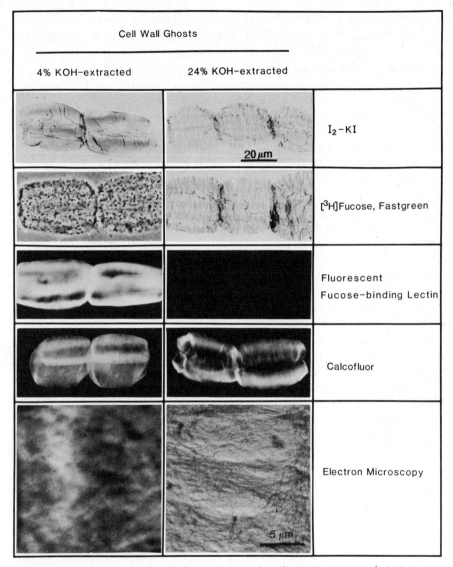

Figure 4.4 Visualization of cell wall ghosts obtained after 4% KOH extraction (xyloglucan and cellulose) and 24% KOH extraction (cellulose only).

REFERENCES

1. T. Hayashi, Y. Kato, and K. Matsuda, *Plant Cell Physiol.,* **21**, 1405 (1980).
2. G. Chambat, F. Barnoud, and J. P. Joseleau, *Plant Physiol.,* **74**, 687 (1984).
3. T. Hayashi and G. Maclachlan, *Plant Physiol.,* **75**, 596 (1984).
4. R. M. Brown, Jr., and D. Montezinos, *Proc. Natl. Acad. Sci. USA,* **73**, 143 (1976).

5. G. Shore and G. A. Maclachlan, *J. Cell Biol.*, **64**, 557 (1975).

6. D. G. Robinson, W. R. Eisinger, and P. M. Ray, *Ber. Dtsch. Bot. Ges.*, **89**, 147 (1976).

7. P. M. Ray, *Biochim. Biophys. Acta*, **629**, 431 (1980).

8. T. Hayashi and K. Matsuda, *J. Biol. Chem.*, **256**, 11117 (1981).

9. T. Hayashi and K. Matsuda, *Plant Cell Physiol.*, **22**, 1571 (1981).

10. J. M. Labavitch and P. M. Ray, *Plant Physiol.*, **53**, 669 (1974).

11. J. M. Labavitch and P. M. Ray, *Plant Physiol.*, **54**, 499 (1974).

12. N. R. Gilkes and M. A. Hall, *New Phytol.*, **78**, 1 (1977).

13. M. E. Terry, R. L. Jones, and B. A. Bonner, *Plant Physiol.*, **68**, 531 (1981).

14. D. P. S. Verma, G. A. Maclachlan, H. Byrne, and D. Ewings, *J. Biol. Chem.*, **250**, 1019 (1975).

15. W. Eisinger, *Annu. Rev. Plant Physiol.*, **34**, 225 (1983).

16. S. P. Burg and E. A. Burg., *Proc. Natl. Acad. Sci. USA*, **55**, 262 (1966).

17. A. Apelbaum and S. P. Burg, *Plant Physiol.*, **48**, 648 (1971).

18. J. M. Lang, W. R. Eisinger, and P. B. Green, *Protoplasma*, **110**, 5 (1982).

19. Y. Raymond, G. B. Fincher, and G. A. Maclachlan, *Plant Physiol.*, **61**, 938 (1978).

20. P. H. Quail, *Annu. Rev. Plant Physiol.*, **30**, 425 (1979).

21. T. Hayashi, Y. Kato, and K. Matsuda, *J. Biochem.*, **89**, 325 (1981).

22. T. Hayashi, Y. S. Wong, and G. A. Maclachlan, *Plant Physiol.*, **75**, 605 (1984).

23. G. A. Maclachlan and Y. S. Wong, *Adv. Chem. Ser.*, **181**, 347 (1979).

24. P. Kooiman, *Rec. Trav. Chim. Pays-Bas*, **79**, 675 (1960).

25. M. Swissa, Y. Aloni, H. Wenhouse, and M. Benziman, *J. Bacteriol.*, **143**, 1142 (1980).

26. N. C. Carpita and D. P. Delmer, *J. Biol. Chem.*, **256**, 308 (1981).

27. T. Hayashi and K. Matsuda, *Agric. Biol. Chem.*, **45**, 2907 (1981).

28. T. Hayashi, Nakajima, and K. Matsuda, *Agric. Biol. Chem.*, **48**, 1023 (1984).

29. G. A. Maclachlan, *Appl. Polym. Symp.*, **28**, 645 (1976).

30. Y. S. Wong, G. B. Fincher, and G. A. Maclachlan, *Science*, **195**, 679 (1977).

Hence the appearance of the *Chlamydomonas* type having a glycoprotein cell wall lattice (12) notably lacking in cellulose and tensile strength, thus condemned to contractile vacuolar osmoregulation. However, once acquired, cell surface glycoproteins underwent rapid exploitation judging from the hydroxyproline-rich sexual agglutinins recently identified in *Chlamydomonas* (13). Indeed throughout eukaryotic evolution cell surface glycoproteins are the dominant paradigm for cell–cell recognition and assembly of the extracellular matrix.

How did cellulose arrive on the scene? Some rather diverse present-day bacteria (*Azotobacter, Rhizobium, Agrobacterium*, etc.) synthesize cellulose, implying a facile origin from bacterial peptidoglycan, which is essentially a 1-4-beta-linked glucan with side chain additions and peptide cross-links. Addition of cellulose to the cell surface probably marked the great divide between a pure crystalline lattice held together by noncovalent bonds (12) and an energetically more efficient (14), stronger covalent cell surface network. This advance led directly to (a) osmoregulation by turgor and wall pressure, (b) morphogenesis and metaphytes, and (c) disease resistance through coevolution of complex host–pathogen interactions at the cell surface.

Although the "invention" of turgor pressure permitted the hydrostatic support of soft tissues and transition from aquatic to land habitat, it also created the problem of regulating cell extension *and* expansion and the raison d'être of a cell wall model! However, a complete cell wall model must also account for its role in differentiation and morphoregulation as well as host–pathogen interactions involving the recently discovered endogenous cell wall elicitors (15).

Almost antithetical properties of plasticity and high tensile strength characterize the primary cell wall. Plasticity is a growth-dependent variable (16). In the most general terms, cell wall growth involves a biochemically controlled creep (17) whose molecular basis is unknown but must be explained by any valid model. Roelofsen (18) made the first step in that direction by introducing the multinet hypothesis suggesting passive reorientation of cellulose microfibrils during cell extension. Bonner (19) had, much earlier (before the discovery of microfibrils), suggested that cellulose was variably cross-linked by "Haftpunkte," whose role since then could be assigned to pectin, xyloglucans, or extensin, depending upon one's favorite model. Pectin was an early favorite, because an apparent auxin-dependent pectin methylesterase fitted in with the calcium bridge hypothesis. But the data were irreproducible, and the hypothesis was discarded. The discovery of xyloglucan hydrogen-bonded to cellulose led to Albersheim's model (20) in which the wall was (with the exception of xyloglucan–cellulose H bonding) one huge, covalently interconnected macromolecule. This model is no longer tenable, chiefly because earlier evidence of a covalent xyloglucan–pectic linkage has softened (the "extent of covalent attachment between hemicelluloses and the pectic substances is unknown," 21) but also because the model predicted easy removal of extensin following major cleavage of glycosidic linkages (22). Other models (including my own, 23, 24) led to similar unfulfilled predictions. The acid test involved specific cleavage of glycosidic linkages leaving others, such as peptide bonds, intact. Thus, anhydrous hydrogen fluoride removed wall

polysaccharides, but *extensin remained insoluble* suggesting possible cross-linkage to itself (25) and therefore the existence of two cell wall networks (25). Two
recent advances support the dual network hypothesis: First, the identification
of the cross-linked amino acid isodityrosine (IDT), and second, the unequivocal
demonstration of two extensin precursors *in muro*, with little or no isodityrosine.

These discoveries led directly to the "warp-weft" model, proposed here, in
which cellulose microfibrils penetrate a glycoprotein network of defined porosity.
The model is strictly a working hypothesis (represented schematically in Fig.
5.1), but attractive because concatenation of polymeric systems by a glycoprotein
network would mechanically couple the load-bearing polymers and hence dis-
tribute stress throughout the cellulose microfibrillar network.

A detailed rationale for the new model involves a discussion of three areas:
(a) the properties of "cross-linked" (firmly bound) extensin *in muro*, (b) the new

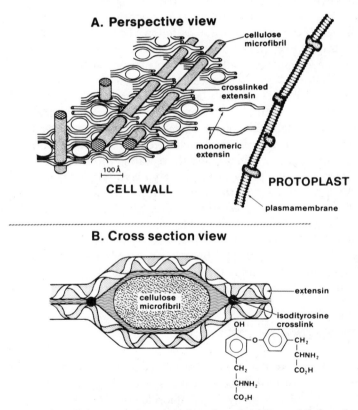

Figure 5.1 Extensin–cellulose network. Artist's three-dimensional illustration shows how extensin
monomers could be cross-linked *in muro* to form a multimeric network of porosity defined by cross-
link frequency and penetrated by cellulose microfibrils. For clarity the illustration omits xyloglucans
linearly adsorbed on the microfibril surface and the pectic substances, some of which may be trapped
in the pores along with the cellulose (42). (Reproduced from Proc. 2nd Annu. Plant Biochem.
Physiol. Symp. Missouri–Columbia.)

cross-link amino acid isodityrosine (IDT), and (c) the isolation and properties of monomeric extensin precursors, including evidence for their *in muro* conversion to multimer.

CROSS-LINKED EXTENSIN *IN MURO*

Although virtually all the evidence is indirect, the name extensin implies a role in growth by cell extension and expansion. But what is extensin? Certainly it is the most abundant hydroxyproline-rich glycoprotein (HRGP) and a serious challenge to "Rubisco" for the title of "most abundant protein on planet earth."

The generic term HRGP also covers arabinogalactan proteins and some solanaceous lectins, notably those from potato and *Datura*, but unlike them extensin is characteristically insoluble (even in anhydrous HF) and therefore presumed to be highly cross-linked . . . but to what?

Cross-linked extensin is a structural component of the primary cell wall. Briefly, the experimental evidence is as follow: First of course the location of extensin outside the cell in a structural "organelle" (the primary cell wall), often as a major component, makes alternative explanations such as enzyme, hormone, lectin, and elicitor difficult. The chemistry of extensin reinforces the *idée fixe*! For example, recent CD spectral data (26) confirm the structural role of the hydroxyproline arabinoside substituents which, according to the wraparound model (8), act as stabilizers of the extended peptide backbone, by hydrogen bonding to it. Peptide periodicity also characterizes structural proteins, the best example being collagen, which contains the repeating tripeptide Gly-Pro-X whose extensin analogue is the pentapeptide Ser-Hyp-Hyp-Hyp-Hyp, although the repeat fidelity is rather less than the collagen tripeptide. Recently obtained peptide sequences of the extensin precursors confirm the Ser-Hyp$_4$ pentapeptide as a major repeat unit and also indicate the occurrence of even longer repeating sequences. These cumulative data show that extensin represents a class of structural glycoproteins. But what is the precise structural role of extensin? The chemical clues are cross-linked tyrosine (isodityrosine) in extensin and the existence of soluble extensin precursors deficient in cross-linked tyrosine.

ISODITYROSINE IN EXTENSIN

Difficulties in studying the structure of extensin derive largely from its insolubility, formerly ascribed to hypothetical cross-linkage by cell wall polysaccharides, but made unlikely by hydrogen fluoride solvolysis (27) of the polysaccharides: Extensin remained insoluble and therefore not cross-linked solely by polysaccharides. Could extensin be cross-linked to itself? If so how? The following lines of evidence identify a promising cross-link candidate.

Acid sodium chlorite oxidation of cell walls released soluble fragments of extensin (28, 29), implying a phenolic cross-link, possibly corresponding to the "unknown tyrosine derivative" observed earlier in some tryptic peptides of extensin (30). Identification of isodityrosine (IDT) first in cell wall hydrolysates (31) and then in tryptic peptides (32) (of firmly bound extensin) showed that the "tyrosine derivative" was indeed a diphenyl ether-linked dityrosine, the phenolic hydroxyl of one tyrosine "donating" the very stable ether link (Fig. 5.1). In principle, such an amino acid could act as an intermolecular or intramolecular linkage and perhaps both. To date, however, only the intramolecular linkage has been identified as an intriguing short internal cross-link encompassing only three amino acids (32). The hypothetical intermolecular linkage remains to be demonstrated, which is hardly surprising considering the technical difficulties involved in the unequivocal isolation of such peptides from a highly cross-linked insoluble substrate!

EXTENSIN PRECURSORS *IN MURO*

Does the cytoplasm transport large prefabricated units to the wall for final assembly, as in some primitive algae (33), or is there insertion of much smaller monomeric polymers followed by extensive *in muro* multimer formation? How much is self-assembly; how much is enzyme directed? The answers vary for each wall component.

Until recently the pathway for assembly of wall protein was unknown, and the status of the small amounts of soluble extensin extractable from the wall was not clear (34). One rapidly growing system even seemed to lack it (35)! However, the recent suggestion (8) that peroxidase might catalyze extensin cross-link formation, together with the even more recent identification of the cross-link candidate as isodityrosine, implied that peroxidase inhibitors might inhibit cross-linking of extensin *in muro*. This idea led to the reexamination of soluble extensin extractable directly from intact cells by salt elution (36); we used a cell column technique developed earlier for elution of cell surface enzymes (23). The yield of soluble extensin from cell suspension cultures, surprisingly, was not dependent on the presence of peroxidase inhibitors, but the yield was highly dependent on species, culture age, and eluant ionic strength. Yields were poor ($\sim$ 30 μg Hyp/g cells dry weight) from sycamore but good from tomato ($\sim$ 300 μg Hyp/g cells dry weight). Highest yields corresponded with rapidly growing cells (5 days) eluted with trivalent cations of aluminum or lanthanum chloride. The elution kinetics clearly demonstrated elution *per se*, rather than secretion (as surmised for an "arabinose-rich" wall component by others (37)); the elution was rapid (within seconds) and temperature independent, pointing to a simple physical mechanism—cationic displacement of extensin from the pectic carboxyl groups (36). (It would be nice to catalog all the wall enzymes also ionically bound to pectin!)

Chemical analysis showed that the elutable hydroxyproline-rich material had an extensin-like composition, and pulse-chase kinetics with [^{3}H]proline showed a turnover flux consistent with wall demand (36). We verified that conclusion

by resuspending salt-eluted cells (pool depletion) in growth medium and following the reappearance (pool repletion) of soluble extensin (Fig. 5.2). In rapidly
growing cells the pool repletion rate exceeded the depletion rate (36). Thus
during rapid growth (pool expansion) the secretion rate of extensin monomers
into the wall exceeds the *in muro* cross-linkage rate (multimer formation),
whereas in aging cells the precursor pool contracts when the rate of multimer
formation exceeds the monomer secretion rate.

Elution of soluble extensin precursors from the cell surface simplified their
purification, as did the excellent yield which, at day 5, was ~ 10% of total wall-
bound extensin. The chemical properties of soluble extensin, which is highly
glycosylated and basic, also assisted its purification; such proteins are generally
soluble in cold trichloroacetic acid (TCA). (It is important to limit exposure to
TCA and especially to avoid hot TCA, which cleaves sensitive linkages such as
Asp-Pro and arabinofuranosides.) After removal of extraneous proteins by TCA
precipitation and chromatography of the dialyzed TCA-soluble fraction on carboxymethylcellulose, we observed two (rather than one as reported by others
(38)) major hydroxyproline-rich fractions, which we designated P1 and P2 from
their chromatographic elution order and their status as precursors to cross-
linked extensin reported earlier (36). These two fractions had "extensin-like"
compositions, and the HF-deglycosylated polypeptides had similar apparent
molecular weights on sodium dodecylsulfate–polyacrylamide gel electro-
phoresis (SDS-PAGE) gradient gels, contained no cystine, and are therefore
probably monomeric (36). Closer inspection of P1 and P2 showed significant
differences in their hydroxyproline arabinoside profiles, and that P1 was histidine-
rich and possibly corresponding to the hydroxyproline-rich "agglutinin" isolated
by others (39, 40).

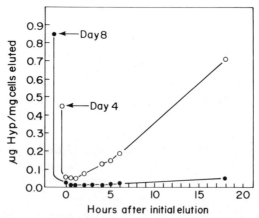

Figure 5.2 Salts elute the *in muro* pool of extensin precursors from cultured cell suspensions of
tomato. After a brief wash and resuspension in growth medium, the eluted cells replenish the
precursor pool (assayed by eluting aliquots taken at various times after the initial elution) at a rate
roughly proportional to the growth rate (36). (Reproduced from Phytochemistry with the permission
of Pergamon Press.)

Peptide mapping in progress (41, 42) also shows fundamental similarities and essential differences between P1 and P2. Although both P1 and P2 generate quite different tryptic peptide maps, all tryptides larger than about five residues contain the $Ser-Hyp_4$ repeat pentapeptide. However, the tryptides occur in two main size classes: In P1 they are $\sim$ 10 and 16 residues (the two major peptides, a 10-mer and a 16-mer, occurring in a 2:1 molar ratio), whereas in P2 the tryptides occur as octapeptides, and a single dipeptide (Tyr-Lys) repeated about 20 times, judging from its high yield. The major P2 octapeptide is $Ser-Hyp_4$-Val-Tyr-Lys. Interestingly, a tryptide previously isolated from covalently wall-bound extensin was

$$Ser-Hyp_4-Val-\overset{\displaystyle \lceil\!-\!-\!o\!-\!-\!\rceil}{Tyr-Lys-Tyr}-Lys$$

(i.e., containing a short intramolecular linkage consisting of two diphenyl ether-linked tyrosine residues (32)). Therefore P2 may well consist of two regularly alternating hexa- and tetrapeptides, $Ser-Hyp_4$-Val and Tyr-Lys-Tyr-Lys. Thus the whole molecule may exist as the highly periodic structure (42):

$$[Ser-Hyp-Hyp-Hyp-Hyp-Val-Tyr-Lys-Tyr-Lys]_n$$

where $n \approx 30$.

Considering the great value of antibodies as immunoreagents, we recently began an immunological approach by raising rabbit antibodies against P1, P2 and their deglycosylated polypeptides. This work in progress yields the following preliminary results and conclusions: Both glycosylated and deglycosylated forms of P1 and P2 were antigenic judging from micro ELISA noncompetitive sandwich assays. Interestingly P2 was more antigenic than P1. Not surprisingly the glycosylated antigens (P1 and P2) elicited polyclonal antibodies (rabbit) that cross-reacted well with those glycosylated antigens (50% for the P1 antibody–P2 antigen combination and 82% for the P2 antibody–P1 antigen). However, these antibodies (raised against glycosylated antigens) reacted much less with the corresponding deglycosylated antigens (ranging from 0% to 30% depending on the antibody–antigen combination (43)).

Likewise the deglycosylated antigens (dP1 and dP2) elicited antibodies that cross-reacted well with those deglycosylated antigens (65% for dP1 antibody–dP2 antigen and 97% for dP2 antibody–dP1 antigen) but much less with the corresponding antigens (range $>$ 10 to 45% depending on the antibody–antigen combination).

Two interactions of lowest cross-reactivity are of special interest: the P1 antibody–dP2 antigen and dP1 antibody–P2 antigen cross-reactions, which ranged from 0% to $<$ 10%. In other words, P1 antibodies did not recognize deglycosylated P2, implying that exposed (i.e., nonglycosylated) domains of P1 are not homologous with any regions of deglycosylated P2. Conversely, dP1 antibodies did not recognize glycosylated P2, implying that the exposed (nonglycosylated) domains of P2 are not homologous with any regions of deglycosylated P1. Yet the high cross-reactivities between deglycosylated antibodies and deglycosylated antigens imply considerable amino acid sequence homology between P1 and P2. This apparent paradox simply means that the nonglycosylated regions of P2 are not common to P1 and vice versa (Fig. 5.3).

Combining the above immunological and sequence data, we conclude that the major epitopes (antigenic determinants) of glycosylated P1 and P2 are the hydroxyproline arabinosides, but in addition glycosylated P2 also contains the major nonglycosylated epitope Val-Tyr-Lys-Tyr-Lys possibly corresponding to a cross-link domain (41). Appropriate immunoaffinity chromatography should thus allow us to prepare a "poor man's monoclonal antibody" against the nonglycosylated epitope, to produce a highly specific immunoreagent.

Use of electron microscopy to characterize P1 and P2 is also underway. To date only the possibly P1 equivalent precursor from carrot cell walls was visualized, by rotary shadowing, as a thin, wormlike structure ~ 80 nm in length (26). Clearly we must develop other more specific methods if we are to realize the eventual aim: to "see" extensin and to determine its porosity and molecular orientation *in muro*.

THE "WARP–WEFT" MODEL

The data reviewed above allowed us to suggest a new model of the primary cell wall, radically different from others, based on two interacting networks of load-

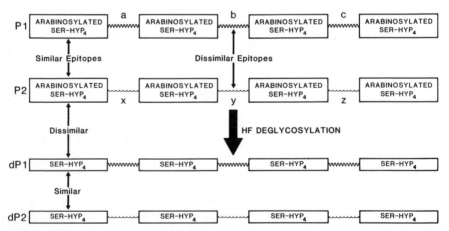

Figure 5.3 Extensin precursors isolated from tomato consist of relatively rigid blocks (arabinosylated Ser-Hyp$_4$) which alternate with more flexible spacers (domains *a, b, c,* in P1 and *x, y, z,* in P2). The Ser-Hyp$_4$ pentapeptide represents an epitope common to P1 and P2. The P2 flexible spacers are not common to both P1 and P2. The P2 flexible spacers seem to consist almost exclusively of the pentapeptide Val-Tyr-Lys-Tyr-Lys (i.e., $x = y = z$), whereas those of P1 (sequence determination in progress) do *not* contain that sequence and probably differ one from the other (i.e., $a \neq b \neq c$).

bearing polymers that are mechanically coupled rather than covalently cross-linked (44). How do the data lead to such a model: What are its implications? Does it lead to useful predictions about growth regulation? What are the pros and cons of the model overall?

Building a Model

The crucial data show that (a) the cell secretes monomeric extensin precursors rather than preformed multimeric network subassemblies, (b) cross-linkage of extensin monomers occurs *in muro* and (c) extensin contains the cross-link IDT.

Network characteristics depend on molecular dimensions (shape and size), cross-link frequency, initial monomer orientation, and site of assembly. C.D. spectra and electron microscopy indicate a rodlike or, perhaps more accurately, wormlike molecule $\sim$ 80 nm in length (26). A width (unhydrated) of 1.5 nm according to molecular models explains in part how such a large molecule passes so rapidly through the relatively small pores of walls impermeable to 68 kD globular proteins (45). The rapidity of elution (36) also argues for intussusception (insertion) of extensin precursors rather than apposition, and one intuits anticlinal orientation of extensin. (This contrasts with the periclinal orientation of cellulose deposited by apposition.) If, by analogy with transmembrane proteins, extensin is a transmural protein, then extensin and cellulose may cross essentially at right angles. Hence, the "warp-weft" analogy, which for a three-dimensional fabric has some limitations! But are the two networks really close enough to interact as suggested? Yes. Cellulose and extensin "copurify" during successive wall extractions (22, 23).

Consider then, two adjacent extensin monomers separated by a single cellulose microfibril; cross-linkage of those monomers across the microfibril would form a pore penetrated by that microfibril (Fig. 5.1). The IDT content certainly indicates a cross-link frequency more than sufficient to create a tight network; for example, the Hyp/IDT molar ratio of 16:1 in wall-bound extensin translates into a cross-link frequency of one every 30 amino acid residues ($\sim$ 10 nm) of P2. But some of the IDT in P2 is undoubtedly intramolecular; intermolecular cross-linkage between P1 and P2 is an assumption. Thus precise porosity remains to be defined.

Such a remarkable concatenation of polymers has precedents ranging from DNA (46) to mechanically coupled glycoprotein networks comprising "Wharton's jelly" of the umbilicus (47). Covalently cross-linked protein networks occur both as elastomers (di- and trityrosine in cross-linked resilin) (48), as the ordered lattice network of fibrin gels (49), and as the peroxidatically-generated fertilization membrane crosslinked by dityrosine in sea urchin eggs (49a).

Implications of the Model

If the cell wall matrix is highly ordered and organized, then the classical description of "cellulose microfibrils embedded in an amorphous matrix" is incorrect. Indeed organized growth involving self-assembly of amorphous structure is a contradiction. On the other hand, the organized cell wall model suggests a highly engineered mechanical coupling of load-bearing polymers leading to stress distribution throughout the wall. Furthermore, even if mechanical failure of the extensin network occurred in older layers of the wall, intussusception of new extensin monomers could repair the fabric and restore mechanical strength. This differs from the *Nitella* cell wall (frequently cited as a general paradigm (50)), where only the inner wall layer bears the major load. However, the *Characeae* represent highly specialized coenocytic green algae of uncertain affinity, possessing almost completely nonesterified pectin, lacking extensin, and therefore probably lacking the "wall repair service" suggested for higher plants! (Perhaps the evolutionary cul-de-sac represented by the aquatic *Characeae* reflects the absence of an extensin network strong enough to provide the support needed for their transition to a land habitat.)

Intussusception of an extensin weft also suggests that extensin is a morphoregulatory protein (51), acting as a matrix organizer by forming a defined network whose molecular properties define the "biochemically controlled creep" involved in cell extension. But these implications are generalities. Can we make more specific predictions?

Predictions from the Model: the Pectin–Extensin Switch

In a woven fabric the warp may slip along its own axis, or it may separate laterally. The model allows microfibril slippage or separation. Slippage corresponds to microfibril movement through an extensin pore, whereas separation involves weft displacement if loosely interconnected or (enzymic?) rupture if highly interconnected. Either warp slippage or separation may predominate depending on the cell type. One suspects that slippage predominates in the mainly isodiametric cells of some suspension cultures (e.g., sycamore), whereas separation predominates in the tubular cells typical of other cultures (e.g., tomato) and extending stems. What regulates these processes? Defined network porosity suggests steric rather than covalent restraints. The adsorbed xyloglucans of the microfibril surface are good candidates because of their enhanced turnover (presumed removal allowing a microfibril to slip through an extensin pore) during auxin-induced growth (52). However extensible tissue responds most rapidly to decreased pH (53) ("acid growth"). These rapid changes in rheological properties suggest a switching mechanism between polymer blocks—for example, between pectin and extensin, which interact strongly judging from the ionic adsorption of extensin precursors (36).

Switching is the essence of any control mechanism (54). The wall pH, poised around the pK of pectic carboxyls, allows a pH-controlled switch: from pectin–extensin interaction at high pH (salt bridge between charged pectic carboxyl and extensin epsilon amino groups) to pectin–pectin interaction—for example, via carboxyl-carboxylate hydrogen bonding (55) at low pH‡ (Fig. 5.4). If a low

---

‡Well understood by our grandmothers, who knew that acid fruits make the best jam!

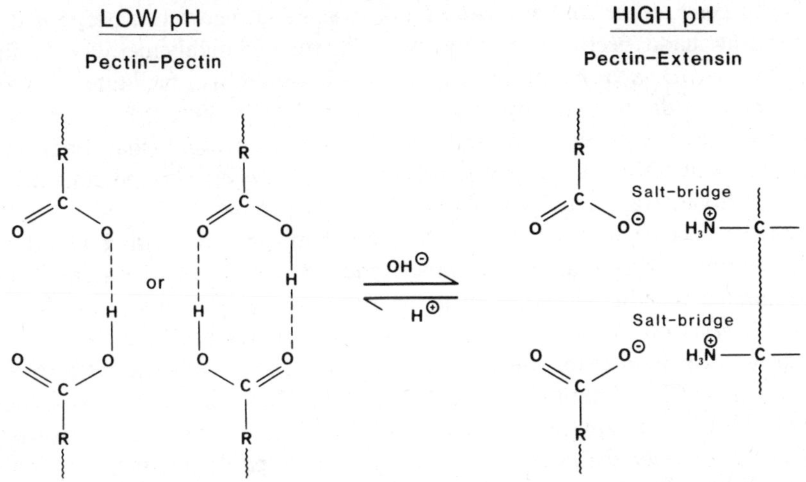

Figure 5.4 pH-Induced pectin–extensin switch. Based on the demonstrated ionic association between pectin and extensin monomers (36), this model shows how a similar interaction between pectin and bound extensin may vary as a simple function of pH and thereby alter the wall rheology. Low pH suppresses pectic carboxyl ionization, and the pectin tends to self-associate via carboxyl–carboxyl and carboxyl–carboxylate hydrogen bonding. Increasing the pH increases the pectin anionic charge, leading to ion pair formation ("salt bridges") between pectic carboxyls and lysine epsilon amino groups. Note that the figure shows carboxyls and epsilon amino groups only.

pH-induced switch from pectin–extensin to pectin–pectin decreases wall viscosity, the resultant cell extension would accord with the "acid growth" hypothesis. But how can increased pectin–pectin (decreased pectin–extensin) association decrease cell wall viscosity? Earlier one could invoke a covalent xyloglucan–pectic linkage (now uncertain) to retard cellulose slippage through the extensin network. However, the principle remains the same: Pectin effectively cross-links cellulose and extensin if extensin pore formation also traps some pectin. Furthermore, such cross-linkage would be pH-dependent. There is NMR evidence for increased pectin association *in muro* at low pH (56) and chemical evidence of a pectin fraction strongly associated with cellulose (57, 58) and solubilized only by cellulase (59).

The model is highly speculative but testable despite obvious limitations. The current model, for example, is based almost exclusively on extensin precursors isolated from two atypical plant systems—cell suspension cultures and excised aging phloem parenchyma of carrots. Both are highly stressed "systems under siege." Current lack of direct evidence for the postulated network porosity is a more serious limitation; intermolecular cross-links and microfibrillar penetration of the network remain to be demonstrated. Then there is the problem of extensin intussusception. What is the driving force? What are the "rules" for self-assembly? Diffusion combined with ionic adsorption is simple but insufficient to account for "deep penetration" of basic extensin monomers into an acidic pectic matrix. On the other hand, pectin of growing tissues is generally highly methyl-esterified, and the low pH *in muro* during rapid growth would also facilitate monomer movement, no doubt aided by turgor pressure. Possibly *in muro* pectin methylesterase creates a gradient of pectin esterification decreasing from the inner to the outer wall. However, extensin precursors themselves showed no intrinsic pectin methylesterase activity (60).

The presence of at least two precursor subunits is an exciting glimpse of future complex molecular design which one assumes involves intermolecular cross-linkage for the assembly of a heteromultimeric network. But that assumption rests on the demonstration of P2 only in tomato, and then only in one growth medium (M6E but not MET) at certain stages of growth (61). Is P2 an artifact? No, because the P2 decapeptide S2Z8 corresponds to tryptide S2A11 isolated in good yield from wall-bound extensin some years ago. However, it is not clear why the sizes of the two precursor pools fluctuate apparently independently. Possibly pool size also reflects extractability, and this might explain why sycamore cell suspensions give such poor yields of (uncharacterized) precursor material.

In some tissues the small amounts of firmly bound extensin might question its mechanical significance; generally one can rationalize those exceptions (e.g., oat coleoptiles) as a loose weft involved in a regulatory rather than a mechanical support role. The clearest example is the extensin-poor pollen tube versus the extensin-rich root hair (6, 23). Both are osmotically protected—one by a controlled environment, the other by a high tensile strength cell wall.

Before succumbing to the danger of ascribing all the mechanical properties of the primary cell wall to a model based solely on a cellulose–extensin network, we should consider the other major wall components: xyloglucans and the pectic substances. How do they fit into the model?

During expansion growth wall polymers play two roles: They bear loads, and they regulate wall rheology. The new model suggests cellulose and extensin as the major tensile load-bearing polymers. Presumably xyloglucans (largely adsorbed to the microfibrillar surface) and pectic substances support compressive loads, but they also regulate wall rheology ("creep") via steric restraints; both xyloglucan turnover and the low pH-induced pectin–extensin switch may release those restraints, thereby enhancing microfibrillar slippage through the extensin network.

Slowdown of cell expansion occurs as restraints return, including covalent cross-linkages generated peroxidatically which result in a general tightening of the extensin network by isodityrosine, cross-linkage of pectic substances by diferulic acid (62), and ultimately, lignification (63). Peroxidases evidently play a major role in growth and development (32). Thus the enzymic complement of the cell wall is now largely explicable, even including ascorbic acid oxidase which, by scavenging any ascorbate released from damaged cells, permits crucial cross-linkage reactions to continue, especially under pathogenic attack (64).

REFERENCES

1. R. D. Preston, E. Nicolai, R. Reed, and A. Millard, *Nature,* **162**, 665 (1948).
2. R. D. Preston and J. Cronshaw, *Nature* **181**, 248 (1958).
3. R. D. Preston, *The Physical Biology of Plant-Cell Walls*, Chapman and Hall, London, 1974, p. 491.
4. R. D. Preston, and J. Hepton, *J. Exp. Bot.,* **11**, 13 (1960).
5. R. D. Preston, *The Molecular Architecture of Plant Cell Walls*, Chapman and Hall, London, 1952, p. 211.
6. D. T. A. Lamport, and J. W. Catt, *Encyclopedia Plant Physiol.* **13B**, 133 (1981).
7. M. F. Mescher and J. L. Strominger, *J. Biol. Chem.,* **251**, 2005 (1976).
8. D. T. A. Lamport, *Biochem. Plants,* **3**, 501 (1980).
9. M. O. Dayhoff, *Precambrian Res.,* **20**, 299 (1983).
10. G. J. Merkel, D. R. Durham, and J. J. Perry, *Can. J. Microbiol.,* **26**, 556 (1980).
11. D. P. Delmer and D. T. A. Lamport, in *Cell Wall Biochemistry Related to Specificity in Host–Plant Pathogen Interactions*, B. Solheim and J. Raa (Eds.), Universitets forlaget, Rodeløkka, Oslo, Norway, 1976, pp. 85–104.
12. K. Roberts, *Philos. Trans. R. Soc. Lond. B,* **268**, 129 (1974).
13. J. B. Cooper, W. S. Adair, R. P. Mecham, J. E. Heuser, and U. W. Goodenough, *Proc. Natl. Acad. Sci. USA,* **80**, 5898 (1983).
14. J. A. Raven, *New Phytol.,* **92**, 1 (1982).
15. M. McNeil, A. G. Darvill, S. C. Fry, and P. Albersheim, *Annu. Rev. Biochem.,* **53**, 625 (1984).
16. A. N. J. Heyn, *Bot. Rev.,* **6**, 515 (1940).
17. R. D. Preston, *Annu. Rev. Plant Physiol.,* **30**, 55 (1979).
18. P. A. Roelofsen, *The Plant Cell Wall*, Borntraeger, Berlin, 1959, p. 335.
19. J. Bonner, *Jb. Wiss. Bot.,* **82**, 377 (1935).
20. K. Keegstra, K. W. Talmadge, W. D. Bauer, and P. Albersheim, *Plant Physiol.,* **51**, 188 (1973).
21. A. Darvill, M. McNeil, P. Albersheim, and D. P. Delmer, *Biochem. Plants,* **1**, 91 (1980).
22. J. A. Monroe, D. Penny, and R. W. Bailey, *Phytochemistry,* **15**, 1193 (1976).
23. D. T. A. Lamport, *Adv. Bot. Res.,* **2**, 151 (1965).
24. D. T. A. Lamport, *Annu. Rev. Plant Physiol.,* **21**, (1970).
25. D. T. A. Lamport, in Proc. 4th Int. Cong. Plant Tissue Cell Culture, Calgary, Canada, 1978, pp. 235.
26. G. J. Van Holst, and J. E. Varner, *Plant Physiol.,* **74**, 247 (1984).
27. A. J. Mort and D. T. A. Lamport, *Anal. Biochem.,* **82**, 289 (1977).
28. A. J. Mort, Ph.D. thesis, Michigan State University, 1978.
29. M. A. O'Neill and R. R. Selvendran, *Biochem. J.,* **187**, 53 (1980).
30. D. T. A. Lamport, *Colloq. Int. CNRS,* **212**, 26 (1973).
31. S. C. Fry, *Biochem. J.,* **204**, 449 (1982).
32. L. Epstein and D. T. A. Lamport, *Phytochemistry,* **23**, 1241 (1984).
33. R. M. Brown, W. W. Franke, H. Kleinig, H. Falk, and P. Sitte, *J. Cell Biol.,* **45**, 246 (1970).
34. M. M. Brysk and M. J. Chrispeels, *Biochim. Biophys. Acta,* **257**, 421 (1972).
35. D. G. Pope, *Plant Physiol.,* **59**, 894 (1977).
36. J. J. Smith, E. P. Muldoon, and D. T. A. Lamport, *Phytochemistry,* **23**, 1233 (1984).
37. M. R. Morris and D. H. Northcote, *Biochem. J.,* **166**, 603 (1977).

38. D. A. Stuart and J. E. Varner, *Plant Physiol.,* **66**, 787 (1980).

39. J. E. Leach, M. A. Cantrell, and L. Sequeira, *Plant Physiol.,* **70**, 1353 (1982).

40. J. E. Mellon and J. P. Helgeson, *Plant Physiol.,* **70**, 401 (1982).

41. J. Smith and D. T. A. Lamport, *Plant Physiol.,* **75**, *Suppl.; Proc. Annu. Mtg.,* U. C. Davis, 1984, p. 62.

42. J. Smith, E. P. Muldoon, J. J. Willard and D. T. A. Lamport, *Phytochemistry,* **25**, (1986) in press.

43. M. Kieliszewski, and D. T. A. Lamport, *Phytochemistry,* **25**, (1986) in press.

44. D. T. A. Lamport and L. Epstein, in *Proc. 2d Annu. Plant Biochem. Physiol. Symp.,* Missouri–Columbia, 1983, p. 73.

45. N. Carpita, *Science,* **218**, 813 (1982).

46. R. Weil and J. Vinograd, *Proc. Natl. Acad. Sci. USA,* **50**, 730 (1963).

47. F. A. Meyer, Z. Laver-Rudich and R. Tanenbaum, *Biochim. Biophys. Acta,* **755**, 376 (1983).

48. S. O. Andersen, *Biochim. Biophys. Acta,* **93**, 213 (1964).

49. B. Blombaeck, M. Okada, B. Forslind, and U. Larsson, *Biorheology,* **21**, 93 (1984).

49a. B. M. Shapiro, in *Fertilization and Embryonic Development In Vitro,* L. Mastroianni, Jr., and J. D. Biggers (Eds.), Plenum, New York, 1981, pp. 233-255.

50. L. Taiz, *Annu. Rev. Plant Physiol.,* **35**, 585 (1984).

51. D. V. Basile and M. R. Basile, *Science,* **220**, 1051 (1983).

52. J. M. Labavitch, *Annu. Rev. Plant Physiol.,* **32**, 385 (1981).

53. R. E. Cleland, in *Plant Growth Substances,* F. Skoog, ed., Springer-Verlag, New York, 1980, p. 71.

54. D. L. D. Caspar, in *Proceedings of the Third John Innes Symposium,* R. Markham and R. W. Horne (Eds.), North-Holland, Amsterdam, 1976, pp. 85-99.

55. L. Sawyer and M. N. G. James, *Nature,* **295**, 79 (1982).

56. I. E. P. Taylor, M. Tepfer, P. T. Callaghan, A. L. Mackay, and M. Bloom, *Appl. Polym. Sci.,* **37**, 377.

57. G. Chambat, J. P. Joseleau, and F. Barnoud, *Phytochemistry,* **20**, 241 (1981).

58. G. Chambat, F. Barnoud, and J. P. Joseleau, *Plant Physiol.,* **74**, 687 (1984).

59. B. J. H. Stevens and R. R. Selvendran, *Carbohydr. Res.,* **128**, 321 (1984).

60. J. Smith and D. T. A. Lamport. (Unpublished results.)

61. J. Willard and D. T. A. Lamport. (Unpublished results.)

62. S. C. Fry, *Planta,* **157**, 111 (1983).

63. T. Higuchi, *Encyclopedia Plant Physiol.,* **13B**, 194 (1981).

64. R. Hammerschmidt, D. T. A. Lamport, and E. P. Muldoon, *Physiol. Plant Pathol.,* **24**, 43 (1984).

6

Structure, Swelling and Bonding of Cellulose Fibers

RAYMOND A. YOUNG

Department of Forestry, School of Natural Resources and School of Family Resources and Consumer Sciences, University of Wisconsin, Madison, Wisconsin

It would be impossible to cover all aspects of the structure, swelling, and bonding of cellulose fibers in this short treatment. The approach is rather to treat the literature that is pertinent to tying together these three related topics on cellulose fibers. Emphasis will be placed on recent developments and/or topics that have not been extensively reviewed in the literature, such as cellulose fiber wettability, hornification, and dry-forming of paper. The swelling and bonding of cellulose fibers are closely related to the unique interaction of cellulose and water. This interaction is due not only to the complex structure and properties of cellulose but also to those of water. The interactions of water at the cellulose interface are further expounded upon by Goring (1) and Caulfield (2, 3). To discuss the swelling and bonding of cellulose fibers properly, it is necessary to give further treatment to cellulose structure and morphology.

FIBER STRUCTURE

There are two basic theories as to how glucose is transformed into a crystalline cellulose polymer. Preston has proposed in Chapter 1 that glucose moieties are added at the microfibrillar tips and simultaneously incorporated into the micro-crystalline structure. The other theory is that the biological formation of cellulose I is through extracellular association of presynthesized $(1-4)$-β-D-glucans by hydrogen bonding which is not controlled enzymatically (4–6).

Colvin (4) has recently purported the latter theory and has gone on to suggest that the extracellular glucan chains hydrogen bond to form long, thin sheets

91

which then superimpose themselves nonenzymatically by London forces to form the nascent microfibril. The lateral (side-by-side) association of the chains is through hydrogen bonds between the oxygens attached to C-3 and C-6 of the glucose residues in adjacent chains. Colvin suggests that the sheets form because the edges of the glucans are hydrophilic and their faces are hydrophobic. There is evidence in the literature for this type of structure both from x-ray diffraction and packing analysis (7–9).

Warwicker and co-workers (10, 11) have also suggested a sheet structure for cellulose based on swelling reactions in caustic soda. These investigators found in their swelling experiments that there was considerable variation of distance between the (101) planes, but that within the (101) planes the interchain distances were virtually invariant. The obvious conclusion was that the true fundamental unit in these reactions was a sheet of chains and not a single chain.

French (12) and Blackwell (Chapter 2) have suggested that there is a perpendicular hydrophobic attraction between the cellulose sheets that holds the structure together. French cites the insolubility of cellulose in strong agents that disrupt hydrogen bonds (such as aqueous NaOH) as evidence for this supposition. Pasteka (13) attributes the cellulose penetration ability of molecules containing planar nonpolar groups (i.e., aqueous quarternary ammonium bases with cyclic nonpolar planar phenyl or benzyl groups) to the hydrophobic portions of the cellulose sheet structure. The curious phenonmenon of cellulose inclusion compounds might also be explained as entrapment in the collapsed sheet structure (14). McConnell (15) found a strong correlation for the percent retention of inclusion compounds such as alcohols, aromatics, and alkanes with a surface tension/solubility ratio parameter of the materials. However, the pore structure of cellulose fibers could also account for entrapment of small molecules as discussed later. Separation of cellulose sheets represents intracrystalline swelling of cellulose, whereas separation of microfibrils is an intercrystalline phenomenon.

Although the precise mechanism of macromolecular biosynthesis of cellulose chains has not yet been elucidated, it is probable that the chains associate mainly by hydrogen bonding in a uniform manner to form characteristic cellulose I crystalline structures based on x-ray diffraction studies. The parallel association of the cellulose chains in the cystallite is currently favored over the antiparallel arrangement that characterized the classic Meyer–Misch unit cell (see Chaps. 2, 3). The main planes of the monoclinic crystal structure of cellulose are visible on x-ray diffractograms; however, the peaks of different intensity from the various lattice planes are set on a diffuse background that has been attributed to a noncrystalline or amorphous structure for cellulose (16). The nature and association of the "amorphous" cellulose with the crystalline structure have been a matter of considerable conjecture and debate for over 50 years.

A number of models have been postulated for the crystalline–amorphous structure of the cellulose. The various models can be summarized into two basic conceptual forms (Fig. 6.1); (a) a single-phase, metastable crystalline structure confined by the microfibril boundary similar to that proposed by Preston in

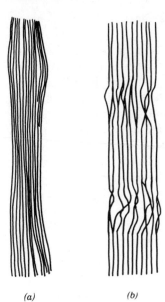

(a) (b)

Figure 6.1. Models of crystalline-amorphous cellulose structure most representative of (a) native cellulose and (b) regenerated cellulose (17).

Chapter 1, and (b) a two-phase fringe-micellar structure encompassing crystalline and noncrystalline features similar to that defined by Frey-Wyssling many years ago (17, 18). The properties of cellulose fibers can be described on the basis of both models.

The basic structural entities beyond the unit cell are the elementary fibril and the microfibril. There is considerable variation in reported dimensions for these basic units. Warwicker et al. (19) summarized the measurements up to 1966, and most recent measurements have just served to more or less confirm previous estimates of sizes (20, 21). It appears that the elementary fibril is about 3.5 nm (2–6 nm) wide. Larger microfibrils are either continuous structures or aggregations of elementary fibrils with dimensions that range up to 3.5–10 nm thick and 10–30 nm wide depending on the source of the cellulose (also see Chap. 1). There can be interfibrillar spaces of 1.2–5 nm wide with water present in the wet state. In the dry state these spaces are only about 1 nm wide (22). The dimensions of the fibrillar elements and the spaces will thus vary with the method and extent of treatment of the fibers and the extent of swelling.

There has never been any microscopic evidence for amorphous materials in native cellulose microfibrils (17). There is overwhelming evidence, however, to suggest that the cellulose microfibrils can be hydrolyzed to relatively uniform micellar fragments (23, 24). Connor et al. (Chap. 15) noted that the readily acid-hydrolyzable fraction accounted for about 10% of the cellulose samples, and Sharples (25) presented mathematical and experimental evidence for endwise depolymerization (rather than transverse attack) in acid hydrolysis of cellulose. Since it is impossible to detect any density variations along the length of the microfibril microscopically, the noncrystalline zones must be less than 5 nm in length (17).

After acid hydrolysis, which presumably cleaves the microfibrils in the weak disoriented zones, the size of the crystalline particles varies in the range of 10–60 nm, with 50 nm the preferred size, although prolonged hydrolysis gave particles 10 nm long (19). Yachi et al. (26) used leveling-off degree of polymerization (LODP), small-angle x-ray scattering, and gel permeation chromatography to investigate particle size of hydrolyzed cellulose samples. They reported a consistent particle size of 20 nm and suggested that periodic variations must occur along the axis of fibrils of native cellulose. Goto et al. (27) published electron micrographs of disintegrated microfibrils of *Valonia* that clearly showed periodic regions of microfibrillar separation after treatment in a homogenizer (Fig. 6.2). As shown, strands run continuously between the separated sections in a rather uniform breakdown pattern. Rånby (28) also reported periodic variations of the length of the elementary fibrils in the range of 10–80 nm. Heyn

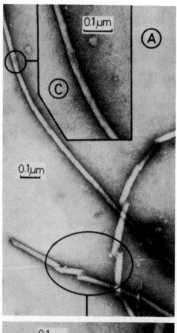

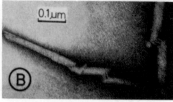

Figure 6.2. Microfibrils of *Valonia* fractured during disintegration with a homogenizer. Negatively stained with uranyl acetate. Photos *B* and *C* are magnifications of photo *A* (27). (Courtesy of T. Gato and Springer-Verlag.)

(29) measured a consistent size (width) of 3.5 nm for elementary fibrils, and Yachi et al. (26) suggested that the elementary fibril had the overall dimensions of $3.5 \times 3.5 \times 20$ nm^3.

Several investigators have suggested that the periodic variations in the crystalline structure of cellulose are due to growth and/or drying stresses. Lofty et al. (30) found a definite dependency of the length of crystallites on the angle of the fibrils and the distance from the core in wood. The crystallite lengths for the experimental material varied from 4.2 nm to 31.1 nm, decreasing with increasing crystallite orientation angle and decreasing radial position. A schematic model of the elementary fibril of cotton with stain-distorted, tilt and twist regions was proposed by Rowland (31, 32).

In his studies on cellulose biosynthesis, Colvin (4) observed that the tips of the microfibrils were not in the shape of blunt rods but rather tapered down to a "chisel" shape. Similarly, after acid hydrolysis of cellulose material, the microcrystalline particles often have tapered ends. This evidence suggests that the native cellulose micelles might have tapered ends which provide the periodic "weak points" in the microfibrils. These areas also would be regions of greater disorder and more accessibility.

An intriguing variation for the structure of the cellulose microfibrils that is compatible with these observations is the hexagonal microfibrillar model proposed by Beall and Murphey in 1970 (33). In Figure 6.3 (a, b), a comparison is made between the generally accepted model of Mühlethaler (34), which has fasciculation in the (101) plane, and the hexagonal model. Both models contain 37 chains, and fasciculation can occur with the Beall and Murphey model in the (101) plane direction through hydrogen bonding of separate hexagonals (Fig. 6.3c). Dislocations would then occur in the depressions between the unit microfibrils. Subsequent layers could be formed on both the (101) and (101) planes. Recent evidence, however, may preclude this model (of Chap. 1, p. 12).

To summarize briefly, it seems highly probable that cellulose microfibrils contain weak zones that occur periodically in the lengthwise direction. However, these are not regions of great disorder as characterized by the fringe–micellar model. Chemical attack can be initiated in these regions, but only small amounts of water sorption can take place. The principal modification reactions are probably on the surface of the microfibrils, especially with cotton. Simple calculations demonstrate that almost 40% of the hydroxyl groups of cellulose would be at the surface if the dimensions of the fibrillar elements were 6×4 nm (ninety-six 8×12 anhydroglucose units in the cross section). It thus becomes very important to understand the complex system of voids present in native cellulose fibers. There is considerable variation in the proposed models and measured sizes of the microfibrillar elements.

Variable Structure of Cellulose Fibers

Cellulose is widespread in nature, occurring in both primitive and highly evolved plants. The highest percentages of cellulose occur in seed hairs such as cotton;

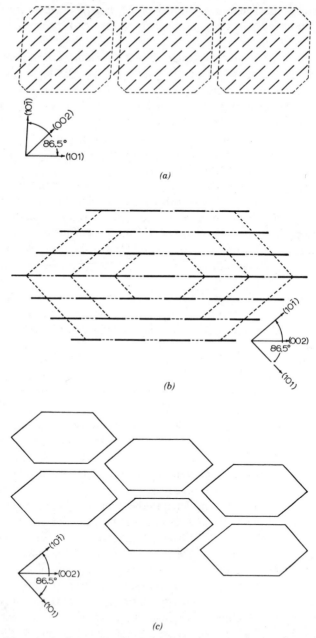

(a)

(b)

(c)

Figure 6.3. Structure of cellulose microfibrils (a) fasciculation of unit microfibrils in the direction of the (101) plane according to Mühlethaler (34); (b) hexagonal unit microfibril configuration according to Beall and Murphey (33); (c) a possible mode of hexagonal unit microfibril fasciculation and layering (33). (Courtesy of Wood and Fiber Science.)

much less is present in the primitive plants (Table 6.1) (35). Commercially, the two most important cellulose fibers are cotton and wood (pulp), and their microfibrillar structures are similar as shown in Figure 6.4 (36, 37). Both fibers contain a primary wall with a rather randomn orientation to the microfibrils and a thin S1 secondary wall layer that is often referred to as the winding layer with cotton fibers. The dominant S2 layer of the secondary wall essentially dictates the properties of both fibers, and the angle of the microfibrils in this layer is only about 20–30°, from the fiber axes. A small S3 layer is also present in both fibers. The near coincidence in the orientation of the chain direction of the tightly packed cellulose macromolecules, the microfibrils in the major S2 layer of the cell wall, and the fiber direction demonstrate the structural continuity developed in nature (17). However, there are some distinct differences in the more detailed structure of wood and cotton fibers.

The crystallinity of different cellulosic materials also varies considerably, as shown in Figure 6.5 (38). Obviously, different methods give different values for the degree of crystallinity, although the relative pattern is consistent, with cotton having the highest crystallinity and regenerated celluloses the lowest. The methods for determination of crystallinity have been divided into three categories; (a) physical methods such as x-ray diffraction, infrared absorption, and density; (b) chemical swelling methods, which include iodine and water sorption, formylation, and acid hydrolysis; and (c) nonswelling chemical methods such as nitrogen sorption and thallation (16, 19, 39). The swelling methods can disrupt interchain hydrogen bonding and penetrate the molecular structure to some extent. The nonswelling methods utilize agents that do not substantially disrupt hydrogen bonding; thus they primarily measure the amount of accessible surfaces such as external surfaces of crystallites and internal voids. Zeronian et al. (40) found that degrees of crystallinity determined from water sorption isotherms fit well with results obtained from other chemical swelling methods if a distinction is made between crystallite surfaces and amorphous regions. In addition to variations in cellulose crystallinity between cotton and wood fibers, there are major differences in composition.

TABLE 6.1   Cellulose Content of
Plant Materials (35)

Plant Material	Cellulose (%)
Cotton	95–99
Ramie	80–90
Wood	40–50
Bark	20–30
Mosses	25–30
Bacteria	20–30
Brown and red algae	1–10

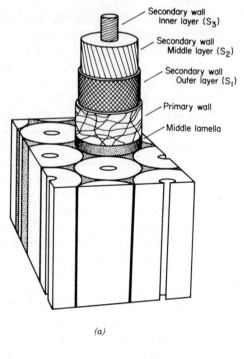

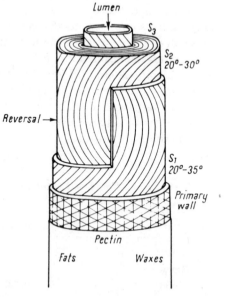

(b)

Figure 6.4. Structure of (*a*) wood and (*b*) cotton fibers showing microfibril orientation and relative size of different layers of the cell wall. (Courtesy of the Shirley Institute, Manchester, England, and John Wiley & Sons.)

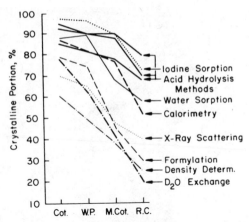

Figure 6.5. Degree of crystallinity of cellulosic materials by different methods according to Mark and Ruck (38). (Cot. = Cotton, W.P. = Wood Pulp, M. Cot. = Mercerized Cotton, R.C. = Regenerated Cellulose.)(Courtesy of Yale University Press.)

Although cotton is 95% cellulose, wood contains only about 42% cellulose, with the remainder of the material composed of mostly hemicelluloses and lignin. There has been considerable effort in recent years to determine the relative distribution of these three main polymeric components in wood fibers. Although the middle lamella is mainly lignin, still over 70% of the total lignin in wood is contained in the secondary wall (41). Electron micrographs taken of wood fibers in which the carbohydrates have been chemically removed clearly demonstrate that the lignin is disbursed throughout the cell wall (Fig. 6.6). A complementary pattern is shown when the lignin is removed to leave a carbohydrate skeleton (20).

The hemicelluloses are also distributed throughout the cell wall in wood fibers, although it has been reported that there is a higher concentration of these low-molecular-weight polysaccharides in the outer layers of the cell wall (S1 and S3) than in the S2 layer (42, 43). It is quite probable that the hemicelluloses are chemically bonded to lignin; it is frequently difficult to obtain a carbohydrate-free lignin preparation (44). Combinations of lignin with xylans, mannans, and pectic type substances have been reported in the literature (45, 46).

Marchassault and co-workers (47–49) have shown by x-ray and infrared spectroscopic techniques that the hemicelluloses are oriented in the direction of the cellulose microfibrils and suggested a strong association of these polysaccharides with the cellulose. Both Parameswaran and Liese (42) and Scott (50) suggested that the glucomannan is more intimately associated with the cellulose than the xylan, based on enzymatic and extraction techniques, respectively. Scott also estimated that about half of the hemicelluloses most rapidly extracted from high-yield pulp were associated with the lumens and large pores. The remaining half, about 5% of the xylan and 2–3% of the glucomannan, was estimated to be the maximum amount potentially, but not necessarily, available for fiber–

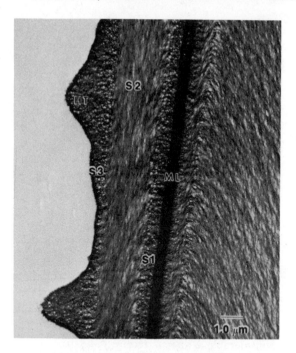

Figure 6.6. Transmission electron micrograph of a lignin skeleton of Douglas fir tracheid wall after removal of polysaccharide components. The S2, S1, and S3 layers as well as the middle lamella (ML) are readily distinguishable (20). (Courtesy of Springer-Verlag.)

fiber bonding. Scott's results did not support the notion that significant amounts of adsorbed xylan remain on the pulp surface after washing. However, other investigators have presented evidence for readsorption of xylan onto the pulp fiber surfaces during alkaline pulping (51, 52).

In the cell wall, hemicelluloses appear to be associated with both the lignin and the cellulose and may form an intermediate layer between the lignin and the cellulose in the microfibrillar structure (35, 53). It is possible that the glucomannan is more closely associated with the cellulose than the xylan, although both these polymers have an easily extractable portion and a more firmly bound fraction (50).

The hydrophilic properties of wood and pulp fibers have been mainly attributed to the presence of hemicelluloses (54), although there are conflicting opinions in the literature (55). Bluhm et al. (56) have discussed the importance of hydration in the crystal structue of these polysaccharides and demonstrated that columnar hydration is characteristic of xylans and sheetlike hydration of glucomannans. They suggested that columnar hydration is more reversible and less morphologically damaging, whereas sheetlike hydration causes uneven expansion or contraction of the material and thereby destroys the structural cohesion. If it is assumed that this is the case for hemicelluloses in wood, the implications are

that the xylans may be more important to reversible swelling and the glucomannans may have a relatively greater structural role. However, the galactose
substituents attached to the 6 position of the glucomannan backbone also serve
to encourage hydration and plasticity. Both these polysaccharides exhibit stress
relaxation under tension as a result of hydration, which is an important structural
adaptation (56).

Various models have been proposed to visualize the complicated cell wall
structure as described above. Those based solely on cellulose crystalline-amorphous or paracrystalline structures have already been presented in Figure 6.1.
Models that take into account the hemicelluloses and lignin in the wood cell
wall have been proposed by Fengel (35) and Goring (57). In Fengel's model
(Fig. 6.7), the elementary fibrils are surrounded by monolayers of hemicelluloses,
with the larger units enclosed by hemicelluloses and lignin. Variable thicknesses
for the hemicellulose layers would explain differences in microfibrillar thickness.
Based on extensive studies on the distribution of lignin in the cell wall by
ultraviolet microscopy, Goring and co-workers (53, 57) proposed an interrupted
lamellar structure for the wood cell wall, as shown in Figure 6.8. The model
shows the cellulose microfibrils as ribbonlike structures consisting of 2–4 pro-
tofibrils bonded on their radial faces with their tangential surfaces coplanar and
parallel to the middle lamella. The lignin was visualized as layers between the
cellulose ribbons. The realization of this model was based partially on the
proposed lamellar structure of cell walls by Scallan and co-workers (59, 60)
described in the next section.

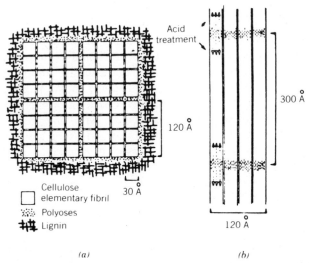

(a)   *(b)*

Figure 6.7. Schematic model of structure of microfibril showing arrangement of cellulose fibrils,
hemicelluloses (polyoses), and lignin in cross section according to Fengel (35). (Courtesy of Tappi
Journal.)

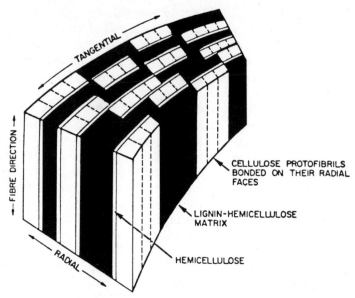

Figure 6.8. Interrupted lamella model for the ultrastructural arrangement of carbohydrates and lignin in the cell wall according to Goring (53, 57). (Courtesy of the American Chemical Society.)

Cell Wall Pore Structure

As Caulfield has pointed out (3), water does not interact with glucose moieties buried within the cellulose crystallite but only on the surfaces of the cellulose crystallites—that is, in the spaces between the cellulose fibrils and microfibrils. Thus the microporous structure of cellulose fibers is of great importance to the properties and reactivity of cellulose fibers in the presence of water. This microstructure influences the penetration of dyes and cellulose-reactive compounds, and with pulp fibers, it influences the extensive imbibition of water, which imparts desirable bonding properties to fibers in wet papermaking.

Pores arise from imperfections in the lateral packing of microstructural elements. Networks of channels from pores do not exist to any great extent in dry fibers, but they are opened up by swelling agents. Measurements of the surface areas of dry cotton fibers by nitrogen adsorption give values in the range of 0.6–0.7 m^2/g, which essentially represents the external microscopic fiber surface. However, if swelling agents are utilized, relatively large internal surfaces are developed in the cotton fiber as follows: (all m^2/g) water, 137; acetic acid, 18.3; and ethanol, 7.3 (58).

With wood fibers the extensive microporous structure is also closely related to the removal of hemicelluloses and lignin in the pulping process. In 1968, Stone and Scallan (59, 60) convincingly demonstrated by a solute exclusion technique that the cumulative pore volume in wood pulp fibers increases with decreasing yield as shown in Figure 6.9. The median pore size increases from

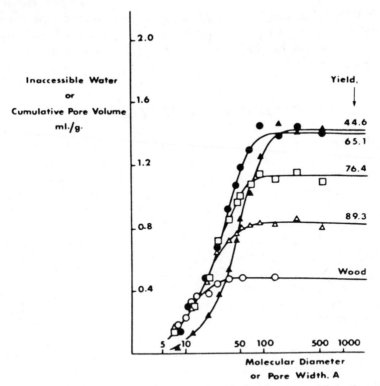

Figure 6.9. Solute exclusion curves of sprucewood and kraft pulps prepared from the same wood (59, 60). (Courtesy of Canadian Pulp and Paper Association.)

1.2 nm to almost 6.0 nm at the lowest yield. The plateau regions of the curves in Figure 6.9 refer to the amount of water inaccessible to a totally excluded macromolecule, which is identical to the fiber saturation point of the fiber. Rowland and Bertoniere (61) have cautioned that the solute exclusion technique may be dependent on the type of solute. Small angle x-ray scattering has also been recently used for determination of the number and size of fiber pores (12, 62, 63).

The pore size distribution varies with the type of cellulosic material. The pulping of wood not only increases the pore volume, but it also creates a wider distribution of pores in the pulp fiber (61, 64). The pore structure of fibers can also considerably influence the reactivity in modification reactions. Nelson and Oliver (65) found that the total volume of micropores in acetic acid swollen fibers correlated closely with acetylation reactivity. They found that the optimum pore size for acetylation was in the range of 2.5–7.5 nm. Young and Fujita (66) have similarly found that the bonding properties of wood are substantially improved by treatment with 3N sodium hydroxide which further expanded the micropore structure in wood. These results are consistent with those of Sawabe and Mori (67), who found that the pore size of holocellulose samples increased with NaOH concentration. They suggested that the change in pore structure with NaOH treatment was due to expansion of pores and unification of adjacent pores, which supports the lamellar pore model proposed by Scallan as discussed in the next section.

BEATING AND SWELLING OF CELLULOSE FIBERS

For papermaking, the cellulose fibers must receive some amount of mechanical treatment in the presence of water to impart the desired physical properties to the resulting paper sheet. The mechanical treatment, referred to as beating or refining, causes a variety of changes in the fiber, but the main effects are short-ening, external fibrillation, and internal fibrillation (68, 69). As a result of beating, the fibers contain much greater quantities of water. For example, bleached low-yield sulfite and kraft pine fibers (never-dried) with an initial water content of 1.42 mL/g and 1.29 mL/g imbibed up to 1.86 mL/g and 1.49 mL/g, respectively, of water when beaten to 250 mL Canadian Standard Freeness (70).

The increased amount of water in the cell wall plasticizes the fiber with the result that the fiber becomes more conformable when randomly laid into a fibrous web by wet-forming processes. The greater conformability and collapse of the fiber lumens give more bonded area at fiber crossover points. Mohlin (71) studied the bonding of single fibers and found that beating had little effect on fiber shear strength but a relatively large effect on fiber conformability, with these effects reflected in the tensile strength of sheets made from the corresponding pulps.

The major changes that occur on beating are at the fibrillar or lamellar level in the cellulose structure. Although the lamellar structure of the cell wall of cotton can be easily seen under the microscope (72), this is not the case with wood fibers. There is, however, good evidence to suggest that the S2 layer of wood fibers is also made up of concentric rings. When mechanically refined, low-yield pulps develop microscopically visible laminar layers, with sulfite pulps producing more lamellae than kraft (73). McIntosh (74) obtained similar evidence of laminar separation on beating of pine holocellulose. Using a layer-expanded technique that involves rapid polymerization of monomer in swollen cell walls, the extensive honeycombed structure of an expanded cell wall can be seen (Fig. 6.10) (3).

Based on his extensive studies of cellulose fiber pore structure by the solute exclusion technique, Scallan (60) proposed a model for the structure of the fiber cell wall. It was composed of sheets of elementary fibrils separated by lamellar spaces of the same order of thickness as the elementary fibril (Fig. 6.11) similar to structures observed under the microscope. Internal fibrillation would then occur by delamination of the concentric layers through intrusion of water and swelling of hemicelluloses in the interlamellar layers. The cleavage occurs mainly in the tangential (101) planes, although some amount of radially oriented de-

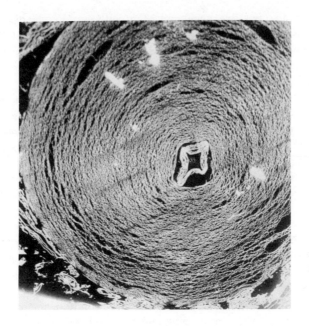

Figure 6.10. Electron micrograph of a cross section of a spruce sulfite fiber after solvent exchange and polymer impregnation (metal shadowed). (Courtesy of USDA Forest Products Laboratory.)

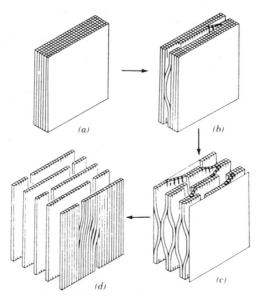

Figure 6.11. Pattern of internal fibrillation of the cell wall with progressive swelling and preferential cleavage of tangentially oriented bonds (zero fibril angle) according to Scallan (60). (Courtesy of Forest Products Research Society.)

lamination is also possible, again consistent with microscopic observations. This extensive disruption and delamination in the cell wall serves to plasticize the fiber, rendering it much more flexible and conformable for production of strongly bonded paper. The cell wall model proposed by Kerr and Goring (Fig. 6.8) is compatible with Scallan's model.

External fibrillation is a more obvious change when viewed under the microscope owing to extensive disruption of the outer layers of the cell wall. Fine fibrillar fragments are left attached to the fiber surface, which results in further hydration (55). Considerable portions of the S1 and even the S2 layers are stripped from the fibers during beating and, together with other cell elements such as parenchyma and fragments, create a fines or crill fraction in the pulp (70). Szwarcztajn and Przybysz (75) have found a strong interdependence of fiber surface fibrillation and crill formation as related to specific surface area and water retention. It is important to note that the physical removal of cell wall layers may also change the chemical characteristics of the fiber surface because of the variable chemical composition of the cell wall. However, it has not been possible to date to sufficiently characterize the chemical composition of the fiber surface so that correlations with fiber bonding strength can be assessed with any degree of certainty. The dissolution and reprecipitation of xylan at the fiber surface in pulping and beating is also still a matter of debate (50, 76).

Although there is extensive damage to the fiber surface with beating, the length of the perimeter of the fiber often remains constant (70). This means that the swelling of the fiber, as a result of imbibition of water, is mainly into the lumen of the fiber. This observation dispels early theories that the outer layers of the cell wall constrained the swelling of the fibers in the beating process (77). Internal fibrillation is thus coincident with, but not dependent on, the extent of external fibrillation.

The importance of external versus internal fibrillation has been discussed by several authors with varying viewpoints as to the relative importance of these two factors (55, 60, 68, 70). Recent evidence strongly indicates that external fibrillation and formation of fines during beating are the main factors affecting drainage and network porosity and that swelling and delamination (internal fibrillation) are critical to sheet consolidation and fiber–fiber bonding (68, 78).

The mechanism and factors controlling the swelling of cellulosic fibers have been a primary concern of both paper and textile chemists for many years. Recent investigations have shown that acidic groups present in the cell wall may have a major influence on the fiber swelling (79). Both Scallan and Lindström and their co-workers (79–86) have found strong correlations between the amount of acidic groups in pulp and the swelling and water retention of the sample. With a higher water retention and fiber saturation point, the fiber bonding improves as previously discussed. Scallan has provided strong evidence to correlate the acidic group content with the breaking length of corresponding sheets (80).

It has been postulated that the acidic groups are mainly associated with the hemicelluloses and lignin in the pulp. The polyelectrolytic character should thus be greater for high-yield pulps. Lindström and Kolman (83) indeed found that bleached kraft pulp was not dramatically affected by pH and electrolyte concentration during beating and sheet forming, whereas the unbleached pulps were affected by the chemical environment. Katz et al. (80) further suggested that the alkali strengthing of mechanical pulps is caused by an increase in acidic group content through alkaline hydrolysis of preexisting esters and lactones in the hemicellulose fraction of the pulp. When ozone treatments were combined with the caustic treatment, the effect on strength was additive. Presumably the ozone introduces additional acidic groups into the lignin component of the high-yield pulp. With high levels of acidic groups it may be possible to get swelling sufficient to break internal cross-links in the cell wall (81).

Cations can also affect fiber swelling and concomitant effects on the strength properties of cellulose fiber webs (87). This effect is closely correlated with the swelling effect of acidic groups, since the ionizable groups present in the pulps are mainly salts of carboxylic acids. The ionized carboxyl group $(-COO^-)$ remains a part of the solid phase while the cation is free in the aqueous solution. Thus the gel swells because of the osmotic pressure differential between the solution of free ions in the gel phase and the aqueous phase outside the gel. Since different cations have different degrees of dissociation and valency, their effect is variable. If the mixture of cations normally present in pulp is removed and replaced with a single species, it is found that the paper made from the pulp is strongest with monovalent species and weakest with trivalent species. The following sequence in order of decreasing strength is typical; $Na^+ > Li^+ > K^+ > NH^{4+} > Rb^+ > Mg^{2+} > Cr^{2+} > Si^{+2} > Ba^{2+} > H^+ > Al^{3+} > Ca^{3+} > Fe^{3+}$ (87, 88). These results have implications for both laboratory experiments and mill operations. A further discussion of the mechanisms of swelling is given in a number of references (79, 89, 90).

HORNIFICATION AND RECYCLING OF CELLULOSE FIBERS

The properties of cellulose fibers in the native state are considerably altered after the fibers are dried. Both cotton and wood fibers have decreased reactivity, and the fiber-bonding properties are generally diminished as a result of drying. This effect is of considerable importance to papermaking, since pulps are often dried for shipment to another papermaking operation and dry wastepaper is the low-cost raw material for recycled papers.

Strength reductions as high as 30% for paper burst and 35% for tensile have been reported for dried versus undried pulp. A linear relationship has been observed between the decrease in burst and the amount of drying, expressed as a percentage of solids of the partially dried pulp (91). Similarly in recycling experiments, there is a progressive reduction in tensile strength with each cycle up to about six cycles (92, 93). However, Horn (92) and McKee (93) have noted an increase in the tearing strength of recycled fiber webs, but Horn found that the tear decreased with subsequent cycles. Fiber breakage as well as hornification influences the strength properties of recycled paper. Klungness (94) evaluated the effect of contaminants on the properties of recycled paper and concluded that the changes were due to the removal process and not to contaminants.

The cause of the strength losses with drying and recycling has generally been attributed to a structural change in the cell wall due to irreversible bonding of cellulose surfaces. Jayme (95) termed this effect "hornification" and later noted that the fibrils aggregated into larger units (strings) on drying, which reduced the area accessible to water and formed a stiffer structure. Akim (96) attributed hornification in dissolving pulps to the amorphous phase of cellulose changing from the pseudo-soft to the glassy state with drying.

Giertz (97), in contrast, stated that the hemicellulose material in pulp fibers is mainly responsible for hornification with drying. Thus, the phenomenon is greatest with hemicellulose-rich pulps. This interpretation does not conflict with other evidence, since the hemicelluloses at the fibril surfaces would serve to draw the fibrils together with drying to form a stiffer structure. Giertz also found that the speed and temperature of drying can affect hornification and that if the drying is not uniform, hornification is noted at an average dry solids content of 60–65%. Generally, hornification increases markedly with drying below 70–75% dry solids.

Stone and Scallan et al. (60, 98) demonstrated by the solvent–solute exclusion technique that dried and reswollen pulp fibers contained much less inaccessible water, had lower fiber saturation points, and had reduced pore sizes in the maximum- and medium-size pore ranges. Apparently with drying the cell wall closes up owing to formation of internal hydrogen bonds, proceeding toward structures *b* and *a* in Figure 6.11. This results in a stiffer and less conformable fiber that cannot bond as well as highly swollen fibers and thus forms sheets with reduced strength properties. The increase in tear properties on recycling is also consistent with stiffer, lesser-bonded fibers in the web.

The strength reductions are of a lesser magnitude for mechanical pulps, and Higgins and McKenzie (91) found that the strength of a neutral sulfite semichemical pulp was less impaired by drying than a kraft pulp. Apparently the lignin in the fiber wall serves to inhibit the internal cross-linking of cellulose fibrils.

Klungness and Caulfield (99) also noted a dramatic reduction in the fiber saturation values for dried compared to never-dried pulps—a 46% reduction for bleached pulp. Drying resulted in loss of both specific surface and specific volume. Although the specific surface area could be easily restored with refining, the specific volume could not be, especially for low-yield pulps. A large increase in the amount of fiber collapse, which indicates an increase in internal bonding, was also reported by these investigators for the dried pulps. An additional effect noted by Klungness and Caulfield was a decrease in fiber wettability upon drying. This effect will be further discussed in the section on wettability of cellulose fibers.

As noted above, partial recovery of the properties of never-dried fibers can be achieved by further refining of the dried fibers. However, strength is recovered at the expense of increased wetness of the pulp. Szwarcztajn and Przybysz (100) have shown that the fines fraction is the major cause of the increase in wetness. It is also possible to use chemical methods to recover the original strength properties of dried pulps. A number of investigators have noted the beneficial effects of a caustic solution treatment of hornified fibers (92, 94). This may be related to the liberation of acidic groups in the cell wall structure, which promotes swelling as described above. Indeed, Lindström and Carlsson (82) have reported that the degree of hornification is less when pulps (which contain carboxyl groups) are dried in the Na^+ form than when they are dried in the H^+ form.

Another approach to the problem of hornification has been that of prevention. This can be accomplished either by incorporation of materials in the pulp that inhibit internal hydrogen bond formation or by chemical modification of the pulp to likewise less bondable structures. Higgins and McKenzie (91) evaluated a wide variety of water-soluble compounds as hydrogen bond inhibitors by incorporation into the pulp before drying. These included glycerol, sucrose, starches, alginate, polyethyleneglycol, guar gum, dyes, urea, borates, and sodium hexametaphosphate. Some success was achieved with sucrose and starches; however, very high concentrations (20%) were necessary, which precluded practical applications. The results for the water-soluble polymers are somewhat clouded because of the inherent adhesive properties of these additives.

In the previous section it was shown that when a pulp was ion-exchanged to contain a single low-valency species such as sodium, the bonding and the strength of the sheet improved. It should be possible to use this concept in the reverse manner—that is, incorporation of high-valency cations in the wet pulp before drying to prevent hornification. Thus trivalent cations such as Al^{3+}, Cr^{3+}, and Fe^{3+} should provide the best protection against hornification followed by divalent ions such as Ca^{2+} and Mg^{2+}, with monovalent ions such as Na^+ and Li^+ offering the least protection. Indeed, Klungness (101) has found that when $Al_2(SO_4)_3$ is present, hornification is inhibited to some extent. This phenomenon warrants further exploration.

Restoration of the original properties of dried pulps can be partially achieved by chemical modification of the cellulose fibers. Carboxymethylation has been utilized by a number of investigators to improve the strength properties of pulps (88, 102). Talwar (102) emphasized the need to have the CMC in the salt form, since all the improvements realized by substitution of the CMC substituent were lost when the product was in the acid form. Ehrnrooth et al. (103) acetylated dried pulps and found that the water retention value (WRV) of the substituted pulp increased up to about 5% acetyl content. The WRV was slightly decreased for the never-dried pulp with acetylation. The pulps were esterified in two different solvents, toluene and pyridine. The WRV for the dried pulp modified in toluene was lower than the WRV for the pulp modified in pyridine. This difference is probably due to inclusion of the pyridine in the swollen cellulose structure, as described in a previous section. The pyridine appears to inhibit internal bond formation by separating sheets or lamallae of cellulose. This observation also illustrates another possible approach to inhibition of hornification—incorporation of cellulose inclusion compounds.

Howarth (104) discussed the fundamental problems in paper recycling in his paper at the Oxford symposium devoted to fiber–water interactions in papermaking. In his concluding remarks he stated that "in order to measure the essential difference between secondary and virgin cellulose fibers we need a technique to measure their surface properties, chemical and physical." In the following section on wettability of cellulose fibers, a technique will be discussed that could well satisfy at least part of this need.

WETTABILITY OF CELLULOSE FIBERS

The previous discussion has emphasized the need of both textile and pulp and paper scientists for information on the surface properties and wettability of cellulose fibers. Wetting influences such important properties as adhesion, interfiber bonding, absorbency, and wicking of fibrous materials and is very important to the manufacturing processes. In addition, wettability information can be valuable for characterization of fiber surfaces, fiber surface modifications, and interaction of fibers with liquids and surfactants.

The importance of fiber wettability to liquid movement in fabrics, paper, and fibrous pads is shown in the basic equation for capillary flow as developed by Lucas and modified by Washburn (105);

$$\frac{dh}{dt} = r\,\gamma_{LV}\,\frac{\cos\theta}{4\eta h} \tag{1}$$

where t is time, h is the length of the capillary, r is the capillary radius, and η is viscosity. (The dominant term in Eq. (1) is $\gamma_{LV}\cos\theta$, which defines the wetting of the fiber.)

Up until recently there was a lack of information on fiber wettability characteristics, because a practical method for measuring this property was not available. Aberson (106) utilized the Lucas–Washburn equation above to obtain fiber contact angles. He calculated r by passing a surfactant solution with a fiber contact angle (θ) of near 0° through a mat of known thickness. This r value was then used to obtain contact angles for different water solutions passed through the mat. Rough estimates for fiber contact angles were obtained by this method. In the 1960s, several scientists proposed the use of direct contact angle measurements on the surface of fibers as a means of assessing the fiber wettability (107, 108). Such techniques are rather unreliable with aniosotropic structures such as filaments, and even further difficulties are encountered with the much smaller diameter cotton and pulp fibers. There are additional limitations to the use of contact angles described in the literature (109, 110).

Because of these difficulties, several investigators (111, 112) have utilized films cast from isolated cellulose and lignin polymers for contact angle measurements.

Luner and Sandell (112) reported water contact angles of 34° for cellulose and 58–60° for kraft lignins. However, the harsh treatments necessary for isolation of the polymers and the altered morphological characteristics cast doubt as to the applicability of this data for practical use.

In the early 1970s several groups attacked the problem of fiber wettability measurement by an alternate approach. Bendure of Procter and Gamble (113), Okagawa and Mason at PPRIC (114), and Young and co-workers at the Textile Research Institute (now at the University of Wisconsin) (109, 115, 116) independently developed similar methods for measurement of fiber wettability based on the Wilhelmy principal (117). This technique avoids problems described above and provides reliable quantitative data for wetting of all types of fibers, including fine native cellulose fibers. The theoretical basis for this approach is described next.

Theory and Methodology

The interaction of a liquid with a solid surface is governed by three interfacial tension forces—γ_{SV}, γ_{SL}, γ_{LV}—related at equilibrium by the Young–Dupré equation:

$$\gamma_{LV} \cos\theta = \gamma_{SV} - \gamma_{SL} \tag{2}$$

where θ is the contact angle and S, L, and V stand for solid, liquid, and vapor, respectively. Dupré also developed the relationship for work of adhesion, WA;

$$WA = \gamma_{LV} + \gamma_{SV} - \gamma_{SL} \tag{3}$$

Combining Eqs. (1) and (2) gives

$$WA = \gamma_{LV} + \gamma_{LV} \cos\theta = \gamma_{LV}(1 + \cos\theta) \tag{4}$$

The work of adhesion is a measure of the interaction of a liquid and a solid. Since γ_{LV} is a property solely of the liquid, the latter term, $\gamma_{LV} \cos\theta$, reflects the direct interaction of a liquid and solid, termed the specific wettability, WS, or adhesion tension. The use of the specific wettability or the work of adhesion is preferred over the contact angle, since the former values have more physical significance. The contact angle is a geometric parameter that complicates quantitative analysis of wetting hysteresis—that is, the difference between the advancing (a) and receding (r) wettability. Thus by use of work of adhesion, the wetting hysteresis, H, can be accurately quantified through the expression (118);

$$H = \frac{WA_r}{WA_a} = \frac{1 + \cos\theta_r}{1 + \cos\theta_a} \tag{5}$$

According to the formula of Wilhelmy (117), the pull exerted on a rod (fiber or filament) inserted into a liquid is expressed by

$$Fw = P\,\gamma_{LV}\cos\theta \qquad (6)$$

where P is the perimeter of the rod. Then the specific wettability is formulated as (109, 115)

$$WS = \frac{\text{force per fiber}}{\text{perimeter of fiber}} = \frac{Fw}{P} = \gamma_{LV}\cos\theta \qquad (7)$$

The wetting force is obtained by measuring the apparent weight increase when the fiber contacts a liquid of known surface tension. This is done with an apparatus of the type shown in Figure 6.12. A small fiber sample is attached to a wire hook and suspended from a microbalance hangdown wire. The liquid is then raised until a force change is observed from contact with the liquid; this marks the zero immersion level. The change in weight is recorded within a few seconds, and the liquid is raised again to obtain a second reading. Depending on the size of the fiber, there will be a counteracting force of buoyancy, Fb, when the fiber is immersed into the liquid beyond the point of contact. Applying this correction gives the expression,

$$F = Fw - Fb \qquad (8)$$

where F is an electrobalance reading corrected for the force reading in air and Fw is the weight of liquid clinging to the fiber or the Wilhelmy wetting force (also designated $Fw(a)$ for advancing). The buoyancy force, Fb, is obtained from the equation,

$$Fb = lad \qquad (9)$$

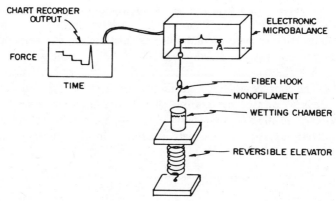

Figure 6.12. Apparatus for measuring the wettability of fibers by the Wilhelmy technique (109). (Courtesy of Textile Research Journal.)

where l = immersed length of fiber, a = area of fiber cross section, and d = density of liquid. Substituting for Fb in Eq. (8),

$$F = Fw - lad \tag{10}$$

Then a plot of F as a function of depth of immersion l should give a straight line with a slope $-ad$ and an intercept Fw. The sign of the slope will depend on whether the wetting force is positive or negative. A typical wetting force plot for a fiber with a contact angle less than 90° is shown in Figure 6.13. Additional features of these plots will be described in a later section. It is possible to calculate a wetted cross-sectional area of the fiber from the slope if the fiber has sufficient buoyancy. Good agreement has been obtained between the area calculated from wetting and that measured by independent methods (109).

The perimeter P must also be known to obtain the specific wettability, WS from Eq. (7). The perimeter can be measured independently by microscopic techniques, or it can be obtained directly from the wetting measurement. In the

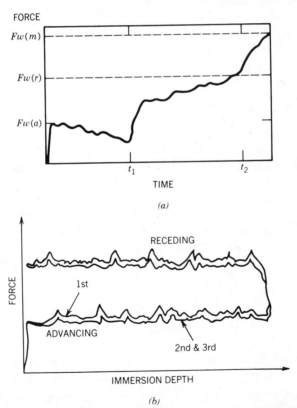

Figure 6.13. Typical force recordings for (a) a buoyant fiber with a contact angle of less than 90° (time to t_1 is for immersion followed by emersion); (b) a nonbuoyant fiber with three cycles of immersion and emersion.

wetting experiment it is possible to measure the maximum wetting force, $Fw(m)$ when a fiber is pulled out of the liquid. Assuming that fiber buoyancy end and wall effects are negligible, the meniscus of liquid contact with the fiber end will be at a contact angle of 0° at the point of fiber pullout. (It is critical that the fiber have a clean cut end.) The equation for fiber perimeter thus reduces to

$$P = \frac{Fw(m)}{\gamma_{LV}} \tag{11}$$

Another approach is to obtain the perimeter from both the advancing and receding wettability measurements. The receding wettability is determined by removing the fiber from the liquid (in practice by moving the liquid down), which gives another, higher force reading, $Fw(r)$ according to the equation,

$$Fw(r) = \gamma_{LV} P \cos\theta_r \tag{12}$$

and for advancing wettability we have,

$$Fw(a) = \gamma_{LV} P \cos\theta_a \tag{13}$$

Then by subtracting Eq. (13) from Eq. (12),

$$P = \frac{1}{\gamma_{LV}}[Fw(r) - Fw(a)]/[\cos\theta_r - \cos\theta_a] \tag{14}$$

The validity of this approach has been demonstrated by Miller et al. (119). With very fine fibers such as those of native celluloses, the measurement is simplified because the buoyancy effect is neglible and, therefore, $F = Fw$ (115). Klungness (120) also demonstrated with wood pulp fibers that the first force indicated on lowering the liquid level after measuring $Fw(a)$ was equivalent to $Fw(m)$ (contact angle = 0°). Thus, the perimeter can be obtained directly for each wetting force along the length of the fiber. This approach was verified by measuring the perimeters with an independent cross-sectioning method. It follows then, for fine fibers that

$$\cos\theta = \frac{Fw}{[\{Fm/\gamma_{LV}\} \times \gamma_{LV}]} = \frac{Fw}{Fm} \tag{15}$$

It is also possible to obtain the contributions to the surface free energy due to dispersion and polar forces. Fowkes (121) proposed that the total free energy at a surface is the sum of the contributions from the different intermolecular forces at the surface. Thus, for a solid,

$$\gamma_S = \gamma_S^d + \gamma_S^p \tag{16}$$

where the superscripts p and d denote dispersion and polar contributions, respectively. Wu (122) further developed the evaluation of dispersion and polar forces as contibutions to work of adhesion (Eq. (3)) by using a reciprocal means approach and suggested the following relationship;

$$\gamma_{LV}\cos\theta = -\gamma_{LV} + \frac{4\,\gamma_S^d\,\gamma_L^d}{\gamma_S^d + \gamma_L^d} + \frac{4\gamma_S^p\,\gamma_L^p}{\gamma_S^p + \gamma_L^p} \qquad (17)$$

Thus, by measuring fiber wettability ($\gamma_{LV}\cos\theta$) in a polar and a nonpolar liquid of known surface tensions, the two simultaneous equations can be solved for γ_S^d and γ_S^p. Water and methylene iodide have been used to obtain the polar and dispersion forces for keratin and glass fibers by the Wilhelmy wettability method through the Wu equation (118, 127), but applications to cellulose fibers have not been reported in the literature. Nguyen and Johns (123) measured contact angles of sessile drops on planar wood surfaces with several different liquids and found that up to 60% of the total surface free energy for Douglas-fir was due to dispersion forces. This is consistent with ESCA data obtained by Young et al. (124) for wood surfaces. The latter investigators found that the surface of wood was rather nonpolar based on oxygen/carbon ratios.

Although cotton, once scoured clean, has a rather consistent cellulose composition, this is not the case with wood pulp fibers, as already discussed in previous sections. Cassie (125) modified the Young–Dupré equation (Eq. (2)) to take into account composite situations such as the surfaces of pulp fibers. He suggested that if two domains of different surface free energy occupy fractions f_1 and f_2 of the surface, then it follows that

$$\gamma_{LV}\cos\theta_c = f_1\,(\gamma_{S_1V} - \gamma_{S_1L}) + f_2\,(\gamma_{S_2V} - \gamma_{S_2L}) \qquad (18)$$

where $f_1 + f_2 = 1$. This equation defines an additional contact angle, θ_c, which represents the composite of the individual equilibrium contact angles θ_1 and θ_2. Equation 18 can therefore also be expressed as

$$\cos\theta_c = f_1\,\cos\theta_1 + f_2\,\cos\theta_2 \qquad (19)$$

However, contributions, f_1, f_2 and so on are difficult to separate and ascertain in practice (115).

Applications

The use of the Wilhelmy principle for measurement of fiber wettability was first described by Collins (126) in 1947. He also recognized that a zero contact angle could be assumed for fiber retreat from the liquid and that a simple and accurate means for measurement of fiber perimeter was thus available. Since cellulose fibers were the predominant fiber of commerce in 1947, all of Collins' original measurements were on cellulosics. By recalculating his data according to Eq. 7

above, wettabilities of 60.47 and 30.23 dynes/cm are obtained for cuprammonium rayon and raw cotton fibers, respectively. This corresponds to respective contact angles of 34° and 65°. The contact angle for the pure cellulose (rayon) fiber obtained by Collins is exactly that obtained by Luner and Sandell (112) on cellulose films! The raw cotton fiber has waxy deposits on the surface, which accounts for the much lower wettability. Values of 105° are expected for contact angles on waxy surfaces, and, indeed, Collins found contact angles up to 94° along the surface of the raw cotton fibers. The hydrophobic surface character of raw cotton fibers was confirmed in subsequent work by Young (116) as shown in Table 6.2.

Several other investigators reported on the use of the Wilhelmy technique for measurement of wettability (127–129). Results for cellulose and a number of other types of fibers, obtained since 1975, are shown in Table 6.2. Although the details of all these experiments cannot be described here, the data demonstrate the variety of applications of this technique for evaluation of many different surface phenomena associated with native and man-made fibers.

For cellulose fibers, several experiments are noteworthy. The effect of scouring raw cotton fibers is shown in Table 6.2 (samples 1 and 2). Cleaning of the fiber surface with mild caustic clearly enhances the fiber wettability by removal of the waxy deposits (116). Delignification also enhances wood pulp fiber wettability by removal of the more hydrophobic lignin, apparently from the fiber surface as well as the interior of the fiber (samples 3–5) (120). Aberson (106) also noted a significant effect of extractives on water contact angles of pulp fibers with his capillary flow technique.

The so-called "self-sizing" effect of pulp fibers (130, 131) can also be evaluated by the Wilhelmy technique. When wood pulp, which contains resins or fatty acids, is heated or aged, the material develops a water-repellent character that can seriously inhibit end-uses dependent on water absorbency. Swanson (130) has suggested that self-sizing is caused by vapor phase migration of resinous materials from the ray cells and resin ducts to the fiber surfaces. This effect was studied by Berg and Hodgson (132, 133) as shown in Table 6.2 (samples 15 and 18). The heating of the fiber decreases the wettability from a contact angle of 30° to 47°. The effect of self-sizing can be counteracted by proper choice of a surfactant as shown for sample 19. However, judicious selection of the surfactant is critical, since some surfactants will only aggrevate the poor wettability (sample 20). Aspler et al. (131) have suggested that the surfactant acts to prevent self-sizing by solubilizing fatty acid molecules and forming a physical barrier to chemical bond formation between cellulose and fatty acid molecules. An additional wetting effect, known as "fugitive sizing," was noted by Klungness and Caulfield (99) for aged pulps (samples 3–8). These investigators found that although the wettability of unbleached pulp fibers decreased with drying, the wettability increased after aging for 1 year (samples 3–5 vs. 6–8). The aging effect was possibly due to surface oxidation of lignin in the high-yield pulps.

The variable nature of wood pulp fibers, both from different processes and from the same process, should be obvious from the data in Table 6.2. Not only

TABLE 6.2 Fiber Wettability with Water by the Wilhelmy Method

Sample	Fiber	Contact Angle (degrees)	Source
1	Cotton, unscoured[a]	107	Young (116)
2	scoured (mild caustic)	0	
	Loblolly pine, oven dried		Klungness and
3	16.2% lignin	49	Caulfield (99,120)
4	4.8% lignin	43	
5	0.2% lignin	21	
	Aged 1 year at 20°C		
6	16.2% lignin	23	
7	4.8% lignin	21	
8	0.2% lignin	22	
9	Sugar maple, bleached kraft[b]	53	Berg and Daugherty (132, 143)
10	Western hemlock, NSSC	82	
11	Wester hemlock, TMP, 50% RH	66	
12	TMP, 100% RH	53	
13	Cationic surfactant treated[c]	57	
14	Anionic surfactant treated[d]	30	
15	Douglas-fir, bleached pulp	30	Berge and Hodgson
16	Surfactant treated, Berocel 584[e]	10	(132, 133)
17	Berocel 564[f]	28	
18	Oven-aged	47	
19	Berocel 584 treated	34	
20	Berocel 564 treated	68	
21	Douglas-fir, kraft pulp	52	Young (115)
22	Polyacrylonitrile grafted	54	
23	Polystryene grafted	72	
24	Aspen, NSSC	0	
25	Polystryene grafted	73	
26	Balsam fir, TMP	43	
27	Wool	112	Young and Miller
28	Polypropylene	86	(109, 110, 134)
29	Polyester	75–80	
30	Nylon	62–71	
31	Carbon	40	
32	Glass	24	
33	Glass, E-glass, dry	79	Wesson and Tarantino
34	heat cleaned	44–49	(127)

[a] 65% Relative humidity (RH).

[b] 50% RH.

[c] Octadecyltrimethylammonium bromide.

[d] Sodium dodecylbenzene sulfonate.

[e] Mixture of 15% cationic, 85% nonionic surfactant.

[f] Partially ethoxylated quarternary ammonium salt.

is there inherent variability due to chemical and morphological differences, but also additional variability is introduced because of different processing conditions, chemical additives, and physical treatment, to name a few. However, detailed analysis of fiber surface changes are possible within a given set of fibers or for similar modifications on different fibers. Young (115) found dramatic differences between the wettability of Douglas fir kraft and aspen NSSC fibers (samples 21 and 24, Table 6.2), but these surface differences were essentially nullified when the fibers were grafted with polystyrene (samples 23 and 25). Additional effects can be noted in Table 6.2 (134). The wettability of wood samples has also been determined using the Wilhelmy principle, although the effect of capillary absorption of water needs further evaluation in this case (135, 136). Additional information can be obtained from the wetting measurement by the Wilhelmy technique, as described in the next section.

Hysteresis and Wetting Profiles

Wetting hysteresis refers to the difference in wetting of a substrate in the advancing and receding modes. With few exceptions hysteresis occurs on all solid surfaces (137). It has been defined quantitatively above (Eq. (5)) as the ratio of the receding WA divided by the advancing WA. A number of reasons have been given for hysteresis including surface roughness, chemical heterogeneity, and changes in adsorption. Johnson and Dettre (138) and Eick et al. (139) have suggested that surface rugosities would need to be larger than a few tenths of a micrometer to cause significant contact angle hysteresis. Rugosity of this magnitude is visible under the microscope, but Oliver et al. (137) found it very difficult to get a strong correlation of contact angle with surface roughness. The effect of roughness is also related to surface topology; that is, the same roughness in the form of grooves gives a different behavior from one in the form of pits (140).

Penn and Miller (141) measured wetting forces for microscopically smooth synthetic fibers and attributed the hysteresis effect to chemical heterogeneity. They found that when the surface tension of the wetting liquid was lowered sufficiently, the difference in intrinsic contact angles was no longer large enough to cause hysteresis. Apparently as the liquid surface tension is lowered, the chemically heterogeneous patches at the fiber surface are not distinquished by the probe liquid. Kamath et al. (142) similarly concluded that hysteresis is eliminated when both the high- and low-energy areas are completely wetted by a low surface tension liquid.

As shown in Figure 6.14, there is an inverse relationship between wetting hysteresis and advancing WA (134). It appears that fibers that are already attractive to the liquid are not altered as much in advancing wetting as fibers that are initially more hydrophobic. It may be that some adsorbed water from the air is already present on the surface of the hydrophilic fibers, which would reduce the differences in the advancing and receding wetting forces in water. Humidity control is therefore important for fiber wetting measurements (also cf. samples 11 and 12, Table 6.2) (132, 143).

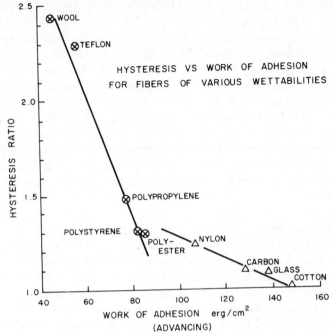

Figure 6.14. Plot of hysteresis ratio versus advancing work of adhesion for various fibers (110, 134).

It is also possible to obtain a wetting profile of a fiber by traversing the liquid slowly across the fiber surface and continuously recording the changes in the wetting forces both in the advancing and receding wetting modes. This slip-stick profile is shown in the idealized plot in Figure 6.13(b) and appears to be a rather random deviation from a mean wetting force. This is not the case, however, since the same profile is produced in subsequent traces. This phenomenon was noted by all three recent developers of the fiber wetting technique (113, 114, 116, 134) and was referred to as capillarography by Okagawa and Mason (114).

Valuable information can be derived from the wetting profile of fibers, since both the chemical and physical heterogeniety along the fiber surface are displayed on the trace. Every force on the wetting curve represents the surface characteristics of the wetted perimeter at that point, and the distribution of the asperities and the hydrophobic and hydrophilic areas is shown by the entire curve. Kamath et al. (142) used the wetting trace to evaluate the deposition, distribution, and removal of surfactants at the surface of human hair fibers, and Miller et al. (144) demonstrated that the force scans could provide information on the number of cavities, the range of cavity shapes, and the transverse length and width of the cavities per scan length. This technique has not been sufficiently exploited for cellulosic fibers and deserves further consideration, since wetting is so important to the absorbent characteristics, recycling, and bonding of cellulose fibers.

DRY FORMING AND PRESS DRYING OF PAPER

Although much has been written about conventional wet-forming methods of papermaking, there is considerably less information available on dry forming of paper. This is because dry forming is a relatively new method of papermaking, and considerable secrecy surrounds commercial applications. The characteristic difference between the two processes is the method of sheet formation; water is the primary fiber carrying vehicle in wet forming, whereas air transports the fibers in dry forming. As described in previous sections, water also serves to soften the fibers and promote interfiber bonding through capillary forces in wet forming of paper. Although water does not participate in web formation with dry-forming methods, it turns out to be critical for sufficient interfiber bonding in the final sheet.

Dry-forming and air-forming technology offer a number of important advantages such as considerably lower capital costs, reduced energy requirements (about 50%), fewer pollution problems, little or no fiber loss, mill locations dependent only on raw materials and customers, and the possibility of a range of unique products by incorporation of man-made fibers. There are several disadvantages such as low strength of the final product and noise and dusting problems, which cause health and safety hazards (145–154).

Dry forming of wood fibers has been used for many years for production of fiberboard, chipboard, and flakeboard. The expertise developed in the building board industry was used to advantage for design of paper and paperboard dry-forming systems (155). Roll pulp or wastepaper is normally the raw material supplied to the dry-forming paper mill. The fibrous material is generally fed to one of several different machines to fluff the dry pulp. Sarkisian et al. (156) have discussed the advantages and disadvantages of four of these machines—the pin cyclinder, the sawtooth cyclinder, the hammer mill, and the disk refiner. The hammermill appears to be the most commonly used apparatus for loosening the fibers. Tostevin and Quimby (157) have shown that the Mullen burst strength is the best indicator of pulp fluffing rquirements. Debonding agents such as quaternary ammonium compounds can also be used in very small amounts to aid defibration and reduce energy consumption; however, these compounds can have a detrimental effect on subsequent interfiber bonding (148, 158).

Individualized fibers are then fed to the forming machine. Swenson (155) has divided the dry-forming methods into four categories: trajectory, aerodyamic, fiberized forming, and sifting processes. All of these methods result in a loosely bonded high-bulk, randomly laid web. To achieve the higher strengths necessary for end uses such as towels, industrial wipes, tablecloths, napkins, and medical products, additional binders are added to the web. The binders are usually latex or starch solutions that are applied by spraying on both sides of the sheet or by immersing the sheet in the polymer solution (size press). For high-bulk, absorbent products, the sheet is dried without excessive compaction.

There has been increasing interest in the use of thermoplastic fibers for thermal bonding of the webs, because the price of the resin binders is now almost equal to the thermoplastic fiber prices (159). Binder-free, thermal-bonded webs can be obtained by the use of bicomponent fibers. These fibers consist of two different polymers arranged in a concentric core sheath formation. The core has a relatively high melting point and acts as the basic structural component while the sheath, with a lower melting point, flows when heated to bond the web (160).

If binders are not added to the web, a very weak sheet is produced that must be consolidated by pressing for development of sufficient mechanical strength. Since water is not used in the formation process, the fibers are hornified, stiff and nonconformable. Even if the fibers are brought into contact, this does not ensure that a strong bond will be formed. Byrd (161), in a study of moisture effects on the strength of dry-formed unbleached kraft softwood paper, found that it was necessary to add more than 60% moisture to a dry-formed web and press (7 MPa, 5 min) to below 20% to obtain strengths approaching conventional linerboard. Young and Dao (162) found that the beneficial effect of moisture (added before pressing) was greater for bleached kraft pulp fibers (southern pine) than for fine denier (1.5) rayon filaments (6.4 mm long) in dry-formed sheets as shown in Figure 6.15. The rayon fibers are weaker and probably do not fibrillate as well as the pulp fibers. Back et al. (163) demonstrated that a strong dry-formed sheet could be produced at low moisture contents if a combination of a high bonding temperature (300°C) and high pressure (2–40 MPa) was used to consolidate the web. Higher densities were noted for these air-laid sheets. Presumably, the heat and pressure caused softening of the fibers and constituent polymers and created a certain amount of "auto-cross-linking" reactions at contact points (164).

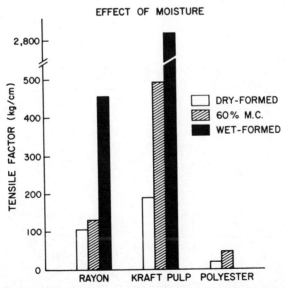

Figure 6.15.   Effect of moisture on strength of sheets pressed at 130°C, 3 MPa for 2 min (162).

Bentley and Pye (165) also investigated bonding of dry-formed webs of pulp fibers. These investigators felt that the fiber preparation stage was critical to production of a strong sheet. They suggested that the low strengths of air-formed sheets was due to fiber cutting and the creation of severe contortions (bends, twists, and curls) in the fibers during the dry defibration of the pulp. Evidence for this was stronger sheets obtained from mechanical pulps compared to those obtained from chemical pulps. The stiff mechanical pulp fibers are not deformed as easily as the chemical pulp fibers. Thus the limitations to dry-formed sheet strength are a consequence of nonuniform stress distribution due to fiber contortions. The lack of fines in dry-formed pulps may also contribute to low strength, although Bentley and Pye felt that their absence mainly affected optical properties of the sheets. These investigators found that the most favorable condition for formation of dry-formed sheets with their furnish was to press the fibers at about 35% moisture content at 150°C with low pressures.

De Ruvo and co-workers (166, 167) also suggested that dry-formed webs were of low strength due to insufficient bonding and fiber contortions. They noted that on a stress–strain curve for a dry-formed sheet, the stress attains a maximum value and then decreases as the elongation proceeds. This was interpreted as due to creation of microcracks, yielding zones of debonded material, or "ripples," as the deformation increases. These microcracks slowly decrease the load-bearing capability of the sheets. The effect of fiber damage was further ascertained by Young and Dao (162) by wet beating of bleached kraft pulp. Whereas beating of pulps is known to greatly enhance the strength of wet-formed sheets, the strength of dry-formed paper is diminished by the mechanical action, as shown in Figure 6.16. The kraft pulp was dried and defibered before dry forming on a laboratory sheet mold. The beneficial effects of fiber fibrillation

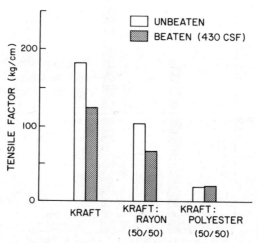

Figure 6.16.   Effect of beating (kraft pulp) on strength of dry-formed sheets (162).

are lost with drying, and only the effect of fiber damage and breakage is noted for the dry-formed sheets.

The future success of dry-formed paper that does not contain binders depends on new innovative approaches, either through fiber modification or process alterations. There has been almost no work published on fiber modification for improved dry-formed products; however, the recent development of press drying for paper and paperboard may be the key to future development of dry forming. According to Setterholm (168), press-drying is a method of drying stiff pulp fibers under a compressive force that induces greater conformability and inter-fiber bonding. There is also more restraint during drying than is now attained by conventional methods of drying paper. This results in a product with higher edgewise compression strength (for linerboard), higher tensile strength, and better dimensional stability. The process also allows for better use of hardwoods, higher pulp yields, and less refining of the pulp.

Some moisture is necessary to realize these effects in press drying; therefore, additional moisture would have to be added to dry-formed sheets to realize the full strength potential. Byrd (169) found that the addition of 40% moisture was sufficient to develop good interfiber bonding with air-laid sheets that were press-dried at 150°C (5.52 MPa). The moisture decreases the softening temperatures of both the hemicelluloses and the lignin, resulting in increased fiber flexibility. However, both Horn (170) and Byrd (171) felt that the major strength development was related to flow of the hemicelluloses in the high-yield pulps. A positive correlation for most stength properties was found for increasing amounts of hemicelluloses in the pulps. The flow of the lignin appears to be important for wet strength, by protecting preexisting bonds against attack by moisture. Atalla (172) also reported increased crystallinity of cellulose for the press-dried paper.

The press-drying process shows great promise for production of linerboard from recycled fiber. At present, the best commercial process utilizes only about 30% of secondary fiber by conventional methods. With press drying many of the problems associated with the common practice of recycling for linerboard production are circumvented. Horn (173) found that the strength properties of the press-dried sheets were superior to the conventionally dried sheets from recycled fibers. Press drying offers a new approach to papermaking and is under intensive investigation in many laboratories throughout the world (174).

REFERENCES

1. D. A. I. Goring, in *Fibre–Water Interactions in Papermaking*, Wm. Clowes, London, 1978.
2. D. F. Caulfield, in *Fibre–Water Interactions in Papermaking*, Wm. Clowes, London, 1978.
3. D. F. Caulfield, in Proceedings of Conference on Paper Science and Technology—The Cutting Edge, Institute of Paper Chemistry, Appleton, WI, 1980.
4. J. R. Colvin, *J. Appl. Polym. Sci. Appl. Polym. Symp.*, **37**, 25 (1983).
5. J. R. Colvin, *Act. Rev. Macromol. Sci.*, **1**, 47 (1972).

6. J. R. Colvin, *Planta*, **149**, 97 (1980).

7. A. D. French, *Carbohydr. Res.*, **61**, 67 (1978).

8. K. H. Gardner and J. Blackwell, *Biopolymers*, **13**, 1975 (1974).

9. A. Sarko and R. Muggli, *Macromolecules*, **7**, 486 (1974).

10. J. O. Warwicker and A. C. Wright, *J. Appl. Polym. Sci.*, **11**, 659 (1967).

11. R. Jeffries and J. O. Warwicker, *Text. Res. J.*, **39**, 548 (1969).

12. A. D. French, in *Cellulose Chemistry and Its Applications*, T. P. Nevell & S. H. Zeronian, eds., Ellis Harwood, Chichester, England, 1985.

13. M. Pasteka, *Cellulose Chem. Technol.*, **18**, 267 (1984).

14. M. V. Merchant, *Tappi*, **40**(9), 771 (1957).

15. W. J. McConnell, unpublished work, University of Washington, Seattle, 1976.

16. L. C. Wadsworth and J. A. Cuculo, in *Modified Cellulosics*, R. M. Rowell & R. A. Young, eds., Academic, New York, 1978.

17. R. A. Marchessault and P. R. Sundararajan, in *Polysaccharides*, G. Aspinall, ed., Academic, New York, 1982.

18. A. Frey-Wyssling, *Protoplasma*, **27**, 372 (1937); *Ber. Deutch Botan. Ges.*, **55**, 119 (1937).

19. J. O. Warwicker, R. Jeffries, R. L. Colbran, and R. N. Robinson, *A Review of the Literature on the Effect of Caustic Soda and Other Swelling Agents on the Fine Structure of Cotton, Silk and Man-Made Fibers*, Man-Made Fibers Research Association, Shirley Institute, Manchester, England, 1966.

20. W. A. Côté, *Wood Sci. Technol.*, **15**, 1 (1981).

21. F. W. Herrick, R. L. Casebier, J. K. Hamilton, and K. R. Sandberg, *J. Appl. Polym. Sci. Appl. Polym. Symp.*, **37**, 797 (1983).

22. J. E. Stone and A. M. Scallan, *J. Polym. Sci.*, *Pt. C*(11), 52 (1965).

23. O. A. Battista, in *Cellulose Technology Research*, A. F. Turbak, ed., ACS Symp Series, 10, American Chemical Society, Washington, D.C., 1975.

24. O. A. Battista and D. A. Smith, *Ind. Eng. Chem.*, **54**(9), 20 (1962).

25. A. Sharples, *Trans. Faraday Soc.*, **54**, 913, 1003, (1958).

26. T. Yachi, J. Hayashi, M. Takai, and Y. Shimizu, *J. Appl. Polym. Sci. Appl. Polym. Symp.*, **37**, 325 (1983).

27. T. Goto, H. Harada, and H. Saiki, *Wood Sci. Technol.*, **12**, 223 (1978).

28. B. G. Rånby, Tappi, *35* (1952) 53; *Makromol. Chem.*, **13** (1954) 40.

29. A. N. J. Heyn, in *The Physics and Chemistry of Wood Pulp Fibers*, D. H. Page, ed., Tappi STAP Series, No. 8, Atlanta, 1970.

30. M. Lotfy, M. El-Osta, R. M. Kellogg, R. O. Foschi, and R. G. Butters, *Wood Fiber*, **6**(1), 36 (1974).

31. S. P. Rowland, *Text. Chem. Color.*, **4**, 204 (1972).

32. S. P. Rowland, in *Applied Fibre Science*, Vol. 2, F. Happey, ed., Academic, New York, 1979.

33. F. C. Beall & W. K. Murphey, *Wood Fiber*, **2**(3), 282 (1970).

34. K. Mühlethaler, in *Cellular Ultrastructure of Woody Plants*, W. A. Côté, ed., Syracuse University Press, Syracuse, NY, 1965.

35. D. Fengel and G. Wegener, *Wood, Chemistry, Ultrastructure, Reactions*, Walter de Gruyter, New York, 1984.

36. R. Jeffries, D. M. Jones, J. G. Roberts, K. Selby, S. C. Simmens, and J. O. Warwicker, *Cell. Chem. Technol.*, **3**, 255 (1969).

37. J. N. McGovern, B. F. Hrutfiord, and R. A. Young, in *Introduction to Forest Science*, R. A. Young, ed., Wiley, New York, 1982.

38. H. Ruck, *Faserforsch. Textiltech.*, **14**(5), 171; **14**(6), 233 (1963); R. E. Mark, *Cell Wall Mechanics of Tracheids,* Yale University Press, New Haven, CT, 1967.

39. S. H. Zeronian, in *Cellulose Chemistry and Its Applications*, T. P. Nevell and S. H. Zeronian, eds., Ellis Horwood, Chichester, England, 1985.

40. S. H. Zeronian, M. L. Cooke, K. W. Alger, and J. M. Chandler, *J. Appl. Polym. Sci. Appl. Polym. Symp.*, **37**, 1053 (1983).

41. B. J. Fergus, A. R. Procter, J. A. N. Scott, and D. A. I. Goring, *Wood Sci. Technol.*, **3**, 117 (1969).

42. N. Parameswaran and W. Liese, in *Proceedings of the International Symposium on Wood Pulping Chemistry*, Vol. 1, SPCI, Stockholm, Sweden, 1981.

43. M. Fujita, T. Takabe, and H. Harada, in *Proceedings of the International Symposium on Wood Pulping Chemistry,* Vol. 1, Tsukuba Science City, Japan, 1983.

44. K. V. Sarkanen and C. H. Ludwig, *Lignins, Occurrence, Formation, Structure and Reactions*, Wiley-Interscience, New York, 1965.

45. J. L. Minor, *J. Wood Chem. Technol.*, **2**, 1 (1982).

46. G. Meshitsuka, Z. Z. Lee, J. Nakano, & S. Eda, *J. Wood Chem. Technol.*, **2**, 251 (1982).

47. H. Chanzy, M. Dube, and R. H. Marchessault, *Tappi*, **61**, 81 (1978).

48. R. H. Marchessault and C. Y. Liang, *J. Polym. Sci.*, **59**, 357 (1962).

49. C. Y. Liang, K. H. Bassett, E. A. McGinnes, and R. A. Marchessault, *Tappi*, **43**, 1017 (1960).

50. R. W. Scott, *J. Wood Chem. Technol.*, **4**, 199 (1984).

51. J. E. Luce, *Pulp Paper Mag. Can.*, **65**, T419 (1964).

52. D. W. Clayton and J. E. Stone, *Pulp Paper Mag. Can.*, **64**, T459 (1963).

53. A. J. Kerr and D. A. I. Goring, *Cell. Chem. Technol.*, **9**, 563 (1975).

54. H. W. Giertz, in *Fundamentals of Papermaking Fibers*, F. Bolam, ed., Br. Board and Papermakers Assn., London, 1958, p. 389.

55. J. d'A. Clark, *Pulp Technology and Treatment for Paper*, Miller Freeman, San Francisco, 1978, p. 156.

56. T. Bluhm, Y. Deslandes, R. A. Marchessault, and P. R. Sundararajan, in *Water in Polymers*, S. P. Rowland, ed., ACS Symp. Series, 127, American Chemical Society, Washington, D.C., 1980.

57. D. A. I. Goring, in *Cellulose Chemistry and Technology*, J. C. Arthur, Jr., ed., ACS Symp. Series, 48, American Chemical Society, Washington, D.C., 1977.

58. S. P. Rowland, in *Textile and Paper Chemistry and Technology*, J. C. Arthur, Jr., ed., ACS Symp. Series, 49, American Chemical Society, Washington, D.C., 1977.

59. J. E. Stone and A. M. Scallan, *Pulp Paper Mag. Can.*, **69**, 288 (1968).

60. A. M. Scallan, in *Fibre–Water Interactions in Papermaking*, Wm. Clowes, London, 1978.

61. S. P. Rowland and N. R. Bertonier, in *Cellulose Chemistry and Its Application*, T. P. Nevell and S. H. Zeronian, eds., Ellis Horwood, Chichester, England, 1985.

62. H. Krassig, F. Phchegger, and G. Faltlhansl, in *Proceedings of the International Dissolving and Specialty Pulps Conference*, 1983.

63. G. Porod, *Makromol. Chem.*, **35**, 1 (1960).

64. E. B. Cowling and W. Brown, in *Cellulases and Their Applications*, G. J. Hajny & E. T. Reese, eds., *Adv. Chem. Ser.*, 95, American Chemical Society, Washington, D.C., 1969.

65. R. Nelson and D. W. Oliver, *J. Polym. Sci.*, Pt. C, **36** 305 (1971).

66. R. A. Young and M. Fujita, *Wood Sci. Technol.*, **19**, 363 1985.

67. O. Sawabe and K. Mori, *J. Jpn. Wood Res. Soc.*, **27**, 409 (1981).

68. H. Corte, in *Handbook of Paper Science*, Vol. 1, H. F. Rance, ed., Elsevier, Amsterdam, 1980, p. 1.

69. H. W. Emerton, in *Handbook of Paper Science*, Vol. 1, H. F. Rance, ed., Elsevier, Amsterdam, 1980, p. 139.

70. D. Atack, in *Fibre–Water Interactions in Papermaking*, Wm. Clowes, London, 1978, p. 261.

71. U.-B. Mohlin, *Sv. Papperstidn.*, **78**, 332 (1975).

72. W. E. Morton and J. W. S. Hearle, *Physical Properties of Textile Fibres*, Wiley, New York, 1975, p. 45.

73. D. H. Page and J. H. DeGrâce, *Tappi*, **50**, 489 (1967).

74. D. C. McIntosh, *Tappi*, **50**, 482 (1967).

75. E. Szwarcsztajn and K. Przybysz, *Cellulose Chem. Technol.*, **9**, 597 (1975).

76. J. C. Roberts and S. Awad El-Karim, *Cellulose Chem. Technol.*, **17**, 379 (1983).

77. H. W. Emerton, *Fundamentals of the Beating Process*, British Paper and Board Industrial Research Association, Kenley, 1957, p. 99.

78. E. Claudio-da-Silva, R. Marton, and S. Granzow, *Tappi*, **65**, 99 (1982).

79. A. M. Scallan, *Tappi*, **66**, 73 (1983).

80. S. Katz, N. Liebergott, and A. M. Scallan, *Tappi*, **64**, 97 (1981).

81. S. Katz and A. M. Scallan, *Tappi*, **66**, 85 (1983).

82. T. Lindström and G. Carlsson, *Sven. Papperstidn.*, **85**, R146 (1982).

83. T. Lindström and M. Kolman, *Sven. Papperstidn.*, **85**, R140 (1982).

84. T. Lindström and G. Carlsson, *Sven. Papperstidn.*, **85**, R14 (1982).

85. T. Lindström, *Das Papier*, **34**, 561 (1980).

86. G. Carlsson, P. Kolseth, and T. Lindström, *Wood Sci. Technol.*, **17**, 69 (1983).

87. A. M. Scallan and J. Grignon, *Sv. Papperstidn.*, **82**, 40 (1979).

88. P. F. Nelson and C. G. Kalkipsakis, *Tappi*, **47**, 107, 170 (1964).

89. J. Grignon and A. M. Scallan, *J. Appl. Polym. Sci.*, **25**, 2829 (1980).

90. R. A. Young, in *Absorbency*, P. K. Chatterjee, ed., Elsevier, Amsterdam, 1985.

91. H. G. Higgins and A. W. McKenzie, *Appita*, **16**, 145 (1963).

92. R. A. Horn, *Paper Trade J.*, **159**(6), 78 (1975).

93. R. C. McKee, *Paper Trade J.*, **155**(21), 34 (1971).

94. J. H. Klungness, *Tappi*, **57**(11), 71 (1974).

95. G. Jayme, *Wochenbl. Papierfabr.*, **42**, 187 (1944).

96. E. L. Akim, *Tappi*, **61**(9), 111 (1978).

97. H. W. Giertz, in *Formation and Structure of Paper*, Vol. 2, British Paper and Board Maker's Association, London, 1962, p. 619.

98. J. E. Stone, A. M. Scallan, & B. Abrahamson, *Sven. Papperstidn.*, **71**(19), 687 (1968).

99. J. H. Klungness and D. F. Caulfield, *Tappi*, **65**, 94 (1982).

100. E. Szwarcsztajn & K. Przybysz, in *Fibre–Water Interactions in Papermaking*, W. Clowes, London, 1978, p. 857.

101. J. H. Klungness, USDA Forest Products Laboratory, Madison, Wisconsin, unpublished results.

102. K. K. Talwar, *Tappi*, **41**(5), 207 (1958).

103. E. Ehrnrooth, M. Htun, & A. DeRuvo, in *Fibre–Water Interactions in Papermaking*, Wm. Clowes, London, 1978, p. 899.

104. P. Howarth, in *Fibre–Water Interactions in Papermaking*, Wm. Clowes, London, 1978, p. 823.

105. E. W. Washburn, *Physiol. Rev.*, **17**, 273 (1921).

106. G. M. Aberson, in *Physics and Chemistry of Wood and Fibers*, D. H. Page, ed., Tappi, Atlanta, 1970, p. 36.

107. T. H. Grindstaff, *Text. Res. J.*, **39**, 958 (1969).

108. W. C. Jones & M. C. Porter, *J. Colloid Interface Sci.*, **24**, 359 (1975).

109. B. Miller & R. A. Young, *Text. Res. J.*, **45**, 359 (1975).

110. B. Miller, in *Surface Characteristics of Fibers and Textiles*, Pt. II, M. J. Schick, ed., Dekker, New York, 1977, p. 417.

111. K. Borgin, *Norsk Skogind.*, **13**, 429 (1959).

112. P. Luner & M. Sandell, *J. Polym. Sci., Pt. C*, **28**, 115 (1969).

113. R. L. Bendure, *J. Colloid Interface Sci.*, **42**, 137 (1973).

114. A. Okagawa & S. G. Mason, in *Fibre–Water Interactions in Papermaking*, Wm. Clowes, London, 1978, p. 581.

115. R. A. Young, *Wood Fiber*, **8**, 120 (1976).

116. R. A. Young, University of Wisconsin–Madison, unpublished results.

117. J. Wilhelmy, *Ann. Phy.*, **119**, 177 (1863).

118. Y. K. Kamath, C. J. Dansizer, H.-D. Weigmann, *J. Soc. Cosmet. Chem.*, **28**, 273 (1977); *J. Appl. Polym. Sci.*, **22**, 2295 (1978).

119. B. Miller, L. S. Penn, & S. Hedvat, *Colloids Surf.*, **6**, 94 (1983).

120. J. H. Klungness, *Tappi*, **64**(12), 65 (1981).

121. F. M. Fowkes, *Ind. Eng. Chem.*, **56**, 40 (1964); *J. Phys. Chem.*, **66**, 382 (1962).

122. S. Wu, *J. Polym. Sci., Pt. C*, **34**, 19 (1971).

123. T. Nguyen & W. E. Johns, *Wood Sci. Technol.*, **12** 63 (1978).

124. R. A. Young, R. M. Rammon, S. S. Kelley, & R. H. Gillespie, *Wood Sci.*, **14**, 110 (1982).

125. A. B. D. Cassie, *Disc. Faraday Soc.*, **3**, 11 (1948).

126. G. E. Collins, *J. Text. Inst.*, **38**, T73 (1947).

127. S. P. Wesson & A. Tarantino, *J. Non-Crystalline Solids*, **38–39**, 619 (1980).

128. P. H. Kaelble, F. H. Dynes, & E. H. Arlin, *J. Adhesion*, **6**, 23 (1974).

129. A. M. Schwartz & F. W. Minor, *J. Colloid Sci.*, **14**, 572 (1959).

130. R. E. Swanson, *Tappi*, **61**(7), 77 (1978).

131. J. S. Aspler, N. Chauret, & M. B. Lyne, *Mechanism of Self-Sizing of Paper*, Fundamental Research Symposium, Oxford, UK, 1985.

132. J. C. Berg, in USA–Sweden Joint Workshop on Composite Systems from Natural and Synthetic Polymers, June 1984.

133. K. T. Hodgson & J. C. Berg, in Fifth International Symposium on Surfactants, Bordeaux, France, July 1984.

134. R. A. Young, TRI results released in ref. 110.

135. R. C. Casilla, S. Chow, & P. R. Steiner, *Wood Sci. Technol.*, **15**, 31 (1981).

136. R. C. Casilla, S. Chow, P. R. Steiner, & S. R. Warren, *Wood Sci. Technol.*, **18**, 87 (1984).

137. J. F. Oliver, C. Hub, & S. G. Mason, *Colloids Surf.*, **1**, 79 (1980).

138. R. E. Johnson & R. H. Dettre, in *Surface and Colloid Science*, Vol. 2, E. Matijevic, ed., Wiley, New York, 1969, p. 85.

139. J. D. Eick, R. J. Good, & A. W. Neumann, *J. Colloid Interface Sci.*, **53**, 235 (1975).

140. A. W. Adamson, *Physical Chemistry of Surfaces*, Wiley-Interscience, New York, 1960.

141. L. S. Penn & B. Miller, *J. Colloid Interface Sci.*, **78**, 238 (1980).

142. Y. K. Kamath, C. J. Dansizer, & H.-D. Weigmann, *J. Appl. Polym. Sci.*, **29**, 1011 (1984).

143. T. H. Daugherty, M.S. thesis, Dept. Chem. Eng., University of Washington, Seattle, 1981.

144. B. Miller, H.-D. Weigmann, & D. Simonetti, in *Physicochemical Aspects of Polymer Surfaces*, Vol. 1, K. L. Mittal, ed., Plenum, New York, 1983.

145. J. P. Casey, *Pulp and Paper Chemistry and Technology*, Vol. 2, Wiley-Interscience, New York, 1980, p. 1111.

146. B. W. Attwood, *Paper Technol. Ind.*, **16**, 44 (T35) (1975).

147. B. W. Attwood & D. G. White, *Tappi*, **62**, 39 (1979); in *Proc. Tappi Eng. Conf.*, Atlanta, 1978.

148. J. Bourka, *Papir a Celluloza*, **29**, 131 (1974).

149. G. Jensen, in *Proc. Tappi Eng. Conf.*, Atlanta, 1978.

150. M. V. Prolov & V. A. Gorbushin, *Bumazh. Prom.*, **8**, 12 (1973).

151. A. Wilson, *Pulp Paper*, **40**, 25 (1966).

152. J. R. Starr, *Tappi*, **64**(9), 87 (1981).

153. J. Mosgaard & K. Rühinen, *Pulp Paper*, March, p. 117 (1983).

154. W. E. Mies, *Pulp Paper*, March, p. 169 (1983).

155. R. S. Swenson, in *Proc. Tappi Eng. Conf.*, Atlanta, 1978.

156. A. Sarkisian, C. F. Murphy, & C. P. Steiner, in *Nonwoven Production Technology Symposium*, INDA, New York, 1975, p. 67.

157. J. E. Tostevin & G. R. Quimby, in *Nonwoven Production Technology Symposium*, INDA, New York 1975, p. 48.

158. S. L. Waahlen, in *Absorbent Products Conference, Insight 83 International Conference*, Marketing/Technology Service, Kalamazoo, MI, Nov. 1983.

159. R. G. Mansfield, in *Advanced Forming/Bonding Conference, Insight 82 International Conference*, Marketing/Technology Service, Kalamazoo, MI, Oct. 1982.

160. R. A. Handon, in *Advanced Forming/Bonding Conference, Insight 83 International Conference*, Marketing/Technology Service, Kalamazoo, MI, Nov. 1983.

161. V. L. Byrd, *Tappi*, **57**(4), 131 (1974).

162. R. A. Young & L. Dao, in *Tappi Progress in Paper Physics Conference*, Institute of Paper Chemistry, Appleton, WI, May 1980.

163. E. L. Back, C. G. Wahlstrom, & R. G. Anderson, *Tappi*, **62**(3), 89 (1979).

164. E. L. Back, *Pulp Paper Mag. Can.*, **68**(4), T165 (1967).

165. S. Bentley & I. T. Pye, *Pulp Paper Mag. Can. Trans.*, **5**, TR77 (1979).

166. A. de Ruvo, H. Hallmark, S. Hartog, & C. Fellers, *Sven. Papperstidn.*, **85**(7), 16 (1982).

167. M. Rigdahl, B. Westerlind, H. Hallmark, & A. de Ruvo, *J. Appl. Polym. Sci.*, **28**, 1599 (1983).

168. V. C. Setterholm, *Tappi*, **62**(3), 45 (1979).

169. V. L. Byrd, *Tappi*, **65**(5), 153 (1982).

170. R. A. Horn, *Tappi*, **62**(7), 77 (1979).

171. V. L. Byrd, *Tappi*, **62**(7), 81 (1979).

172. R. A. Atalla, in *Proceedings of the Press Dry Conference*, USDA Forest Products Laboratory, Madison, WI, Aug. 1983.

173. R. A. Horn, in *Proceedings of the Press Dry Conference*, USDA Forest Products Laboratory, Madison, WI, Aug. 1983.

174. V. C. Setterholm, in *Proceedings of the Press Dry Conference*, USDA Forest Products Laboratory, Madison, WI, Aug. 1983.

PART 2

Cellulose Modification

7

Cellulose Carbamate

KURT EKMAN, VIDAR EKLUND, JAN FORS,
JOUKO I. HUTTUNEN, JOHAN-FREDRIK SELIN,
and OLLI T. TURUNEN
Technology Research Centre, Neste Oy, Kulloo, Finland

Cellulose has three reactive hydroxyl groups per anhydroglucose repeating unit that form inter- and intramolecular hydrogen bonds. These bonds strongly influence the chemical reactivity and solubility of cellulose. As a result, cellulose is virtually insoluble in most common solvents, and the successful preparation of derivatives, mostly ethers and esters, requires a special approach.

When urea is heated to 135°C, it starts to decompose into ammonia and isocyanic acid (Fig. 7.1) (1). The latter reacts with alcoholic hydroxyl groups to form carbamates. Under suitable conditions cellulose and urea combine together in a heterogeneous reaction to form cellulose carbamate:

$$\text{Cell—OH} + \text{NH}_2\text{CO} \cdot \text{NH}_2 \rightarrow \text{Cell—O—CO} \cdot \text{NH}_2 + \text{NH}_3 \qquad (1)$$

When Hill and Jacobson (2) first reacted urea with cellulose in the mid 1930s, they were not fully aware of the chemical nature of the product but called it "urea–cellulose" or only a "nitrogen-containing derivative of cellulose." They noted, however, that this product could be dissolved in dilute caustic soda and precipitated from solution by acids to form films or filaments. The nitrogen content ranged from 1% to 3.5%.

The reaction of cellulose with urea was studied in more detail by Segal and Eggerton in 1961 (3). They suggested that the product was a true chemical derivative. This conclusion they based on the following experimental facts: (a) the treated cellulose was insoluble in cupriethylenediamine (CED); (b) there were changes in the infrared (IR) spectra; (c) the nitrogen content was not affected by treatment in boiling water; (d) the dyeing properties of the cellulose

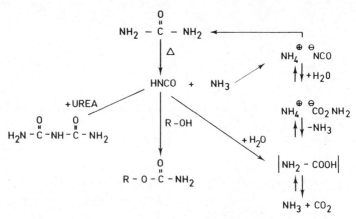

Figure 7.1 Schematic representation of the reactions of urea.

were modified; and (e) the amount of moisture regained by the cellulose was changed. From these facts they established the existence of cellulose carbamate. From the insolubility in CED they further concluded that the main reaction took place with the hydroxyl groups in positions 2 and 3 of the anhydroglucose unit.

Further work was done in 1981 by Nozawa and Higashide (4). They studied the IR spectra of cellulose carbamate and correlated the increase in nitrogen content with specific peaks in the spectra. Recently Hebeish and co-workers in Egypt performed a series of investigations in which they treated cotton cloth with urea in alkaline solutions (5–9). After curing to get the carbamates they evaluated the textile properties of these samples. They also used the carbamate groups for further modifications including grafting and dye attachment. These attempts to modify textiles met with only moderate success, however.

For two decades Neste Oy has been engaged in research into wood chemistry and the combination of the best properties of wood and petrochemicals. The most recent such project was a study of the reaction between urea and cellulose and an investigation into the potential applications of cellulose carbamate (10, 11).

ACCESSIBILITY AND REACTIVITY OF CELLULOSE

Both natural cellulose, as in cotton, and the cellulose isolated from wood by pulping have a high degree of crystallinity. Typical values range from 50% to 90%. In the crystalline regions the cellulose molecules are arranged in ordered lattices, in which the hydroxyl groups are bonded by strong secondary forces. To form derivatives we have to proceed in one of two ways. Either the lattice structure must be expanded sufficiently to let the reactants penetrate into the crystals or the whole structure must be broken down to a state of solution before a reaction can take place. Examples of the first are the syntheses of cellulose nitrate or carboxymethyl cellulose (CMC), whereas a complete opening or solubilization is required for the preparation of cellulose xanthate and cellulose acetate, for example (12).

Cellulose can be made to swell in several ways. Strong bases and acids as well as strong solutions of salts are usually used for this purpose. These same types of agents may also be used to dissolve cellulose. The most effective solvents are certain basic metal hydroxide complexes. It should be stressed that cellulose does not dissolve as such, but as soluble addition complexes. The indirect way to dissolve cellulose is by means of soluble derivatives (13).

Liquid ammonia causes cellulose to swell markedly. It not only expands the crystal lattice, but transforms the crystalline regions into an essentially amorphous configuration (Fig. 7.2). This form of cellulose is usually called cellulose III. By this process the accessibility of cellulose to various reagents is greatly increased (14–16).

Disregarding the end groups, which in a polymer are of relatively minor importance, cellulose has three reactive hydroxyl groups per repeating unit. They are in positions 2, 3, and 6. The hydroxyl groups in positions 2 and 3 are secondary hydroxyl groups, whereas the one in position 6 is a primary hydroxyl group. The acidity of these hydroxyl groups increases in the order OH-6 < OH-3 < OH-2. It should be noted, however, that when OH-2 is ionized or substituted, the acidity in position 3 is lowered, and the hydroxyl group in position 6 may then be the more acidic one.

For etherification reactions the hydroxyl group in position 2 is most reactive since it is the most acidic. The primary hydroxyl group in position 6 is the first to react in esterification reactions. Steric hindrance is an important consideration, particularly in the case of bulky reacting species; position 6 is least sterically

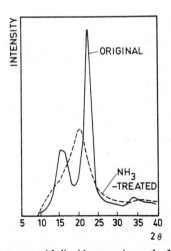

Figure 7.2 Effects of treatment with liquid ammonia on the X-ray structure of cellulose.

hindered (13). But, as already pointed out, it is the inter- and intramolecular forces that most influence the reactivity, and a cellulose that is not swollen will react only on the surface of the molecular bundles.

Reactions in Suspension

Hill and Jacobson (2) and later workers used alkali cellulose or cellulose impregnated with alkaline solutions of urea. The addition of alkali played an active part in those reactions. Right from the start of our experiments we used no alkali but tried to make the cellulose swell by other means. First we impregnated the cellulose with aqueous solutions of urea and different salts. After pressing and drying, the uptake of salt and urea was estimated, and the reaction was started by heating the cellulose in an inert medium, usually xylene, to 135°C. At this temperature urea decomposes to form isocyanate and ammonia. Typical results from these early experiments are seen in Table 7.1.

In the examples the reaction time was 3 hr, and fairly high ratios of urea to anhydroglucose units (AGU) were used. The nitrogen contents of the products were usually high, between 3.5 and 4.5%, which means that the degree of substitution approached 0.6.

When these carbamates were dissolved in cold aqueous sodium hydroxide, the quality of the solutions was determined by the so-called clogging value, K_w. This value can be calculated from Eq. 2 (17, 18):

$$K_{w_{20,60}} = 2 \times 10^5 \times \frac{\dfrac{60}{p_{60}} - \dfrac{20}{p_{20}}}{60 - 20}$$

$$= \frac{1}{2} \times 10^4 \times \left(\frac{60}{p_{60}} - \frac{20}{p_{20}} \right)$$

(2)

TABLE 7.1   Synthesis of Cellulose Carbamate from Cellulose Impregnated with Different Salts and Urea (Reaction Time 3 hr at 135°C in Xylene)

Salt	Concentration in Impregnation Solution (%)	Urea/AGU[a] (mole ratio)	Nitrogen Content of Product (%)	$K_{w_{20,60}}$[b]
KOCN	5.0	3.0	4.8	1,380
	2.5	3.0	4.5	1,770
	1.5	2.5	3.3	11,300
NaOCN	10.0	3.0	3.4	3,620
K₂CO₃	5.0	4.5	4.1	7,250
Na₂CO₃	5.0	4.5	3.9	3,820

[a] AGU = anhydroglucose units.

[b] Clogging value of solution in 10% NaOH at −5°C

where p_{20} = mass in grams of solution that filtered in 20 min and p_{60} = mass in grams of solution that filtered in 60 min.

The clogging value is a measure of the clogging of a filter due to solid particles and gels in a solution. It is of special importance for the fiber industry, where clean solutions are required to get strong fibers and to avoid clogging of the spinnerets. Clogging values depend on the filter used in the test apparatus, on the viscosity of the solutions, and on many other factors. It is very difficult to compare values obtained at different laboratories using somewhat different methods. However, clogging values give useful information about the quality of a solution or solute when compared with others measured using the same standardized method in the same laboratory.

Clogging values of the cellulose carbamates mentioned above show variations depending on the reaction conditions and salt concentrations, but they are all high, and this means that the cellulose carbamates do not dissolve completely. We interpret the results thus: The distribution of the carbamate groups along the cellulose backbone was not uniform, although the degree of substitution was fairly high. The crystalline regions of the cellulose had not opened up enough for the urea to penetrate. Another possibility is that a certain amount of crosslinking had occurred, which would not necessarily be reflected by an increase in the measured degree of polymerization (DP).

Impregnation with Liquid Ammonia

As the salts were not satisfactory in opening up the cellulose structure we tried liquid ammonia. Near its boiling point at $-33°C$ liquid ammonia dissolves about 10% urea, and because cellulose absorbs roughly three times its own weight of the solution, we get between 20 and 30% urea into the cellulose with one impregnation.

Examples of cellulose carbamates prepared by impregnating the cellulose with urea from a solution in liquid ammonia are shown in Table 7.2. The temperature during the impregnation was $-35°C$, and we used a 10% (w/w) solution of urea. After a brief immersion in the urea solution, the cellulose was drained and allowed to dry, first at room temperature and finally by a short treatment at 100°C. This heating removed the last ammonia bound to the cellulose. The cellulose–urea mixture was then allowed to react in the dry state by heating in an oven to 135°C for 3 hr. Higher reaction temperatures will naturally shorten the reaction time, and under proper conditions no adverse effects occur.

The samples in Table 7.2 dissolve well in sodium hydroxide, and the filterability of the solutions is good, as can be seen from the clogging values. Compared with the samples mentioned in Table 7.1, the degree of substitution is less, but judging from the solubility behavior, the uniformity of the substitution is superior. Treatment in liquid ammonia allowed the reagents to diffuse into the crystalline regions of the cellulose matrix. Good solutions are usually obtained when the degree of substitution ranges between 0.15 and 0.25. In terms of nitrogen contents this corresponds to between 1.2 and 2.0% nitrogen in the carbamates.

Reaction Rate

The progress of the reaction with time measured in terms of the increase in nitrogen content is shown in Figure 7.3. The reactions were carried out at 155°C, and three urea-to-cellulose ratios were used. At the two lower ratios the reactions are essentially complete after about 2 hr, and the nitrogen content levels off at 1.8–2.0%. When the cellulose contains 25% by weight of urea, or about 1 mole of urea per anhydroglucose unit, the amount of nitrogen introduced rises above 2% after a reaction time of 3 hr.

To investigate the mechanism of the formation of cellulose carbamate, we determined the reaction velocity of the deamination reactions for urea *per se* and for urea added to cellulose by means of liquid ammonia. The reactions are first order, and for pure urea the reaction velocity soon slows down (probably because the biuret is formed). For urea impregnated into cellulose, the reaction rate is constant up to the degree of substitution (DS) which is of interest in the carbamate synthesis, (i.e., DS = 0.3). In this case the formation of biuret was negligible.

By comparing the rate constants for urea and urea impregnated into cellulose, we concluded that urea and cellulose form a transition complex leading to the formation of cellulose carbamate without the formation of free isocyanic acid. The heat of reaction, as calculated from the heats of combustion, was small; the reaction is endothermic with about 80 kJ per substituted carbamate group.

The main by-product in this reaction is biuret, which is always formed when urea is heated (1). When the temperature is raised above 190°C, the biuret starts to decompose and thus can take part in the carbamoylation reaction. In fact, we have synthesized good-quality cellulose carbamates starting from biuret and cellulose (19).

PRODUCT CHARACTERIZATION

Hill and Jacobson published no data on the chemical characterization of their reaction products obtained from cellulose and urea (2). The first to do so were

TABLE 7.2 Synthesis of Cellulose Carbamate from Liquid Ammonia–Urea Impregnated Cellulose (Reaction for 3 hr at 135°C in the Dry State)

Urea/AGU (mole ratio)	Nitrogen Content of Product (%)	DP[a]	$K_{w_{20,60}}$ [b]
1.0	1.2	420	515
2.0	1.2	410	475
3.0	1.1	400	685

[a] Initial DP of cellulose 430.

[b] Clogging value of solution in 10% NaOH at −5°C.

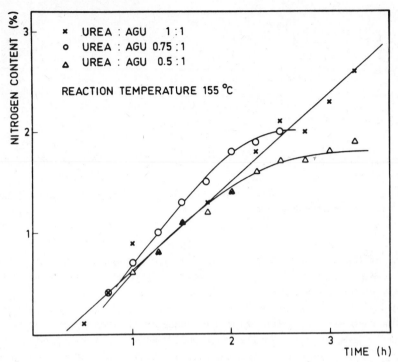

Figure 7.3 Nitrogen content of cellulose carbamate as a function of time at three urea: AGU ratios. (AGU = anhydroglucose units)

Segal and Eggerton in 1961 (3), followed by Nozawa and Higashide in 1981 (4). Because the carbamates prepared by the present method probably differ from those of the earlier researchers, it is justified to report facts and conclusions we have drawn although they are similar to those of the above authors.

The easiest way to follow the course of the reaction is to measure the increase in nitrogen content of thoroughly washed samples as a function of time. Estimates can be made from the nitrogen content of the number of carbamate groups, and thus the degree of substitution can be ascertained. Figure 7.3 shows examples of the increase in nitrogen content with reaction time.

Confirmation of the chemical nature of the nitrogen bound to cellulose in the reaction — that is, the existence of carbamate groups – has been obtained in several ways. Figure 7.4 shows the IR- spectra of cellulose and of the reaction product, cellulose carbamate. The most prominent new feature in the carbamate spectrum is the strong peak at wavenumber 1715 cm^{-1}. This peak can be ascribed to the stretching vibrations of the carbonyl group, $C=O$, in the carbamate. Other new peaks, though lesser ones, are seen at wavenumbers 1230 and 770 cm^{-1}. Both peaks are typical of derivatives of cellulose and especially of cellulose esters (20).

The extent of the carbonyl absorption can be directly correlated with the nitrogen content of the samples. This good correlation, shown in Figure 7.5, is also an indication that no other types of nitrogen-containing structures are

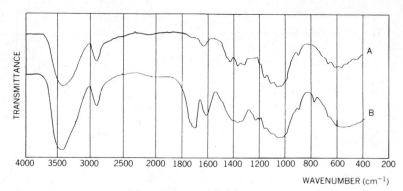

Figure 7.4 Infrared spectra of (A) cellulose and (B) cellulose carbamate. (Copyright 1983. Tappi. Reprinted with permission from reference 11.)

formed in the carbamoylation reaction. The high nitrogen contents, more than 3%, shown in Figure 7.5, have been achieved by repeated carbamoylation of already synthesized cellulose carbamates.

Susceptibility of the nitrogen-containing groups in the reacted celluloses to alkaline hydrolysis constitutes further evidence for the identity of these as carbamate groups. In Figure 7.6 the progress of the hydrolysis, measured in terms of the nitrogen content of the treated samples, is shown at a few different temperatures.

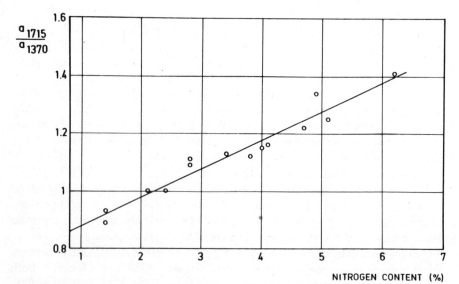

Figure 7.5 Ratio of absorbances at 1715 cm^{-1} and 1370 cm^{-1} as a function of nitrogen content of some cellulose carbamates. The correlation factor $r = 0.971$. (Copyright 1983. Tappi. Reprinted with permission from reference 11.)

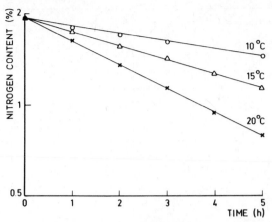

Figure 7.6 Hydrolysis as a function of time for cellulose carbamates in solution (7% in 9% NaOH) at selected temperatures. The figure shows residual nitrogen in the samples. Note the logarithmic scale of the y axis.

The alkaline hydrolysis was also followed by recording the IR spectra at various intervals. This is shown in Figure 7.7. The cellulose carbamate sample was dissolved in alkali, and parts thereof were precipitated in methanol and their spectra were recorded. The size of the carbonyl peak at 1715 cm^{-1} decreases with time along with a decrease in the nitrogen content. Also seen from the spectra is the appearance and later disappearance of a strong peak at 2155 cm^{-1}. This peak stems from cyanate, the first breakdown product of the hydrolyzed carbamate group. In alkaline environments cyanate degrades further to carbon dioxide, or carbonate, and ammonia (Eq.(3)). This explains the later disappearance of the peak. Semiquantitative calculations and determinations of carbonate formed in the sodium hydroxide solutions supported this theory.

$$Cell-O-\overset{\overset{\displaystyle O}{\displaystyle \|}}{C}-NH_2 + OH^- \longrightarrow Cell-OH + CNO^- + H_2O$$

$$CNO^- + H_2O + OH^- \longrightarrow CO_3^{2-} + NH_3 \tag{3}$$

Because of the different reactivities of the hydroxyl groups in cellulose we expect the reaction to take place in position 6. Segal and Eggerton (3), however, concluded that because the carbamate derivative they synthesized did not dissolve in CED, the reaction probably took place at positions 2 and 3. Our carbamate derivatives dissolve readily in CED. Because of differences in our methods, in particular that we do not use alkali cellulose or alkali addition during the reaction, it is possible that the derivatives are different.

We have used ^{13}C-NMR to localize the carbamate groups introduced into the cellulose. The spectra of Figure 7.8 show that the only differences upon

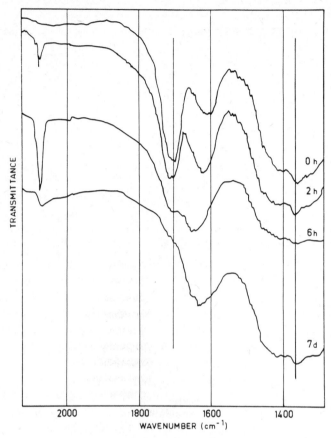

Figure 7.7 IR Spectra of the carbonyl peak from the carbamate group during alkaline hydrolysis of cellulose carbamate. (Copyright 1983. Tappi. Reprinted with permission from reference 11.)

synthesis of the carbamate occur in the peak of carbon atom 6. Hence we conclude that this is the principal site of carbamate groups in the cellulose carbamates we have prepared.

Solution Properties

Cellulose carbamate dissolves readily in cold sodium hydroxide solutions. The practical ranges of concentration for both the solute and the solvent are determined by the viscosity of the solutions. The viscosity, on the other hand, is a function of both the degree of polymerization and the degree of substitution of the samples. For most applications, 5–10% by weight of cellulose carbamate in solutions with 7–9% by weight of sodium hydroxide would be the preferred range.

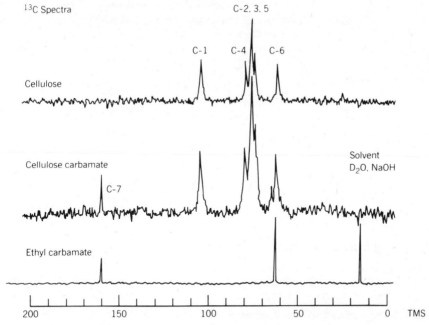

Figure 7.8   NMR spectra of cellulose and cellulose carbamate.

As with most cellulose derivatives, the dissolution is carried out at low temperatures, preferably at or below 0°C. The effect of the dissolving temperature on the filterability of cellulose carbamate samples is shown in Table 7.3.

To avoid the detrimental effects of hydrolysis of the carbamate groups, solutions of cellulose carbamates should be kept cold. Properly stored, their shelf life is several days. Addition of zincate (ZnO) or urea (21) may be used, partly to enhance the dissolution and filterability of the solutions and partly to prolong the storage time before gelation starts. The use of ZnO considerably increases the viscosity of the solutions. Examples of this are shown in Table 7.4.

TABLE 7.3   Dissolution of Cellulose Carbamate at Different Temperatures[a]

Temperature (°C)	$K_{w_{20,60}}$	Viscosity (Pa·sec)
10	∞	7.6
0	2060	15.4
−5	280	95.4

[a] Effects on filterability and viscosity of solutions of 7% cellulose carbamate in 9% NaOH.

TABLE 7.4 Effect of Zincate on the
Viscosity and Filterability of a Solution of
Cellulose Carbamate in 9% NaOH at
$-5°C$

ZnO (%)	Viscosity (Pa·sec)	$K_{w_{20,60}}$
0	11.4	6,110
0.1	14.2	2,350
0.5	16.9	1,590
1.0	18.5	1,300
1.5	22.8	870
3.0	35.0	740

Degree of Polymerization

For almost all purposes where cellulose is reacted, either to get a stable derivative or for the preparation of regenerated cellulose, the DP has to be reduced to obtain better reactivity or suitable properties of the products. This adjustment of the molecular mass is usually done by making alkali cellulose and by letting this age until it reaches the desired maturity.

In the work by Hill and Jacobson (2), as well as in most subsequent carbamoylations, sodium hydroxide was used in the steeping liquors when the cellulose was impregnated with urea solutions. An alkali was thus present during the reactions. No mention is made of the molecular masses before or after the reactions, but it may be assumed that the cellulose in some cases degraded extensively during the treatment.

We impregnated the cellulose with urea in a solution of liquid ammonia and did not get any alkaline degradation reactions during the synthesis. DPs of the reaction products differ very little from those of the starting materials.

In place of a chemical depolymerization we adjusted the molecular mass of the cellulose before the reaction or of the carbamate after the synthesis by means of irradiation. Both accelerated electrons and γ-radiation are suitable for this purpose. For small-scale applications we have routinely used γ-radiation from a ^{60}Co source. For irradiation of larger quantities of samples, or when whole cellulose webs have been irradiated, the use of accelerated electrons was preferred.

Radiation-induced degradation of cellulose is well documented (22–25). The DP is readily reduced in the samples regardless of the initial degree of crystallinity. The radiation primarily produces radicals, which combine with oxygen to form peroxyradicals. These disintegrate to form carbonyls, leading to a random breaking up of the molecules and an opening of the glucose rings. However, most of the cited work was with high-energy radiation. In the case of accelerated electrons, energies of several MeV were used. These radiation sources require extensive shielding and safety precautions. We have therefore looked into the use of electrons accelerated in fields of 250–400 keV. This type of accelerator needs no additional shielding, and commercial equipment is available. Because of the limited ability of these electrons to penetrate deeply into the cellulose, irradiation from both sides of a cellulose sheet or web is required. An example of the DP distribution in a layered cellulose sheet upon treatment with accelerated electrons from one side or from both sides is shown in Figure 7.9 (26).

APPLICATIONS

Over 3 million tons of man-made cellulosic fibers, viscose or rayon, are manufactured annually. Many plants are old, and cuts in capacity as well as ra-

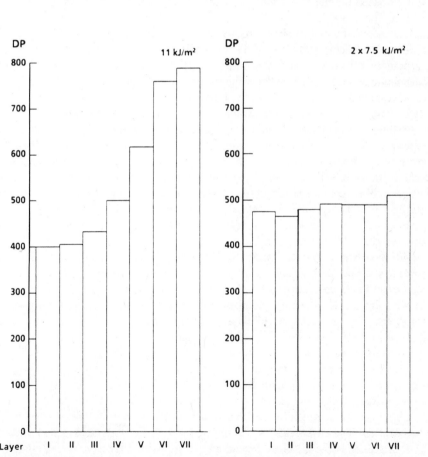

Figure 7.9 DP of different layers in sheets of cellulose irradiated from one or both sides with accelerated electrons of 250 keV energy.

tionalizations have been carried out. In the long term, however, this type of fiber will continue to fill a need that synthetic fibers cannot meet. With the introduction of the high wet modulus (HWM) or modal fibers, of which we already have the "second generation," these fibers correspond to existing quality requirements. Yet their manufacture is connected with environmental and hygiene problems, and new production methods have therefore been sought.

Considerable effort has been put into seeking organic solvents for cellulose. One method of preparing high-quality fibers from such solutions is to use N-methylmorpholine-N-oxide or similar amine oxides (27, 28; also cf. Chap. 14). This process has not reached commercialization, however, probably because of problems with solvent recovery and recycling. Other such solvents, or solvent systems, include paraformaldehyde-DMSO (29), LiCl-N,N-dimethyl-acetamide (27), and liquid ammonia–ammonium thiocyanate (27,30). For all of these the use of expensive chemicals places high demands on recovery and recycling. A system based on cheap inorganic chemicals and water-based solvents is still the best choice.

The most obvious applications for cellulose carbamate are the same as for cellulose xanthate, and one of the main aims during the development of the carbamate derivative was to find an alternative for the viscose process. Cellulose carbamate would then have the significant advantage of not requiring highly toxic chemicals for its preparation. Furthermore, in the dry state it is a stable derivative, and it is reasonably stable in properly stored solutions.

As mentioned in the Introduction, Hebeish and co-workers (5–9) introduced carbamate groups into cotton textiles in order to give reactive sites either for grafting reactions or for reactive dyes. The results published do not seem very promising, however. Hill and Jacobson (2) prepared fibers and films from solutions of their cellulose carbamates. They gave very few figures for the properties of their products, and it is difficult to assess their method of preparing the carbamate. Attempts in our laboratory to repeat their results have met with considerable difficulty.

We have spun fibers on a small pilot spinning machine. The cellulose carbamate samples were dissolved in aqueous sodium hydroxide of between 6 and 9% (w/w) to concentrations of 7–10% by weight. In some cases ZnO was added to the solutions to speed up the dissolution and to yield better and more stable solutions. The solutions of cellulose carbamate were then filtered if necessary and de-aerated. Fibers were formed by pumping the solutions through spinnerets into coagulation baths. The tows formed were picked up on reels, stretched to various degrees, washed, and dried either as filaments or cut into staple lengths. The cellulose was regenerated by alkaline hydrolysis of the carbamate groups either in the process immediately after precipitation or at a later stage.

Cellulose can be precipitated from solutions in several different media. Unlike viscose, cellulose carbamate is precipitated with the substituent groups intact, and only later is the cellulose regenerated. The baths that have so far given the best results are acidic baths with sulfuric acid and sodium sulfate, thus resembling those of the viscose industry. Aluminum sulfate may be added to the coagulation baths to increase the dry matter content of the fresh fibers.

It is also possible to precipitate cellulose carbamate in baths containing other salts, or even lower alcohols like methanol and ethanol. Table 7.5 shows some results from the precipitation of fibers in an acidic coagulation bath. The fibers were spun on a small laboratory apparatus with an uptake speed of about 30 m/min. The diameter of the spinneret orifices was 50 μm. The best stretched fibers have properties of the same class as regular viscose.

As mentioned above, cellulose retains its carbamate groups during normal precipitation in acidic solutions. Some hydrolysis of the groups occurs when the cellulose is in solution. The extent of this hydrolysis depends on the temperature and length of time in solution, but usually the fibers still contain about half of the original carbamate groups. Because of this the alkali solubility is high, and for most purposes the fibers must therefore be treated to remove the remaining carbamate groups. Table 7.6 shows the influence of alkaline hydrolysis of the remaining carbamate groups on the fiber properties. We have recently discussed in more detail some aspects on the spinning of fibers from cellulose carbamate (31).

Besides fibers, many other products can be made from cellulose carbamate—for instance, films and sponges with good properties. The possibility of either retaining the carbamate groups or of regenerating to cellulose increases the versatility of this derivative, as the final properties of a product will depend on the amount of carbamate groups it contains. Although work by the Egyptian group did not point to any specific advantage as regards the chemical reactivity of the carbamate groups, this possibility must not be excluded (5–9). Generally, the carbamate derivative of cellulose is a conveniently made compound of a natural polymer. It can be stored in the dry state, dissolved in water-based

TABLE 7.5   Regenerated Cellulosic Fibers from Cellulose Carbamate (Acid Precipitation)

	Stretch		
Property	50%	75%	100%
Titer (dtex)	1.8	1.8	1.9
Tenacity (cN/dtex)			
Conditioned	2.0	2.5	3.0
Wet	1.1	1.5	1.7
Wet modulus	6	8	12
Elongation (%)			
Conditioned	20	17	15
Wet	19	16	12

TABLE 7.6 Effect of Post-treatment at 95°C on Fibers from Cellulose Carbamate

Treatment	Tenacity (cN/dtex)		Elongation (%)		Wet Modulus (cN/dtex)	Nitrogen Content (%)	Alkali Solubility (%)	DP
	Conditioned	Wet	Conditioned	Wet				
Untreated	2.4	0.9	10	15	8	1.2	90	310
A = 3 min in 0.75% NaOH	2.4	1.2	11	15	8	0.4	55	—
A + 3 min in 3% NaOH	2.3	1.2	15	14	7	0.18	11	305
A + 3 min in 4% NaOH	2.2	1.1	15	14	7	0.14	5	—

systems, precipitated unchanged from solutions, and finally regenerated back to the original polymer.

REFERENCES

1. C. E. Reddmann, F. C. Riesenfeld, and F. S. La Viola, *Ind. Eng. Chem.*, **50**, 633 (1958).

2. J. W. Hill and R. A. Jacobson, *U.S. Pat.* 2,134,825 (1938).

3. L. Segal and F. V. Eggerton, *Text. Res. J.*, **31**, 460 (1961).

4. Y. Nozawa and F. Higashide, *J. Appl. Polym. Sci.*, **26**, 2103 (1981).

5. A. Hebeish, A. Waly, N. Y. Abou-Zied, and E. A. El-Alfy, *Text. Res. J.*, **48**, 468 (1978).

6. A. Hebeish, N. Y. Abou-Zeid, E. A. El-Alfy, and A. Waly, *Cellulose Chem. Technol.*, **12**, 671 (1978).

7. A. Hebeish, A. Waly, E. A. El-Alfy, N. Y. Abou-Zeid, A. T. El-Aref, and M. H. El-Rafie, *Cellulose Chem. Technol.*, **13**, 327 (1979).

8. A. Hebeish, A. E. El-Alfy, A. Waly, and N. Y. Abou-Zeid, *J. Appl. Polym. Sci.*, **25**, 223 (1980).

9. M. I. Khalil, E. El-Alfy, M. H. El-Rafie, and A. Hebeish, *Cellulose Chem. Technol.*, **16**, 465 (1982).

10. J. Huttunen, O. Turunen, L. Mandell, V. Eklund, and K. Ekman, *U.S. Pat.* 4,404,369 (1983).

11. K. Ekman, V. Eklund, J. Fors, J. I. Huttunen, J.-F. Selin, and O. T. Turunen, in *International Dissolving and Specialty Pulps Conference*, Tappi Press, Atlanta, Georgia, 1983, p. 99.

12. D. F. Durso, in *Wood and Agricultural Residues: Research on Use for Feed, Fuels and Chemicals*, E. Soltes, Ed., Academic, New York, 1980, p. 79.

13. E. Sjöström, *Wood Chemistry. Fundamentals and Applications*, Academic, New York, 1981.

14. K. Bredereck and A. Saafan, *Melliand Textilber.*, **7**, 510 (1982).

15. L. Cheek and H. Struszcyk, *Cellulose Chem. Technol.*, **14**, 893 (1980).

16. F. W. Herrick, *J. Appl. Polym. Sci. Appl. Polym. Symp.*, **37**, 993 (1983).

17. K. Götze, *Chemiefasern nach dem Viskoseverfahren*, Springer-Verlag, Berlin, 1967, p. 495.

18. H. Sihtola, B. Nizovsky, and E. Kaila, *Pap. Puu*, **44**, 295 (1962).

19. V. Eklund, K. Ekman, O. T. Turunen, J. I. Huttunen, J.-F. Selin, and J. Fors, *Finn. Pat. Appl.* 840, 999 (1984).

20. R. G. Zhbankov, *Infrared Spectra of Cellulose and Its Derivatives*, Consultants Bureau, New York, 1966.

21. J.-F. Selin, J. I. Huttunen, O. T. Turunen, V. Eklund, and K. Ekman, *Pat. Pending*.

22. E. J. Lawton, W. D. Bellamy, R. E. Hungate, M. P. Bryant, and E. Hall, *Tappi*, **34**, 113A (1951).

23. J. L. Neal and H. A. Krässig, *Tappi*, **46**, 70 (1963).

24. J. F. Saeman, M. A. Millett, and E. J. Lawton, *Ind. Eng. Chem.*, **44**, 2848 (1952).

25. R. Imamura, T. Ueno, and K. Murakami, *Bull. Inst. Chem. Res. Kyoto Univ.*, **50**, 51 (1972).

26. K. Ekman, V. Eklund, and J. I. Huttunen, *J. Ind. Irr. Tech.*, **2**, 345 (1984).

27. A. F. Turbak, in *International Dissolving and Specialty Pulps Conference*, Tappi Press, Atlanta, Georgia, 1983, p. 105.

28. M. Dubé and R. H. Blackwell, in *International Dissolving and Specialty Pulps Conference*, Tappi Press, 1983, p. 111.

29. R. B. Hammer, A. F. Turbak, R. E. Davies, and N. A. Portnoy, ITT Rayonier Inc., Eastern Res. Div. Contr. No. 163.

30. S. M. Hudson, J. A. Cuculo, and L. C. Wadsworth, *J. Polym. Sci. Polym. Chem. Edn.*, **21**, 651 (1983).

31. O. T. Turunen, J. Fors, and J. I. Huttunen, *23rd International Man-Made Fibres Congress,* Dornbirn, Austria, Sept. 26–28, 1984.

8

Description and Analysis of Cellulose Ethers

JACQUES REUBEN

Research Center, Hercules Incorporated, Wilmington, Delaware

The reactivity of the three hydroxyl groups at positions 2, 3, and 6 on the glucosyl unit of cellulose offers a variety of possibilities for making useful derivatives. Cellulose-based polymers with predetermined properties are made by appropriate derivatization. In most cases the desired properties are achieved by making a partially substituted derivative—that is, by substituting on the average only one or two of the three hydroxyl hydrogens. Thus, the original homopolymer is transformed into a copolymer of eight or more monomers randomly distributed along the polymer chain. The description and characterization of polymers of such complexity have been challenging problems of long–standing interest. The theoretical framework for studying the distribution of substituents in partially substituted cellulose esters and ethers was presented by Spurlin in 1939 (1). It was only recently, however, that the validity and utility of his statistical models were clearly demonstrated (2, 3). As expected (1), cellulose nitrate esters are consistent with the equilibrium model (2), whereas (carboxymethyl)cellulose ethers conform to the kinetic model (3). The kinetic model has been extended to include cases such as (hydroxyethyl)cellulose and (hydroxypropyl)cellulose, in which a new hydroxyl group introduced by derivatization is also susceptible to reaction and side-chain extension (4). This chapter provides a summary of the mathematical framework for the description of cellulose ethers and a discussion of the analytical strategies that can be used to obtain such a description.

MATHEMATICAL DESCRIPTION

Spurlin realized that statistical models would be most suitable for a simple yet accurate description of polymers of such molecular complexity as that of cellulose

149

ethers (1). The statistical kinetic model involves the following assumptions:

1. All glucosyl units in the cellulose chain are equally accessible for reaction.
2. The relative rate constants (k_i and k_x) of reaction of the hydroxyls remain unchanged throughout the process.
3. Substitution within a given unit does not affect the reactivity of the remaining hydroxyls.
4. The effects of end groups are negligible.

For any given position i on the glucosyl unit one can write the following set of reactions

$$-C_i-OH + RX \xrightarrow{k_i} -C_i-OR + HX \qquad (1)$$
$$\underset{p_i}{} \qquad \underset{x_i}{}$$

The degree of substitution at each position, x_i, is given by

$$x_i = 1 - \exp(-Bk_i), \qquad i = 2, 3, \text{ or } 6 \qquad (2)$$

where B is a factor with the dimension of time. Thus, the probability, p_i, of having an unsubstituted hydroxyl at position i is

$$p_i = \exp(-Bk_i) \qquad (3)$$

The total (average) degree of substitution of the glucosyl residues is

$$D = x_2 + x_3 + x_6 \qquad (4)$$

The other quantities of interest may now be derived from p_i and x_i on the basis of simple statistical considerations. Thus, the mole fraction of unsubstituted glucose residues is given by the product of the probabilities of having unsubstituted hydroxyl groups at each of the three positions:

$$s_0 = p_2 p_3 p_6 \qquad (5)$$

The mole fraction of residues monosubstituted at position i is given by the product of the probabilities of having a substituent at that position (x_i) and the probabilities of having unsubstituted hydroxyl groups at the remaining positions:

$$s_i = x_i p_j p_k \qquad (6)$$

Similarly, for the disubstituted residues:

$$s_{ij} = x_i x_j p_k \qquad (7)$$

Finally, the mole fraction of trisubstituted glucosyl residues is given by the product of the probabilities of having a substituent at each one of the three positions:

$$s_{236} = x_2 x_3 x_6 \tag{8}$$

A complete description of cellulose ethers produced in the reaction of Eq. (1) can be obtained if the values of the three relative rate constants and the factor B are known. Examples of such materials include methylcellulose, ethylcellulose, and (carboxymethyl)cellulose.

A more complex situation arises when a new hydroxyl group is created in the substitution reaction. Thus, in the reaction with ethylene oxide (EO), a pendent hydroxyethyl group is produced. The latter can also react with EO leading to an oxyethylene side chain:

$$-C_i-OH + EO \xrightarrow{k_i} -C_i-OEOH \tag{9}$$

$$-C_i-OEOH + EO \xrightarrow{k_x} -C_i-OEOEOH \tag{10}$$

Here we have a fourth relative rate constant, k_x. Also, in addition to the degree of substitution of hydroxyl groups, D, one has the molar substitution, M, which is the average number of oxyethylene units per glucosyl residue. It has been shown (4) that the molar substitution is given by the closed form expression

$$M = 3Bk_x + D - k_x \sum_i \frac{x_i}{k_i} \tag{11}$$

The molar substitution, M_i, at position i is given by

$$M_i = Bk_x + x_i(1 - \frac{k_x}{k_i}) \tag{12}$$

For the complete description of materials obtained in reactions of the type of Eqs. (9) and (10), one needs the values of four relative rate constants and the factor B.

It is emphasized that the above expressions describe statistical distributions and related quantities characteristic of a final product rather than the actual kinetics of its formation.

CARBOXYMETHYLCELLULOSE

Because of the inherent intra- and interchain heterogeneity of the polymer, the usual analytical procedures for the determination of the composition of carboxymethylcellulose (CMC) involve the acid hydrolysis of the material (3, 5–8). In this way a mixture of 16 monosaccharide species is obtained, which includes the α and β pyranose forms of each of the eight monomeric sugars. Buytenhuys and Bonn (7) analyzed such mixtures by gas chromatography and mass-spectrometric identification of the silylated monosaccharides. Only the mole fraction of 2,3- and 3,6-di-O-(carboxymethyl)glucoses were not resolved from one another. Reuben and Conner (3) were able to obtain the complete monomer composition from the carbon-13 NMR spectrum. Ho and Klosiewicz (8) assigned the resolved proton NMR signals of the carboxymethyl methylenes and obtained the positional degrees of substitution (x_i).

With data on the monomer composition one can readily test the validity and applicability to CMC of the statistical model. Thus, for example, combining Eqs. (5) and (6) and using Eqs. (2) and (3) one obtains

$$p_i = \frac{s_0}{s_0 + s_i} \tag{13}$$

The relative rate constants are calculated as

$$\frac{k_i}{k_j} = \frac{\ln p_i}{\ln p_j} \tag{14}$$

Note that only the relative values of the rate constants can be obtained, since in the pertinent equations k_i appears as a product with the factor B. Defining $k_3 = 1$, a value for B can be obtained:

$$B = -\ln p_3 \tag{15}$$

With a set of relative rate constants one can use Eqs. (5)–(8) to calculate values for the mole fractions of each of the monosaccharides. A comparison between experimental and calculated values is shown in Figure 8.1. The agreement between the two sets is excellent.

The positional degrees of substitution can be obtained from the monomer composition, namely:

$$x_2 = s_2 + s_{23} + s_{26} + s_{236} \tag{16}$$

$$x_3 = s_3 + s_{23} + s_{36} + s_{236} \tag{17}$$

$$x_6 = s_6 + s_{26} + s_{36} + s_{236} \tag{18}$$

and their values used to calculate p_i:

$$p_i = 1 - x_i \tag{19}$$

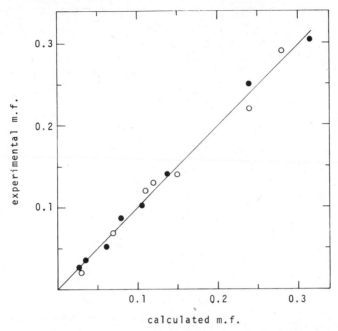

Figure 8.1   A plot of experimental versus calculated values of the mole fractions of the CMC monomers determined by gas chromatography (7) (open circles) and by carbon-13 NMR (3) (solid circles). The line is drawn with a slope of unity.

Using Eqs. (3) and (5) and taking the logarithms one obtains:

$$\ln(1 - x_i) = \frac{(\ln s_0)k_i}{k_2 + k_3 + k_6} \qquad (20)$$

A plot of $-\ln(1 - x_i)$ versus $-\ln s_0$ for a series of CMC samples of different degrees of substitution is shown in Figure 8.2. The linearity of the plots in Figure 8.2 demonstrates the applicability of the statistical model to treat data for a series of samples differing only in their degree of substitution.

Conformity of CMC to a simple statistical model can serve as a guide in the search for analytical strategies other than determination of the complete monomer composition. Since only three parameters are necessary for the description of the material, three properly chosen analytical quantities should suffice. Positional degrees of substitution obtained by proton NMR analysis of hydrolyzed CMC afford a complete description (8).

Values of the relative rate constants determined in recent studies (3, 7, 8) are in good mutual agreement. The ranges are 2.0–2.5 for k_2/k_3 and 1.5–1.8 for k_6/k_3. Note, however, that for CMC prepared using a water/cellulose mole ratio of 14, Buytenhuys and Bonn (7) report relative constants of $k_2/k_3 = 4.6$

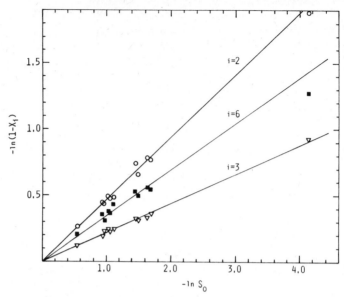

Figure 8.2 Logarithmic plots of the mole fractions of unsubstituted hydroxyl groups in CMC against the mole fraction of unsubstituted glucose (see Eq. (20)). (Reproduced with permission from ref. (3); copyright 1983 Elsevier Science Publishers.)

and $k_6/k_3 = 3.6$, which are higher than values obtained with a mole ratio of 7.

HYDROXYETHYLCELLULOSE

The monomer composition and substituent distribution of hydroxyethylcellulose (HEC) can be fully described by five parameters: the four rate constants k_2, k_3, k_6, and k_x, and the factor B. Because the latter parameter appears in the equations as a product with a rate constant, only relative rate constants can be determined. Thus, one is left with four unknown parameters. At least four independent observations are needed for their determination. Croon and Lindberg (9) analyzed the hydrolysis mixture of HEC by fractionation on a carbon column. However, the substitution levels of their samples were too low to be of practical interest. Results on molar substitution (M), unsubstituted glucosyl units (s_0), and vicinal diols (V_{23}) were reported by Glass et al. (10). They used a stochastic model and computer simulations to obtain a description of the material. An alternative approach to the analysis of such data is summarized below (4).

The probability of having vicinal diols (i.e., unsubstituted hydroxyls at positions 2 and 3 of the glucosyl unit) is given by the product of probabilities of having such hydroxyls:

$$V_{23} = p_2 p_3 \tag{21}$$

In combination with Eqs. (3) and (5) and taking the logarithms one obtains:

$$\ln V_{23} = \frac{(\ln s_0)(k_2 + k_3)}{k_2 + k_3 + k_6} \tag{22}$$

$$\ln\left(\frac{s_0}{V_{23}}\right) = \frac{(\ln s_0)k_6}{k_2 + k_3 + k_6} \tag{23}$$

Plots of the data of Glass et al. (Fig. 1 of ref. 10) according to Eqs. (22) and (23) are shown in Figure 8.3. The ratio of slopes of the two lines gives $(k_2 + k_3)/k_6 = 0.702$. Assuming $k_6 = 1$, a value for B can be calculated from $\ln(s_0/V_{23})$. With these, the data on M can be analyzed with Eq. (11) (and Eq. (4)) and a set of equations of the type of Eq. (2) to yield values of k_x for different ratios of k_2/k_3. As could be anticipated, k_x is rather insensitive to the value of k_2/k_3, since none of the measured quantities separate k_2 and k_3. The result of this analysis is $k_x/k_6 = 1.471$.

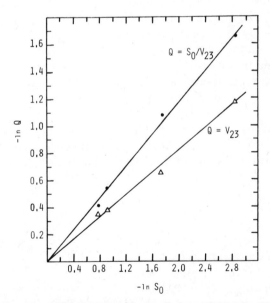

Figure 8.3 Graphical analysis according to Eqs. (22) and (23) of the HEC data from ref. (10) on unsubstituted glucosyl units (S_0) and vicinal diols (V_{23}). (Reproduced with permission from ref. (4); copyright 1984 American Chemical Society.)

Wirick (11) reported data on the molar substitution (M), unsubstituted glucosyl units (s_0), and number of enzymatically effected chain breaks (C) for a series of commercial samples. Klop and Kooiman (12) have shown that the cellulases catalyze the hydrolysis between unsubstituted glucosyl residues as well as between residues with the sequence β-glucosyl$(1 \rightarrow 4)$–6–O–(substituted)glucosyl. It can be shown that the total number of such chain breaks (per monomer unit) is given by (4)

$$C = s_0{}^2(1 - s_0) + \frac{1}{2} s_0{}^2(1 - s_0)^2 \left(\frac{1}{p_6} - 1\right) . \qquad (24)$$

With data on C and s_0, one can solve Eq. (24) for p_6 and use the result to calculate the ratio

$$\frac{k_2 + k_3}{k_6} = \frac{\ln s_0}{\ln p_6} - 1 \qquad (25)$$

The ratio k_x/k_6 is then calculated from the experimental value of M as described above. The results obtained from Wirick's data are $(k_2 + k_3)/k_6 = 0.404$ and $k_x/k_6 = 1.006$. It is of interest to note that the mole fraction of vicinal diols (V_{23}) and the number of enzymatically effected polymer chain breaks (C) are related quantities and contain similar, rather than complementary, information.

A complete description of HEC requires a set of four parameters. Therefore, at least four analytical quantities are necessary for the determination of these parameters. With only three quantities such as, the molar substitution (M), the fraction of unsubstituted glucosyl units (s_0), and the fraction of vicinal diols (V_{23}), or M, s_0, and the number of enzymatically effected chain breaks (C), one can obtain a partial description. Specifically, in these examples, the ratios $(k_2 + k_3)/k_6$ and k_x/k_6 can be determined. The ratio k_2/k_3 is needed to complete the description.

The potential of carbon-13 NMR spectroscopy for the characterization of HEC (13) has not yet been realized. However, the problems one may anticipate have been illustrated in a study of (hydroxypropyl)cellulose (14).

REFERENCES

1. H. M. Spurlin, *J. Am. Chem. Soc.*, **61**, 2222 (1939).
2. T. K. Wu, *Macromolecules*, **13**, 74 (1980).
3. J. Reuben and H. T. Conner, *Carbohydr. Res.*, **115**, 1 (1983).
4. J. Reuben, *Macromolecules*, **17**, 156 (1984).
5. T. E. Timell and H. M. Spurlin, *Sven. Papperstidn.*, **55**, 700 (1952).
6. I. Croon and C. B. Purves, *Sven. Papperstidn.*, **62**, 876 (1959).
7. F. A. Buytenhuys and R. Bonn, *Papier*, **31**, 525 (1977).

8. F. F. L. Ho and D. W. Klosiewicz, *Anal. Chem.*, **52**, 913 (1980).

9. I. Croon and B. Lindberg, *Sven. Papperstidn.*, **59**, 794 (1956).

10. J. E. Glass, A. M. Buettner, R. G. Lowther, C. S. Young, and L. A. Cosby, *Carbohydr. Res.*, **87**, 245 (1980).

11. M. G. Wirick, *J. Polym. Sci., Part A-1*, **6**, 1705 (1968).

12. W. Klop and P. Kooiman, *Biochim. Biophys. Acta*, **99**, 102 (1965).

13. A Parfondry and A. S. Perlin, *Carbohydr. Res.*, **57**, 39 (1977).

14. D. S. Lee and A. S. Perlin, *Carbohydr. Res.*, **106**, 1 (1982).

9

Novel Methods for Accelerating Photografting of Monomers to Cellulose

J. L. GARNETT
School of Chemistry, University of New South Wales, Kensington, New South Wales, Australia

Procedures for modifying cellulose, particularly one-step methods, are extremely valuable since they produce products that are of use in a wide variety of applications. Grafting processes, involving monomers, are specific examples of such procedures, especially those systems that are initiated by ultraviolet light (1–6). In this respect, UV is a complementary tool to ionizing radiation (7–12) as an initiator for these copolymerization reactions. The lower-energy UV process does possess certain advantages in these studies, in particular, for minimizing degradation in the system especially of the backbone polymer. In addition, small, relatively nonhazardous, and inexpensive laboratory-scale UV sources are readily available for this type of work.

In this chapter methods for optimizing the yields of copolymer in the photografting of monomers to cellulose are discussed. Styrene will be used as representative monomer in the initial treatment which will subsequently be extended to include a variety of monomers, particularly the acrylates. The parameters influencing the grafting yield to be discussed include sensitizer effects, the nature of the solvent, and the role of unique additives in accelerating the copolymerization reaction. The relevance of this current work in UV rapid-cure surface polymerization processes will also be considered.

GRAFTING METHOD

The UV source used for the work was a Philips 93110E 90W medium-pressure mercury vapor lamp mounted in the center of an enclosed cylindrical metal box and fitted with a rotating rack with space for the irradiation of 36 samples simultaneously. The cellulose samples (5 × 4 cm) consisted of Whatman No. 41 filter paper that had been extracted in appropriate hot solvent for 72 h prior to irradiation. These strips were conditioned at 21°C and 65% relative humidity (RH), before and after grafting, to minimize errors in weighing due to the natural water content of cellulose. For the actual copolymerization runs, the cellulose samples were fully immersed in a solution (25 mL) of monomer in solvent containing, where relevant, the appropriate additive including photosensitizer. Lightly stoppered Pyrex tubes were used for simplicity in these studies. Although evacuated quartz tubes give higher grafting efficiencies, reasonable rates of reaction were still obtained in the Pyrex vessels. Actinometry was performed with the uranyl nitrate–oxalic acid system. At the completion of the reaction, the cellulose strips were quickly removed from the tubes to avoid post-irradiation effects, washed in the appropriate solvents, then extracted in these same solvents in a Soxhlet for 72 h to remove homopolymer. The percentage graft was the percentage increase in weight of the cellulose strip.

Homopolymerization was determined by the following modification of the Kline method (13). At the completion of the irradiation, the grafting solution was poured into methanol (200 mL) to precipitate the homopolymer, and the sample tube was rinsed with methanol (50 mL). Homopolymer that adhered to both polymer film and tube was dissolved in dioxane (20 mL), and the dioxane solution was added to the methanol, in a beaker, together with benzene washings from the extraction of the original film. The beaker was heated to coagulate all polymer. The mixture was cooled, filtered through a sintered glass crucible, and washed with methanol (3 × 100 mL), and the crucible was dried to constant weight of 60°C. The percentage homopolymer was calculated from the weight of homopolymer divided by the weight of monomer in solution.

EFFECT OF SENSITIZER

From preliminary studies it was observed that a monomer concentration of 30% (v/v) in a solvent such as methanol was a suitable grafting solution to evaluate the efficiencies of various sensitizers. Grafting does occur in the absence of sensitizer (2, 6); however, for reasonable rates of reaction, such as for the irradiation times shown in Table 9.1, sensitizer is required. Otherwise no copolymerization is observed under these conditions. Of the sensitizers studied in Table 9.1, uranyl nitrate is the most efficient for the styrene monomer system examined, followed by biacetyl and benzoin ethyl ether. The anthraquinone dye is similar in structure to the compound used by other workers (1) but is not as effective as the previously mentioned materials. Manganese carbonyl, although useful in photopolymerization work (14), is not very efficient for grafting under the experimental conditions reported in Table 9.1. The data for the benzophenone charge-transfer complexes of dimethylaminoethanol and triethylamine are in-

TABLE 9.1 **Effect of Sensitizer in UV Grafting of Styrene in Methanol to Cellulose**[a]

Run	Sensitizer	Distance from UV Lamp (cm)	Graft
1	Nil	24	0.0
2	Nil	24	0.0[c]
3	Nil	12	0.0[c]
4	Uranyl nitrate	24	1.6[c]
5	Uranyl nitrate	24	15.2
6	Uranyl nitrate	24	3.4
7	Uranyl nitrate	12	10.9
8	ADS[b]	24	1.8
9	ADS[b]	12	3.7
10	Biacetyl	12	0.0[c]
11	Biacetyl	24	3.8
12	Biacetyl	12	11.0
13	Benzoin ethyl ether	24	0.0[c]
14	Benzoin ethyl ether	24	6.7
15	Benzoin ethyl ether	24	1.4
16	Benzoin ethyl ether	12	5.6
17	Manganese carbonyl	24	0.9[c]
18	Manganese carbonyl	24	1.5
19	DMAE:BPO[d]	24	0.0
20	TEA:BPO[e]	24	0.0

[a] Solutions of styrene (30% v/v) in methanol contained in Pyrex tubes with cellulose, irradiated for 3 hr with UV lamp except runs 4, 5, 13, and 14, where quartz tubes were used, and runs 17 and 18, where styrene (25% v/v) solutions were utilized. Sensitizer concentrations (1% w/v).

[b] ADS = anthraquinone-2,6-disulfonic acid, disodium salt.

[c] Ultraviolet light excluded.

[d] Dimethylaminoethanol:benzophenone (3:2; 2:1).

[e] Triethylamine:benzophenone (3:2; 2:1).

teresting, since neither combination sensitizes the styrene grafting reaction, although both are used extensively in UV rapid-cure processing of oligomer/acrylate mixtures. As later work will show, both sensitize methyl methacrylate copolymerization to cellulose. Thus the mode of action of the sensitizer is very closely related to the structure of the monomer being grafted. The final aspect of the data in Table 9.1 is that the copolymerization in quartz is more efficient than grafting in Pyrex; however, in the latter vessels, rates of reaction are still satisfactory.

EFFECT OF SOLVENT

Alcohols

The results in Figure 9.1 show that those solvents that wet and swell cellulose are the most efficient for grafting styrene. Copolymerization yields in the lower-molecular-weight alcohols are relatively high and decrease with increasing length of alcohol chain such that at n-octanol grafting is low at all monomer concentrations studied. If a long-chain, low-activity alcohol (octanol) is mixed with a short-chain active alcohol (methanol), there is significant enhancement in the original methanol graft. The remaining important feature of the data in Figure 9.1 is the peak in grafting observed at monomer concentrations of 70–90%, depending on the solvent system used. This Trommsdorff peak is not only of interest mechanistically but is also of practical value, since, in a preparative context, if copolymer yield is the sole property required from the reaction, then the photolytic conditions required to yield a Trommsdorff peak are important.

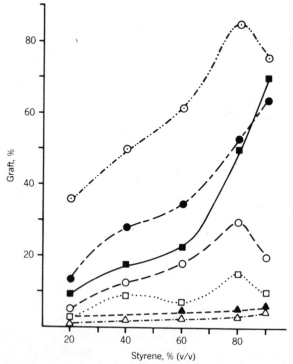

Figure 9.1. Photosensitized grafting of styrene to cellulose using uranyl nitrate as sensitizer (1% w/v) and irradiated for 24 hr at 24 cm from UV lamp. ● methanol; ■ ethanol; ○ n-propanol; □ isopropanol; ▲ n-butanol; △ n-octanol; ⊙ n-octanol/methanol (1:1).

Other Solvents

In the nonalcohol solvents, the grafting trend is the same as previously discussed for the alcohols where wetting and swelling properties are important in determining the efficiency of grafting. From the data in Figure 9.2, dimethyl sulfoxide is easily the most efficient grafting solvent, being marginally better than the best of the alcohols. These results reflect the well-known sensitizing properties of dimethyl sulfoxide in the UV. Of the remaining solvents in this group, dimethyl formamide, dioxane and acetic acid give reasonable grafting yields, as predicted, whereas with benzene and hexane, virtually no grafting of styrene is observed at any concentration studied (this property reflecting the poor cellulose swelling characteristics of these last two solvents).

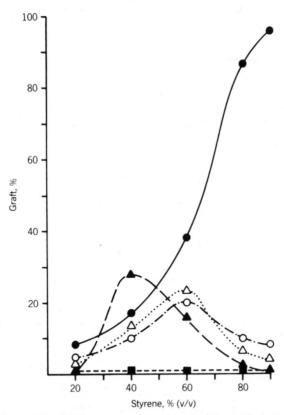

Figure 9.2. Photosensitized grafting of styrene to cellulose using uranyl nitrate as sensitizer (1% w/v) except for benzene and hexane (benzoin ethyl ether) and irradiated for 24 hr at 24 cm from UV lamp. ● dimethyl sulfoxide; ○ dimethyl formamide; △ dioxan; ▲ acetic acid; ■ benzene, hexane. Data for neat styrene: benzoin ethyl ether 3 hr, 1.7%; 24 hr, 37.9%; biacetyl 3 hr, 1.3%; 24 hr, 9.8%.

EFFECT OF ACID

Inclusion of either organic or inorganic acids in the grafting solution enhances the copolymerization yields significantly at particular concentrations of monomer in solvent (Table 9.2). With uranyl nitrate as a sensitizer, formic acid increases the grafting yield at higher monomer concentrations, whereas both inorganic acids show grafting enhancement at all monomer concentrations examined. When the organic sensitizer benzoin ethyl ether (BEE) is used, inclusion of mineral acid leads to grafting enhancement only at the lower styrene concentrations studied. The magnitude of the copolymerization yield is also significantly lower in the acid solutions containing BEE than when uranyl nitrate is used. In addition, the presence of acid increases the size of the Trommsdorff peak (Table 9.2, uranyl nitrate runs) and also induces the formation of such peaks (Table 9.2, BEE runs).

Homopolymerization

Concurrent with grafting significant amounts of homopolymer are formed at all monomer concentrations studied, the homopolymer yield decreasing, relative to grafting, as the concentration of monomer is increased in neutral solution (Table 9.3). A similar trend in reactivity is observed in the acid solutions. The effect is more significant at the higher monomer concentrations, the grafting efficiency (essentially the ratio of graft to homopolymer) being particularly high in the 90% styrene solutions containing nitric acid.

TABLE 9.2 Enhancement Effect of Organic and Inorganic Acids in Photografting Styrene to Cellulose Using Organic and Inorganic Sensitizers[a]

Styrene in Methanol (% v/v)	Graft (%)					
		UO_2^{2+}			BEE	
	Neutral	HCOOH (1% w/v)	H_2SO_4 (1 M)	HCl (2 M)	Neutral	H_2SO_4 (0.1 M)
10	—	—	—	—	5	8
20	13	7	54	34	11	23
40	18	21	73	47	32	18
60	34	34	77	72	38	15
80	53	79	95	127	—	—

[a] Irradiations at 20°C using uranyl nitrate (UO_2^{2+}, 1% w/v) for 24 hr at 24 cm from 90-W source and benzoin ethyl ether (BEE, 1% w/w) for 15 hr at 24 cm.

TABLE 9.3   Effect of Acid on Homopolymerization during Photografting of Styrene in Methanol to Cellulose[a]

Styrene (% v/v)	Graft (%)		Homopolymer (%)	
	Neutral	HNO$_3$ (1 M)	Neutral	HNO$_3$ (1 M)
20	13	20	10	18
40	28	23	18	19
60	34	37	24	24
80	53	77	24	24
90	64	107	34	32

[a] Irradiation conditions as in Table 9.2 with uranyl nitrate (1% w/v) as sensitizer.

GRAFTING OF METHYL METHACRYLATE

All data in Tables 9.1–9.3 and Figures 9.1 and 9.2 are for styrene as the monomer in these photografting reactions. Other monomers, particularly the acrylates, are very useful in this field. Methyl methacrylate (MMA) is a representative acrylate monomer of value in these studies. Previous results with styrene are a convenient guide for the reactivity expected of MMA in photografting, although, as the following data show, MMA is generally more reactive in photografting processes than styrene.

Role of Sensitizer in MMA Grafting

The results of typical sensitizers used in the photografting of MMA in methanol to cellulose are shown in Table 9.4. As for styrene grafting, BEE and uranyl nitrate are both efficient sensitizers for photografting MMA, the former being particularly effective at the higher monomer concentrations. By contrast, the amine–benzophenone (BPO) (dimethylaminoethanol, DMAE; triethylamine, TEA) charge-transfer complexes do not readily sensitize photografting of styrene to cellulose under the conditions reported in Table 9.4 (<0.3%), whereas TEA/BPO, particularly at lower monomer concentrations, is effective in photografting MMA. In addition to demonstrating that the amine–benzophenone initiators can be used to prepare acrylate graft copolymers, the present result is also of importance in radiation rapid-cure (RRC) polymerization work with UV sources where films of oligomer–monomer mixtures, mainly acrylates, are cured onto cellulose substrates with UV light. The possibility that concurrent grafting can occur with film formation is valuable, since the adhesive properties of the cured film are improved if some grafting is achieved. The amine–benzophenone result

TABLE 9.4 Photografting of Methyl Methacrylate (MMA) in Methanol to Cellulose Using Various Sensitizers[a]

MMA (% v/v)	UO_2^{2+}	BEE	Graft (%)	
			DMAE/BPO (3/1)[b]	TEA/BPO (6/1)
30	29	74	1.2	20
60	—	165	0.8	7
80	—	124	0.9	3

[a] Irradiation conditions as in Table 9.3, using sensitizer (1% w/v). Time of irradiation 6 hr except UO_2^{2+} (1 hr). DMAE = dimethylaminoethanol; TEA = triethylamine; BPO = benzophenone. Graft < 1.0% when amine or benzophenone used separately.

[b] With this sensitizer under these UV conditions using MMA (30% v/v) grafting yields in dimethyl sulfoxide and dimethyl formamide were 135% and 111%, respectively.

involving MMA is significant in this respect, since these sensitizers are much cheaper than BEE and are widely used in RRC acrylate work because of the economics. The data in Table 9.4 indicate that with certain amines, concurrent photografting with film formation may be achieved in RRC acrylate processes.

Acids and Polyfunctional Monomers as Additives in MMA Grafting

Consistent with the styrene data in Table 9.2, inclusion of sulfuric acid in the MMA grafting solution leads to a characteristic enhancement reaction (Table 9.5). The magnitude of this acid effect is very large, especially when the relative irradiation times of the uranyl nitrate solution with (6 hr) and without (16 hr) acid are considered. Inclusion of acid also leads to a shift in the Trommsdorff

TABLE 9.5 Comparison of Acid with Polyfunctional Monomers as Additives for Enhancing Photografting of MMA to Cellulose[a]

MMA (% v/v)	Graft (%)		
	UO_2^{2+}[b]	$UO_2^{2+} + H_2SO_4$	$UO_2^{2+} + TMPTA$[b]
30	51	186	3
40	95	228	160
60	67	373	160
80	10	227	32

[a] Irradiation conditions as in Table 9.4 for 6 hr; H_2SO_4 (0.2 M); TMPTA = trimethylol propane triacrylate (1% v/v); MMA = methyl methacrylate; solvent methanol.

[b] Irradiation time 16 hr.

peak in grafting from 40% MMA without acid to 60% MMA in the presence of this additive.

The other significant feature of the data in Table 9.5 is the use of polyfunctional monomers (PFMs) such as trimethylol propane triacrylate (TMPTA) in additive amounts (1% v/v) to accelerate the MMA photografting reaction to cellulose. PMFs have previously been used for an analogous purpose with the polyolefins (15) and also for accelerating the polymerization efficiency of MMA (16) and styrene (17) in RRC work. From the results in Table 9.5, the modes of action of acids and PFMs appear to be different, since the shapes of the enhancement curves are different (i.e., the TMPTA monomer dependence grafting curve is very much sharper than the corresponding graph for acid). TMPTA is also not as efficient in grafting enhancement as sulfuric acid.

Synergistic Effect of Acids and Polyfunctional Monomers in MMA Grafting

Because the shapes of the grafting enhancement curves with acids and PFMs are different, the mechanism by which each additive operates is probably different, and thus synergistic effects would be expected from grafting solutions containing both additives. These predictions are realized from the data in Table 9.6, where extremely large enhancement yields are observed when sulfuric acid and TMPTA or divinylbenzene (DVB) is present during grafting. TMPTA is particularly effective, especially at higher MMA concentrations. The actual high values for grafting reported in Table 9.6 (e.g., the data for 60 and 80% solutions containing acid and PFM) should be treated with reservation, since at this level of MMA graft, homopolymerization is also appreciable and can lead to the presence of turbidity in the grafting solution because of phase separation problems. With the onset of turbidity, grafting can cease, and this property may contribute to the differences in actual graft percentages reported between TMPTA and DVB in Table 9.6.

TABLE 9.6 Synergistic Effect of Acid with Polyfunctional Monomers as Additives for Enhancing Radiation Grafting of MMA to Cellulose[a]

MMA (% v/v)	Graft (%)		
	UO_2^{2+} [b]	UO_2^{2+} + H_2SO_4 + TMPTA	UO_2^{2+} + H_2SO_4 + DVB
30	51	433	83
40	95	580	195
60	67	1043	380
80	10	1417	485

[a] Irradiation conditions as in Table 9.5; UO_2^{2+} = uranyl nitrate (1% w/v); TMPTA = trimethylol propane triacrylate (1% w/v); DVB = divinylbenzene (1% w/v).

[b] Irradiation time 16 hr.

STYRENE COMONOMER TECHNIQUE FOR MMA GRAFTING

Efficiency of Sensitizer

The difficulty previously discussed with MMA photografting is the problem of competing homopolymerization. In earlier ionizing radiation work involving grafting MMA to wool (18), homopolymerization was controlled by a styrene comonomer procedure. Not only did the addition of 25% styrene to the total monomer in solution minimize homopolymer formation, but the resulting graft contained predominantly MMA with the properties of that polymer. The results in Table 9.7 show that the styrene comonomer technique is also applicable to the photografting of MMA to cellulose. Copolymerization yields are high when uranyl nitrate is used as sensitizer, whereas with TEA/BPO, grafting was low. However, the solutions remained clear, and thus turbidity problems were not the reason for the poor graft with this latter sensitizer.

Effect of Acid and Polyfunctional Monomers in Styrene Comonomer Technique

Inclusion of acid and PFMs in the monomer solution enhanced the MMA photografting yields in the styrene comonomer technique (Table 9.8) using the TEA/BPO sensitizer system. This result is consistent with the previous data discussed for the photografting of MMA (Table 9.6). Significant enhancement in MMA grafting is found when acid is used alone (Table 9.8), whereas TMPTA alone has little effect. When the two additives are combined, a significant synergistic effect is observed giving significant grafting yields that could be of

TABLE 9.7   Effect of Sensitizer on Styrene Comonomer Technique for Photografting MMA to Cellulose[a]

Total Monomer (% v/v)	Graft (%)		
	UO_2^{2+}	TEA/BPO	
		A	B
30	70	1.4	—
40	—	1.6	—
60	102	1.3	$2.9(2.9,^b\ 3.9^c)$
80	111	1.1	—

[a] Irradiation for 6 hr in methanol except column B (57 hr); UO_2^{2+} = uranyl nitrate (1% w/v); TEA/BPO = triethylamine/benzo-phenone (6/1, 1% w/v); total monomer = MMA/styrene (3/1, v/v).

[b] MMA/styrene (5/1, v/v).

[c] MMA/styrene (7/1, v/v).

TABLE 9.8 Effect of Acid and Polyfunctional Monomers on Styrene Comonomer Technique for Photografting MMA to Cellulose with TEA/BPO Sensitizer[a]

Total Monomer (% v/v)	Graft (%)			
	No Additive	H_2SO_4 (0.2 M)	TMPTA (1% v/v)	H_2SO_4 + TMPTA (0.2 M) (1% v/v)
40	1.6	7.8	1.2	7.3
60	1.3	8.4	1.2	8.0
80	1.1	11.4	1.4	12.2

[a] Irradiation conditions as in Table 9.7; TEA/BPO = triethylamine/benzophenone (6/1, 1% w/v); TMPTA = trimethylol propane triacrylate; MMA/styrene (3/1, v/v).

value in practical applications. Grafts of up to 12% are readily obtained after 6 hr of irradiation. The solutions remained clear at this time, and thus longer exposure to UV would be expected to give significantly higher yields. These additive results are also of importance in RRC work where the TEA/BPO sensitizer is frequently used, since they demonstrate that TEA/BPO is capable of sensitizing grafting as well as polymerization of films.

NEW SALT ADDITIVES FOR ENHANCING PHOTOGRAFTING

In experiments designed to discern the mechanism of the acid enhancement effect in the photografting of styrene to cellulose (Table 9.2), a new series of salt additives that increase grafting yields has been discovered (19, 20). Of the inorganic salts studied, certain lithium salts (perchlorate and nitrate) are particularly effective. The data in Table 9.9 indicate that lithium perchlorate (0.2

TABLE 9.9 Comparison of Acid with Lithium Salt as Additives in UV Grafting of Styrene in Methanol to Cellulose[a]

Styrene (% v/v)	Graft (%)		
	No Additive	H_2SO_4 (0.1 M)	$LiClO_4$ (0.2 M)
10	5	8	11
20	11	23	38
30	25	39	41
40	32	18	38
60	38	15	40

[a] Irradiations in sealed Pyrex vessels for 15 hr at 24 cm from 90-W source at 20°C. Solutions contained benzoin ethyl ether (1% w/w) as sensitizer. Solutions without sensitizer showed no graft.

M) is more efficient than sulfuric acid in enhancing photografting of styrene in methanol to cellulose under the experimental conditions reported. The presence of lithium perchlorate leads to an increase in grafting yield at all monomer concentrations studied, whereas with sulfuric acid the enhancement is observed only at the lower monomer concentrations. The parameters that define the efficiency of the inorganic salt in the grafting enhancement depend on the nature of the anion, the solubility of the salt in the grafting solution, and the polarity of the backbone polymer (21).

The significant feature of the lithium perchlorate results is that the trends are similar to those observed with acid, particularly when the data are examined from the number average molecular weights $(\bar{M}_n)$ of the copolymers from the photografting of styrene to cellulose (Table 9.10). The corresponding $\bar{M}_n$ values for both acid and lithium perchlorate samples are similar and consistently higher than for the copolymers without additive.

MECHANISM OF PHOTOGRAFTING

The data in the figures and tables show that styrene and MMA monomers will photograft to cellulose, but only very slowly in the absence of sensitizers. The mechanism of such processes is free radical and involves absorption of incident UV by the backbone polymer yielding radicals that act as sites for grafting as depicted in Eq. (1) (CeOH = cellulose).

$$\text{CeOH} \xrightarrow{h\nu} \text{CeO}^{\cdot} + \text{H}^{\cdot} \tag{1}$$

Monomers (MH) may also absorb UV to yield radicals that then abstract H atoms from the backbone polymer to give additional sites for grafting:

$$\text{MH} \xrightarrow{h\nu} \text{M}^{\cdot} + \text{H}^{\cdot} \tag{2}$$

$$\text{M}^{\cdot} + \text{CeOH} \longrightarrow \text{MH} + \text{CeO}^{\cdot} \tag{3}$$

TABLE 9.10   Effect of Acid and Lithium Salt on Number Average Molecular Weights $(\bar{M}_n)$ of Copolymers from UV Grafting of Styrene to Cellulose[a]

Styrene (% v/v)	Graft (%)		
	No Additive	H_2SO_4 (0.1 M)	$LiClO_4$ (0.2 M)
20	3	3.8	4
30	2.9	5.4	4.5
40	3.6	5.3	5
60	3.3	4.9	5.3

[a] Irradiation conditions as in Table 9.7. Molecular weight determinations performed on a Waters Associates model ALG/GPS201 g.p.c. instrument.

The inclusion of sensitizers leads to rapid photografting to cellulose. Typical of the organic photosensitizers used is BEE where free radical reactions again are involved (12, 22) initiated by absorption of energy in BEE:

$$C_6H_5-\overset{\overset{O}{\|}}{C}-\underset{\underset{H}{|}}{\overset{\overset{OEt}{|}}{C}}-C_6H_5 \xrightarrow{h\nu} C_6H_5\overset{\cdot}{C} + C_6H_5-\underset{\underset{H}{|}}{\overset{\overset{O}{\|}}{C}}-OEt \qquad (4)$$

With inorganic sensitizers, such as uranyl nitrate, the mechanism of the process is depicted in Eqs. (5)–(7) where radical reactions even involving the solvent methanol can be involved (22, 23).

$$UO_2^{2+} + h\nu \rightarrow (UO_2^{2+})^* \qquad (5)$$

$$(UO_2^{2+})^* + CH_3OH \rightarrow CH_3O\cdot + H^+ + UO_2^+ \qquad (6)$$

$$(UO_2^{2+})^* + CeOH \rightarrow UO_2^+ + H^+ + CeO^{\cdot} \qquad (7)$$

By contrast amine–benzophenone (amine/BPO) complexes sensitize the reaction by forming exiplexes from reaction with the triplet state of the benzophenone (Eqs. (8) and (9), where S and T denote the lowest excited singlet and triplet states, respectively). Especially with tertiary amines, such as triethylamine as hydrogen donors, the following charge–transfer complexes are formed as the excited states (Eq. (9)):

$$C_6H_5-\overset{\overset{O}{\|}}{C}-C_6H_5 \xrightarrow{h\nu} C_6H_5-\overset{\overset{O}{\|}}{C}-C_6H_5^* \,(S) \xrightarrow{h\nu} C_6H_5-\overset{\overset{O}{\|}}{C}-C_6H_5^* \,(T)(8)$$

$$C_6H_5-\overset{\overset{O}{\|}}{C}-C_6H_5^* \,(T) + \underset{\underset{Et}{|}}{\overset{\overset{Et}{|}}{:N}}-Et \longrightarrow \left[\underset{\underset{C_6H_5}{|}}{\overset{\overset{C_6H_5}{|}}{O=C^{\cdot}}}{}^{\ominus} \quad {}^{\oplus}\underset{\underset{Et}{|}}{\overset{\overset{Et}{|}}{\cdot N}}-Et \right]^* \qquad (9)$$

From current studies, uranyl and BEE type sensitizers are more efficient in photografting than amine/BPO complexes. The size of these complexes may sterically hinder the formation of the transition states involved in photografting such as when monomer, solvent, or cellulose substrate is the reagent being attacked. In the presence of styrene, an additional explanation for the poor sensitizer efficiency of the amine/BPO is available. Thus, with styrene alone or in the styrene/MMA comonomer runs, grafting is low, and this result is attributed to styrene preferentially quenching the triplet state of BPO leading to the poorer yields.

As predicted from previous ionizing radiation grafting studies with cellulose (12), solvents that wet and swell cellulose are the most useful for photografting. DMSO is particularly effective with both styrene and MMA in the current UV grafting. This result suggests that DMSO also possesses sensitizing properties, since its effectiveness in UV is significantly higher than other solvents when compared with analogous results from ionizing radiation initiation systems.

MECHANISM OF ADDITIVE EFFECTS

Organic and Inorganic Acids

Both organic and inorganic acids enhance photografting reactions, as shown by the present studies. Organic acids, such as formic, contain functional groups that absorb strongly in the UV, thus providing additional sensitization to the copolymerization reaction. With inorganic acids, the mechanism of acid enhancement can be due to several processes. Residual water in cellulose ($\sim$ 7% at ambient conditions) can lead to acid enhancement in photosensitized grafting by processes similar to the sulfuric acid catalyzed photopolymerization of monomers in aqueous solutions where, in the presence of acid (e.g., H_2SO_4), additional radicals are formed. These may subsequently react with trunk polymer by abstraction reactions to give increased numbers of grafting sites (Eqs. (10) and (11)). Changes in the physical parameters of cellulose by the inclusion of acid can also advantageously affect grafting. Thus acid can (a) produce both intercrystalline and intracrystalline swelling, which loosens the "order–disorder" structure of the cellulose, making it more accessible to reagents, and (b) act as a catalyst in the hydrolysis of the cellulose, leading to uncoiling of the chains and improved monomer accessibility. Such hydrolysis reactions involve an intermediate complex between the glycosidic oxygen and the proton. In the presence of UV, breakage of the glycosidic bond of the complex is facilitated, leading to additional grafting sites.

$$SO_4^{2+} + H_2O \xrightarrow{h\nu} SO_4^- + H^{\cdot} + OH^- \qquad (10)$$

$$SO_4^- + H_3O^+ \longrightarrow H_2SO_4 + OH^{\cdot} \qquad (11)$$

Although the above mechanisms do contribute to the overall acid enhancement in photografting to cellulose, these schemes fail to explain several further observations concerning the phenomenon including (a) the occurrence of enhancement only at certain monomer and acid concentrations and (b) the variation in acid efficiencies under different experimental conditions. A further model is thus required to explain the predominant mechanism by which acid increases grafting yields in these systems. The discovery of novel salts (Tables 9.9 and

9.10) as enhancement additives has enabled the mechanism of the acid effect to be clarified further.

Salt Effect

The fact that nonprotic salts such as $LiClO_4$ enhance photografting to cellulose (Tables 9.9 and 9.10) in a manner similar to acid is important, since these salts do not exert the same physical effects on the cellulose structure as acids. Thus mechanisms already proposed based on the effect of additive on cellulose structure are not relevant to the salt system. A variety of salts, in addition to $LiClO_4$, have been found to accelerate the photografting (19, 20). The molecular weight data ($\bar{M}_n$ values, Table 9.10) are also of importance in this respect. Thus the grafted celluloses reported in Table 9.9 were hydrolyzed, and the $\bar{M}_n$ values of the graft copolymer that was separated from the cellulose trunk polymer were determined by gel permeation chromatography. The $\bar{M}_n$ values of samples from additive-free solutions are independent of monomer concentration. Samples produced in the presence of acid show a significant increase in $\bar{M}_n$ over reference values and reach a maximum at 30–40% monomer. $\bar{M}_n$ values of copolymers produced in the presence of salt are markedly higher than those of the reference and increase with increasing monomer concentration.

Considering the above results and more extensive evidence to be published elsewhere (20, 21), a new model is proposed to explain the effect of acids and metal salts such as $LiClO_4$ in enhancing photografting in the present studies.

In any grafting system at any one time, there is an equilibrium concentration of monomer absorbed within the grafting region of the backbone polymer. This grafting region may be continually changing as grafting proceeds. Thus in grafting styrene to cellulose, during the initial part of the reaction the grafting region will be essentially cellulosic in nature; however, as reaction proceeds, the grafting region will become more involved with styrene. The degree to which the monomer will be absorbed by this grafting region will therefore depend on the chemical structure of the region at the specific time of grafting. Experimental data now available indicate that increased partitioning of monomer occurs in the graft region when ionic solutes are dissolved in the bulk grafting solution. Thus higher concentrations of monomer are available for grafting at a particular backbone polymer site in the presence of these salt additives. The extent of this improved monomer partitioning depends on the polarities of monomer, substrate, and solvent and also on the concentration of ionic solute. Metal salts such as $LiClO_4$ are more efficient than acids in enhancing grafting because of the overall monomer partitioning effect. It is thus the effect of these ionic solutes on partitioning that is essentially responsible for the observed increase in photografting yields in the presence of these additives. Because of all the parameters involved, the enhancement process is mechanistically complicated. It is obvious that mechanisms based on changes in physical parameters of the cellulose also contribute to the enhancement reaction, but not as the predominant pathway, which is essentially due to the partitioning process.

Polyfunctional Monomer Effect

The mechanism of the enhancement with the polyfunctional monomers in additive amounts is obviously different from that of acid and salt, since synergistic effects are observed when both acid and PFM additives are included in the same solution. With a PFM such as TMPTA, branching of the grafted polymer can occur when one end of the PFM, immobilized during grafting, is bonded to the growing chain. The other end is unsaturated and free to initiate a new chain growth via scavenging reactions. The new branched polymer may eventually terminate, cross-linked by reacting with an adjacent polystyrene chain or immobilized PFM radical. Grafting is thus enhanced mainly through branching of the grafted polymer chain.

SIGNIFICANCE OF CURRENT RESULTS IN PHOTOGRAFTING APPLICATIONS

Photografting can be used as a direct, one-step technique to improve the properties of cellulosics. Thus, in a batch process, styrene or MMA can be photografted to improve such properties as water resistance, solvent repellency, soil release, and so on (1–7, 24, 25). By photografting monomers containing appropriate functional groups, cellulose can also be used as a backbone polymer for immobilization of enzymes (12), anchoring catalytically active metal complexes (12), and forming new ion exchange resins (26). In most of these applications, the yield of graft copolymer required is usually not high (15%). Any process—presence of solvent, type of sensitizer, inclusion of acid, salt, or PFM additives—to improve yields is extremely valuable, since any graft enhancement will lead to the utilization of lower UV doses with the obvious advantages such a step possesses.

The other general field where enhanced photografting can be of importance is in the radiation rapid curing (RRC) of films on to cellulose using UV as initiator (22). The RRC technique involves essentially polymerizing a film containing oligomer–monomer (usually acrylate) onto a substrate such as cellulose under influence of UV. In this work competing photografting can be useful, since a true carbon–carbon bond between film and substrate is formed in the photografting process, leading to improved adhesion and flexibility. In this respect, the results of the amine/BPO sensitizers reported in the present work are significant. Sensitizers of this type are frequently used in RRC formulations because of economic efficiency; however, it is obvious that only certain of these sensitizers catalyze significant photografting even though most catalyze fast polymerization rates and therefore rapid film formation in RRC work. Inclusion of additives such as acids and PFMs to mixtures containing those amine/BPO sensitizers leads to an enhancement in photografting, and this could be a practical advantage in RRC treatments. The potential applications of these additives in RRC work are currently being evaluated.

Overall, the discovery of the novel salt additives reported here for enhancing photografting has proved to be of value both mechanistically and practically.

In terms of the acid additive effect, the salt work has enabled a new, more plausible mechanism to be proposed for explaining the acid enhancement in photografting. The availability of salt additives for increasing photografting yields means that, in terms of practical considerations, these additives can be used to replace acid in those applications where the trunk polymer is acid-sensitive and thus not amenable to acid enhancement techniques.

REFERENCES

1. V. Stannett, in ACS Symp. Ser. No. 187, *Graft Copolymerization of Lignocellulose Fibers*, D. N.-S Hon, ed., American Chemical Society, Washington, D.C., 1982, p. 3.

2. J. C. Arthur, Jr., ACS Symp. Ser. No. 187, *Graft Copolymerization of Lignocellulose Fibres*, in D. N.-S. Hon, ed., American Chemical Society, Washington, D.C., 1982, p. 21.

3. H. L. Needles and W. L. Wasley, *Text. Res. J.*, **39**, 97 (1969).

4. H. Kubota, T. Murata, and T. Ogiwara, *J. Polym. Sci.*, **11**, 485 (1973).

5. D. N.-S. Hon and H.-C. Chan, in ACS Symp. Ser. No. 187, *Graft Copolymerization of Lignocellulose Fibres*, D. N.-S. Hon, American Chemical Society, Washington, D.C., 1982, p. 101.

6. N. P. Davis, J. L. Garnett, C. H. Ang, and L. Geldard, in *Polymer Applications of Renewable Resource Materials*, C. E. Carraher, Jr., and L. H. Sperling, eds., Plenum, New York, 1983, p. 323.

7. A. Hebeish and J. T. Guthrie, *The Chemistry and Technology of Cellulosic Copolymers*, Springer-Verlag, Berlin, 1980.

8. A. Chapiro, *Radiation Chemistry of Polymeric Systems*, Interscience, New York, 1962.

9. R. J. Demint, J. C. Arthur, Jr., A. R. Markevich, and W. F. McSherry, *Text. Res. J.*, **32**, 918 (1962).

10. H. A. Krassig and V. Stannett, *Adv. Polym. Sci.*, **4**, 111 (1963).

11. A. Charlesby, *Atomic Radiation and Polymers*, Pergamon, Oxford, 1960.

12. J. L. Garnett, *Rad. Phys. Chem.*, **14**, 79 (1979).

13. G. M. Kline, *Analytical Chemistry of Polymers*, Interscience, New York, 1966, Part I.

14. G. Oster and N. L. Yang, *Chem. Rev.*, **68**, 125 (1968).

15. C. H. Ang, J. L. Garnett, R. Levot, and M. A. Long, *J. Polym. Sci. Polym. Lett.*, **21**, 257 (1983).

16. M. M. Micko and L. Paszner, *J. Rad. Curing*, **1**(3), 3 (1974).

17. M. M. Micko and L. Paszner, *J. Rad. Curing*, **1**(4), 2 (1974).

18. J. L. Garnett and J. D. Leeder, in ACS Symp. Ser. No. 49, *Textile and Paper Chemistry and Technology*, J. C. Arthur, Jr., ed., American Chemical Society, Washington, D.C., 1977, p. 197.

19. J. L. Garnett, S. V. Jankiewicz, R. Levot, and D. F. Sangster, *Rad. Phys. Chem.*, in press.

20. J. L. Garnett, S. V. Jankiewicz, M. A. Long, and D. F. Sangster, *J. Polym. Sci.*, submitted.

21. J. L. Garnett, S. V. Jankiewicz, M. A. Long, and D. F. Sangster, in preparation.

22. J. L. Garnett and G. Major, *J. Rad. Curing*, **9**, 4 (1982).

23. H. D. Burrows and T.J. Kemp, *Chem. Rev.*, **3**, 139 (1974).

24. J. P. Fouassier, in ACS Symp. Ser. No. 187, *Graft Copolymerization of Lignocellulose Fibers*, D. N.-S. Hon, ed., American Chemical Society, Washington, 1982, p. 83.

25. A. Takahashi and S. Takahashi, *Kobunshi Ronbunshu*, **37**, 151 (1980).

26. C. H. Ang, J. L. Garnett, R. C. Levot, and M. A. Long, in ACS Symp. Seri. No. 212, *Initiation of Polymerization*, F. E. Baily, Jr., ed., American Chemical Society, Washington, 1983, p. 209.

10

Anionic Graft Polymerization onto Cellulose

RAMANI NARAYAN and GEORGE T. TSAO

Laboratory of Renewable Resources Engineering, Purdue University, West Lafayette, Indiana

The grafting of a synthetic vinyl polymer to cellulose and other natural materials like wood offers the potential of preparing a new class of engineering materials with specific and improved properties for a wide range of applications. Almost all the work done to date in the preparation of cellulosic graft polymers involves radical polymerization methods (1–3, Chap. 9). Using these methods, the grafting of only a few high-molecular-weight molecules involving a low level of graft substitution has been obtained. There is no control of the molecular weight and molecular weight distribution of the grafted side chains. A lot of homopolymer formation occurs, and there exist poor reproducibility and little control of the graft process, graft yields, properties, and other features of the graft polymers. These problems have been reviewed in detail by Stannett (4).

If the tremendous applications potential of the cellulosic graft polymers is to be realized, new and better processes for grafting onto cellulose are needed. The new processes should provide for (a) control of the molecular weight and molecular weight distribution of the graft side chains, (b) elimination or at least minimization of concurrent homopolymer formation, (c) grafting onto both the crystalline and amorphous regions of the cellulose, and (d) better reproducibility and control of the grafting process, graft yields, properties, and other features of the graft polymers.

We have recently developed a new process for the preparation of cellulosic graft polymers (5, 6) which would take into account the process conditions outlined above. Our process for the preparation of cellulosic graft polymers involves (a) introduction of "electrophilic groups" or "leaving groups" on to the cellulose backbone by chemical modification, (b) Preparation of the "living" synthetic polymer of desired molecular weight by anionic polymerization techniques, and (c) Reaction of the "living polymer" with the modified cellulose under homogeneous reaction conditions.

A few reports have appeared in the literature on the preparation of cellulosic graft polymers using anionic polymerization techniques (7–10). However, these involved the use of sodium cellulosate as an initiator for grafting vinyl monomers onto cellulose under heterogeneous conditions.

GRAFTING OF POLYSTYRENE ONTO CELLULOSE ACETATE AND TOSYLATED CELLULOSE ACETATE

Cellulose Acetate

Acetone-soluble cellulose acetate 3 (Scheme 10.1; degree of substitution 2.4) was used as the chemically modified cellulosic substrate. "Living" polystyryl carbanion 2 of desired molecular weight was prepared in tetrahydrofuran (THF) using n-butyl lithium as the initiator, following standard anionic polymerization techniques.

The cellulose acetate 3 was dissolved in THF and reacted with the polystyryl carbanion 2. The acetate group functions as a leaving group in a nucleophilic displacement reaction with the polystyryl carbanion resulting in the formation of the cellulose-polystyrene graft polymer 4 (Scheme 10.1). The infrared (IR) spectra of 4 (Fig. 10.1b) supported the formation of a polystyrene graft. The spectrum showed the characteristic aromatic -CH stretching vibrations around 3100 cm^{-1}. The carbonyl band corresponding to the unreacted acetate groups on the cellulose backbone is seen at 1720 cm^{-1}.

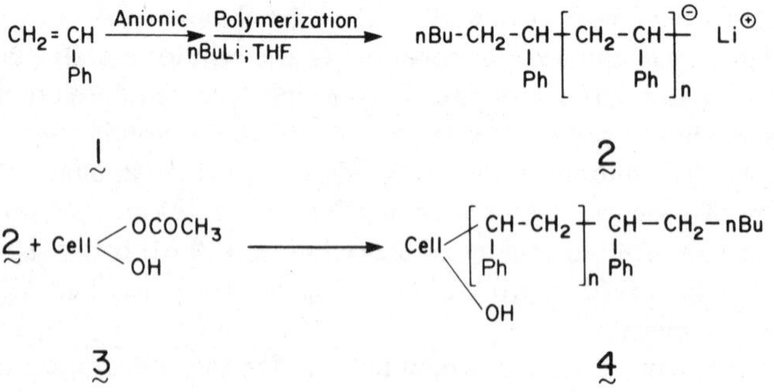

Scheme 10.1

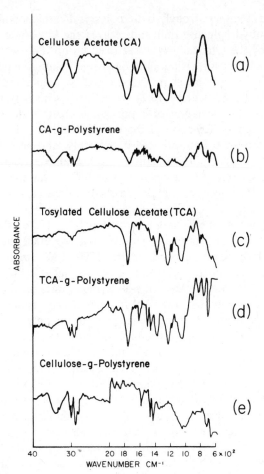

Figure 10.1 Infrared spectra of (*a*) cellulose acetate (CA); (*b*) CA–g–PS; (*c*) TCA; (*d*) TCA–g–PS; and (*e*) cell–g–PS.

Tosylated Cellulose Acetate

The acetone-soluble cellulose acetate used in the earlier experiment was further modified by tosylation of the unesterified hydroxyl groups following the method of Craemer and Purves (11). The effect of this modification was twofold: introduction of the tosyl group provided a much better leaving group than the acetate, in nucleophilic displacement reactions, and, since there are no more free hydroxyl groups left, any homopolymer formation, due to quenching of the polystyryl carbanion by the hydroxyl protons, was eliminated. However, when the tosylated cellulose acetate (TCA) was reacted with the polystyryl carbanion, no grafting occurred. A lot of homopolymer (polystyrene) was formed with concurrent degradation of the cellulose chain.

It appears that the very strongly basic polystyryl carbanion abstracts the C-1 proton of the anhydroglucose unit and cleaves the glycosidic bond, which is rendered labile by the introduction of the tosyl group. This would result in the observed homopolymer formation and concurrent cellulose chain degradation.

To circumvent this problem, it was decided to reduce the basicity of the polystyryl carbanion **2** (Scheme 10.2) by reaction with 1,1-diphenylethylene, which resulted in the formation of a polystyryl chain with a diphenylethane carbanion end group **6**. Reaction of the modified polystyryl carbanion **6** with the tosylated cellulose acetate proceeded smoothly, and excellent yield of the graft product was obtained. The weight of product after extraction with toluene to remove any polystyrene homopolymer was 90% of the theoretical. Proof of the formation of the polystyrene grafts on the cellulose backbone was obtained by IR spectroscopy.

The IR spectrum of the TCA-g-polystyrene (Fig. 10.1d) showed the characteristic aromatic —CH stretching vibrations above 3000 cm^{-1} and also the sharp aromatic ring vibration bands in the 1500–1600 cm^{-1} region, both of which are not present in the spectrum of the tosylated cellulose acetate (Fig. 10.1c). This supports the presence of polystyrene grafts on the cellulose backbone. The strong carbonyl band at 1720 cm^{-1} is clearly visible in both the TCA and TCA-g-PS. To further confirm the presence of the polystyrene grafts on the cellulose backbone, the TCA-g-PS product was subjected to mild base hydrolysis to remove the acetate and tosyl groups, and the IR spectrum of the hydrolysis product was again recorded (Fig. 10.1e). The absence of the — C=O stretching vibrations at 1720 cm^{-1} and the appearance of the —OH stretching vibrations at 3500 cm^{-1} show clearly that the acetate groups have been hydrolyzed to the hydroxyl groups. The absence of any sulfur based on elemental analysis shows that any unreacted tosyl groups present have also been hydrolyzed. The IR spectrum (Fig 10.1e) still revealed the characteristic aromatic —CH stretching vibrations just above 3000 cm^{-1}, and the bands due to the aromatic ring vibrations in 1440–1600 cm^{-1}, thus confirming the presence of polystyrene grafts on the cellulose backbone. Since there are no more acetate or tosyl groups on

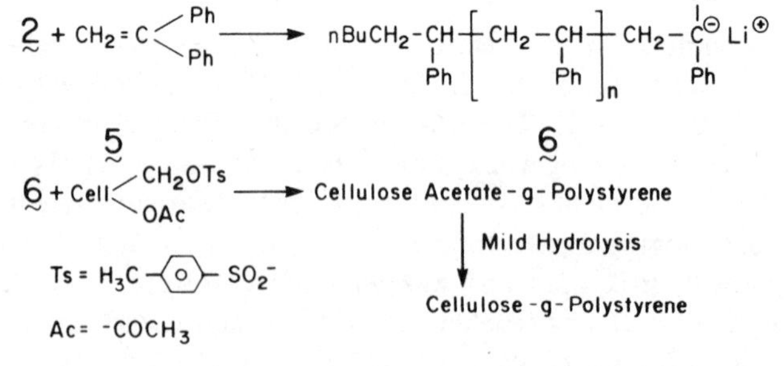

Scheme 10.2

the cellulose backbone, this hydrolysis product represents a true cellulose–polystyrene graft polymer.

GRAFTING OF POLYACRYLONITRILE ONTO CELLULOSE ACETATE

The polyacrylonitrile (PAN) carbanion was prepared using sodium napthalene as initiator in a 1:1 mixture of THF-DMF at $-78°C$ following the procedure of Tsukamoto (12). While the exact structure of the PAN carbanion is not very clear, Tsukamoto (12) proposes a highly branched structure for the anionically polymerized acrylonitrile based on molecular weight studies. This branched structure is attributed to a chain transfer process to existing polymer. Both carbanionic species **7** (arising from chain transfer to polymer) and **8** (propagating species) could be present and would be capable of reacting with the cellulose acetate (see Scheme 10.3).

The "living polyacrylonitrile" was reacted with cellulose acetate, and the crude polymer product was isolated by quenching with methanol and washed with acetone to remove any unreacted cellulose acetate. The IR spectrum of the graft polymer (Fig. 10.2b) showed the characteristic carbonyl band of the acetate groups at 1720 cm^{-1} and the typical nitrile stretching vibrations at 2200 cm^{-1} indicating the formation of a cellulose acetate–PAN graft polymer. To get rid of any PAN homopolymer and to obtain a true cellulose–PAN graft polymer, the crude reaction product was subjected to mild hydrolysis with 15% ammonia to remove the unreacted acetate groups. The hydrolysis product was extracted with DMF to remove any PAN homopolymer, and the IR spectrum was recorded (Fig. 10.2c). The spectrum revealed the typical nitrile stretching vibrations at 2200 cm^{-1}, the —OH stretching vibrations at 3500 cm^{-1}, and the absence of the strong carbonyl band at 1720 cm^{-1}. This confirmed that the product was, indeed, a true cellulose–PAN graft polymer.

If the hydrolysis is done with dilute NaOH, some of the nitrile groups on the cellulose backbone are hydrolyzed to a carboxylic acid functionality. The IR spectrum (Fig. 10.2d) shows the weak nitrile stretching vibrations at 2200 cm^{-1}, but more importantly, all the characteristic stretching vibrations of the —COOH grouping are present. The broad strong band extending from 3500–2200 cm^{-1} is due to the —O—H of the —COOH group and the strong band at 1720 cm^{-1} is due to the —C=O.

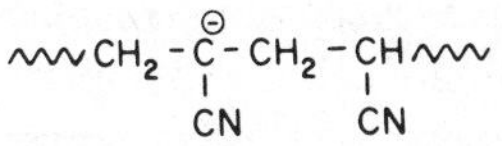

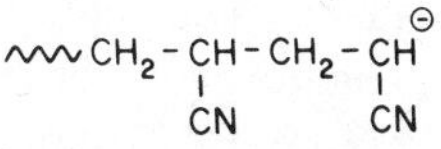

Scheme 10.3

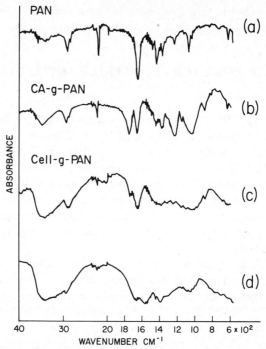

Figure 10.2 Infrared spectra of (*a*) polyacrylonitrile; (*b*) cellulose acetate–g–PAN; (*c*) cellulose–g–PAN; and (*d*) NaOH hydrolysis product of CA–g–PAN.

GRAFTING OF POLYSTYRENE AND POLYACRYLONITRILE ONTO WOOD

The new synthetic methodology developed for the preparation of cellulosic graft polymers has been extended to preparation of graft polymers from wood. Acetylation of wood chips with acetic anhydride in the presence of concentrated H_2SO_4 furnished a triacetate which was partially hydrolyzed to give a secondary acetate (13). IR spectrum of this product showed the characteristic carbonyl stretching at 1750 cm^{-1}, confirming the introduction of acetate groups onto the wood backbone.

Using anionic polymerization techniques, polystyryl carbanion and PAN carbanion were prepared. Each of these "living polymers" was reacted with the acetylated wood chips. The highly reactive carbanions displaced the acetate groups in a SN$_2$ reaction with the formation of the wood–polystyrene (W-PS) and the wood–PAN (W-PAN) graft polymers. Any homopolymer formed was removed by extraction with toluene for the W-PS graft polymer and dimethyformamide for the W-PAN graft polymer. Proof for the formation of these graft polymers was obtained from IR and thermogravimetric analysis. The IR spectrum of the W-PS graft polymer showed the presence of the aromatic —CH

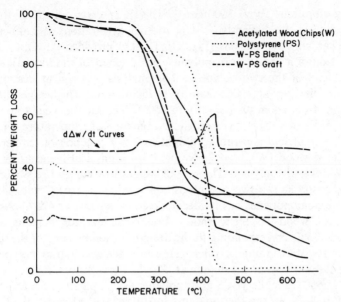

Figure 10.3 Thermogravimetric analysis—grafting of polystyrene onto wood.

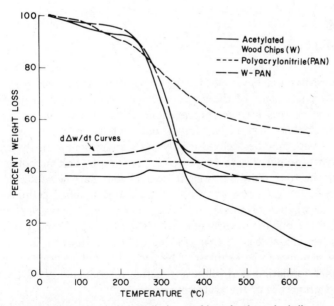

Figure 10.4 Thermogravimetric analysis—grafting of polyacrylonitrile onto wood.

stretching vibrations above 3000 cm^{-1}. The IR spectrum of the W-PAN graft polymer showed the distinctive —CN stretching vibrations at 2200 cm^{-1}.

Thermogravimetric analysis (Figs. 10.3, 10.4) provided strong evidence for the formation of a "true" graft polymer and discounted the formation of mere physical blend of the two polymers. The analysis was run at heating rate of 6°C/min in nitrogen at a flow rate of ∼75 mL/min. The temperature range covered was from room temperature to 650°C. The sample weight and the rate of change of the weight were recorded as a function of temperature. Figure 10.3 shows the weight loss and rate of weight loss as a function of temperature for (a) acetylated wood (W),(b) polystyrene, (c) 1:1 physical blend of W and PS, and (d) W-PS graft polymer. The rate of weight loss curves (dΔW)/(dt) of the acetylated wood shows two decomposition maxima, at 265°C and 345°C. The polystyrene homopolymer has its decomposition maxima at 425°C. As is to be expected, the wood–PS blend shows three decomposition maxima at 265°C, 345°C, and 425°C, corresponding to the decomposition maxima of the individual wood and polystyrene hompolymers. However, the W-PS graft polymer shows only one decomposition maximum at 300°C. Furthermore, a 25% residue is obtained at 650°C for the W-PS graft polymer as compared to 5% for the blend.

Figure 10.4 shows the thermograms for (a) acetylated wood, (b) PAN, and (c) W-PAN graft polymer. Again, the W-PAN graft polymer shows only one decomposition maximum at 325°C. However, a higher amount of residue (35%) is obtained as compared to the W-PS graft polymer. Since the PAN homopolymer leaves as much as 50% residue at 650°C, the increased residue obtained for the W-PAN graft polymer is not surprising and can be taken as additional evidence for the formation of W-PAN graft polymer.

SUMMARY

A new process has been developed for the preparation of cellulosic graft polymers that can overcome some of the major problems encountered in grafting onto cellulose. The process involves preparation of the "living synthetic polymer" using anionic polymerization techniques and reacting the polymeric mono- or dicarbanions with modified cellulosic substrates under homogeneous conditions. Since the synthetic polymer to be grafted is prepared by anionic polymerization techniques, effective molecular weight control and narrow molecular weight distribution of the graft can be achieved. By using modified celluloses, which are soluble in organic solvents, the displacement reaction between the modified cellulose and the polymeric carbanion can be done under homogeneous conditions. This would also ensure that grafting occurs on both the crystalline and amorphous regions of the cellulose. The process should eliminate concurrent homopolymer formation and allow for better reproducibility of the grafting process and properties of the graft polymers.

REFERENCES

1. D. J. McDowall, B. S. Gupta, and V. T. Stannett, *Prog. Polym. Sci.*, **10**, 1 (1984).
2. A. Hebeish and J. T. Guthrie, *The Chemistry and Technology of Cellulosic Copolymers*, Springer-Verlag, Berlin, 1981.
3. J. C. Arthur, Jr., *Adv. Macromol. Chem.*, **2**, 1 (1970).
4. V. T. Stannett, *ACS Symp. Ser.* **187**, 1 (1982).
5. R. Narayan, "Abstracts of Papers," 186th National Meeting of the American Chemical Society, Washington, 1983.
6. R. Narayan and G. T. Tsao, *Polym. Preprints*, **25**(2), 29 (1984).
7. R. F. Schwenker, Jr. and E. Pacsu, *Tappi*, **46**, 665 (1963).
8. B.-A. Feit, A. Bar-Nun, M. Lahav, and A. Zilkha, *J. Appl. Polym. Sci.*, **8**, 1869 (1964).
9. Y. Avny, B. Yom-Tov, and A. Zilkha, *J. Appl. Polym. Sci.*, **9**, 3737 (1965).
10. Y. Avny and L. Rebenfield, *Text. Res. J.*, **38**, 684 (1968).
11. F. B. Craemer and C. B. Purves, *J. Am. Chem. Soc.*, **61**, 3458 (1939).
12. A. Tsukamoto, *J. Polym. Sci.*, **3**,2767 (1965).
13. B. L. Browning, *Methods of Wood Chemistry*, Vol. 2, Interscience, New York, 1967.

11

Economic Analysis of an Innovative Process for Cellulose Acetate Production

W. H. KLAUSMEIER

EES Division, Argonne National Laboratory, Argonne, Illinois *

The objective of this chapter is to address recent innovations in the production of dissolving grade cellulose and cellulose acetate. Much of the analysis to be reported is the result of an in-depth evaluation of the novel biomass conversion processes performed by Argonne National Laboratory (ANL) with assistance from Chem Systems, Inc. (1). In that study, organosolv fractionation was compared to kraft pulping as a means of separating cellulose from the other wood components. Results indicated that organosolv pulping with alcohol–water mixtures had a number of attractive features but was still somewhat more expensive and energy-intensive than kraft pulping as currently conceived. However, there are many ways to greatly improve the economics and energy efficiency of organosolv processes.

One of the most notable developments in cellulose derivative production has been the impending commercial introduction of the Eastman process for manufacture of the reagent, acetic anhydride, needed for making cellulose acetate. The Eastman process is based on coal gasification to supply methanol and carbon monoxide reactants. ANL examined the possibility of employing wood gasification and developed the economics for cellulose acetate manufacture based entirely on wood feedstock. The process economics were very promising compared to those for making cellulose acetate with purchased acetic anhydride made by the conventional ketene route. Energy requirements were also very low

*Present address: Sylvatex Corporation, Burke, Virginia.

for the wood gasification route. However, favorable economics depend on the availability of acetic acid from acetate production for recycle to the Eastman process. Unless acetic acid and acetic anhydride were transported back and forth by rail, both pulp and acetic anhydride manufacture would have to be integrated within the same facility. To supply enough acetic anhydride to derivatize the entire pulp output of a pulp mill, an equivalent amount of wood would have to be gasified. Such an additional wood demand would severly strain wood supplies if the pulp mill employed 500 dry ton/day of feedstock or more. Nevertheless, further analysis reveals a plausible strategy for integrated cellulose acetate manufacture using purchased methanol, and this rationale is developed in the subsequent discussion. The potential exists for vertical integration of a pulp mill to produce cellulose acetate at a greatly reduced cost of production compared to present practice.

SOLVENT PULPING

The idea of using solvents to dissolve lignin was first explored in 1893 (2). In the 1930s, Kleinert was the first to develop and patent a process using 50% aqueous ethanol (3, 4). There was little incentive to pursue substitutes to the kraft process further until recently. Energy and chemicals were inexpensive, there were few environmental regulations, and there was little interest in by-products. New environmental regulations and rising energy and raw material costs have prompted many to examine lignocellulose as a potential source of glucose for fermentation to ethanol and other chemicals. Enzymes were found that could saccharify (hydrolyze) cellulose to glucose in high yield, but saccharification rates were very low with untreated lignocellulosic materials. However, if the feedstock was pretreated to remove lignin, saccharification was greatly facilitated, and organosolv delignification was identified as a preferred method (5). This has stimulated a renewed interest in solvent fractionation. Curiously, several firms that initiated solvent fracctionation process development as a pretreatment for enzymatic hydrolysis have decided to focus instead on pulping process development because of the greater value of pulp compared to the value of fermentable sugar (6). Current research activity for both applications is intense, and at least three new organosolv processes have reached the pilot plant stage within the past few years (Battelle-Geneva, BEC, and MD-Organosolv processes). The basic elements of organosolv processing are depicted in Figure 11.1. Organosolv process will hereafter be defined as a fractionation process using solvents for the purpose of delignification for cellulose pulp production.

Basic Elements of Organosolv Process

Subdivided feedstock such as wood chips is contacted with an aqueous solvent mixture and cooked (digested) for a period of time at elevated temperatures. Digestion pressures range from atmospheric to 40 psig. Cooking initially releases

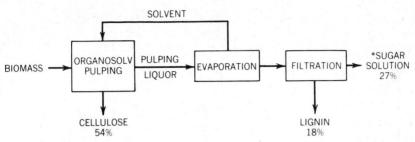

Figure 11.1 Organosolv biomass fractionation.*Sugars can also be obtained by dilute acid pre-hydrolysis of the biomass before pulping.

acetic and formic acid from natural ester functionalities in the feedstock that then promote the hydrolysis of hemicellulose and lignin to low-molecular-weight species. Catalysts such as mineral acids (HCl), organic carboxylic acids (acetic, oxalic), sulfonic acids, and Lewis acids and bases ($AlCl_3$, $Fe_2(SO_4)_3$, $Mg(SO_4)_2$, $CaCl_2$. . . .) are often employed to accelerate the hydrolysis process. Anthra-quinone has also been shown to promote delignification by promoting the frag-mentation of polymeric lignin. Cellulose pulp is filtered from the liquor, and solvent is then recovered from the liquor, usually by distillation. Lignin often becomes insoluble at this point and precipitates from the liquor, leaving an aqueous solution of primarily hemicellulosic sugars. Conditions for a variety of organosolv processes are compared in Table 11.1. The processes fall into two general categories: Aqueous solutions of organic acids, and non-acidic, volatile solvents. Because the organic solvents in the first five processes are acids, hy-drolysis of hemicellulose and lignin is promoted, and lower digestion tempera-tures can be employed. The lower tempereature and lower volatility of many organic acids often allow digestion to be carried out at atmospheric pressure, which simplifies process design. Recovery of acidic organic solvents, however, can be extremely difficult. Purification of acetic acid requires much energy, although formic acid forms a lower boiling azeotrope with water (8, 18). The Battelle-Geneva process uses one of the simplest solvent recovery methods in which the phenol becomes immiscible with the aqueous phase upon cooling of the pulping liquor, allowing simple recycling of the pulping solvent. Unfortu-nately, the aqueous phase still contains 8% phenol, and this must also be extracted from the aqueous phase with another organic solvent or other means to avoid undue solvent losses and environmental risk (19).

The organosolv processes employ nonacidic, volatile solvents of intermediate to high polarity such as acetone, methanol, and ethanol. Since these solvents are not acidic, higher digestion temperatures are needed to autohydrolyze hemi-cellulose and lignin. Reaction temperatures are usually kept just below the temperature at which pentoses begin to decompose to furfural (205–210°C) to obtain maximum rates without by-product formation. If furfural forms, it can react with lignin and induce repolymerization and deposition onto the pulp product. Low digestion rates mean high capital costs per unit of product. At

TABLE 11.1 Solvent Pulping Process Characteristics

Process Name Developer	Solvent Used	Additional Reactants	Digestion Temperature	Digestion Time	Reference
(Young)	50–75% Acetic Acid[a]	Ethyl Acetate[e]	170°C	1 hr	(7)
(Herdle)	90% Acetic acid[a]	0.2% H_2SO_4	150°C	4 hr	(7a)
(Jordan)	80% Formic acid[a]	Unspecified organic catalyst	105°C	45 min	(8)
Hydrotropic Pulping (Springer)	2.0 M Xylene-sulfonic acid[a]	None	100°C	60 min	(9)
Battelle-Geneva Process (Sachetto)	Aqueous phenol[b]	Acid catalyst	100°C	30–60 min	(10)
(Clermont)	50% Sulfolane	None	175°C	2 hr	(11)
(DeHaas and Lang)	Aqueous acetone[b]	NH_3/Na_2S	160–190°C	1–2 hr	(12)
CASPL (Paszner) MD Organosolv	Aqueous methanol[b]	Alkali earth salt catalysts	195–210°C	25–45 min	(13–15)
Kleinert Process Organosolv Process APR Process BEC Process	20–75% Ethanol[a,c] usually 50%	Uncatalyzed or metal salt catalysts	160–195°C	30–60 min	(3, 4, 16)
(Nayak)	Ethylene diamine[d]	Na_2S	170–180°C	90 min	(18)

[a]Other component is water.
[b]Concentration not reported.
[c]n-Butanol has also been employed.
[d]Ethanolamine and ethylene glycol also studied.
[e]A three-component (ethyl acetate/acetic acid/water) pulping process has recently been developed with phase separation for solvent recovery (7).

temperatures of 205–210°C, aqueous ethanol and methanol digestions develop pressures of 38–41 psig (14). The use of volatile solvents at high temperatures also requires that digestion be done in pressurized reactors. In addition to the need for higher digestion temperatures, acid and Lewis acid catalysts are almost always needed to obtain sufficiently rapid rates of delignification. Only a few processes do not employ catalysts, and the issue of whether any uncatalyzed process can be viable is still hotly debated. Although catalyst cost does not usually appear to be addressed in evaluating process alternatives, there is evidence that acids and even Lewis acids will attack the cellulose to a limited extent. This is reflected in a reduction in degree of polymerization (determined by measuring viscosity), which, in turn, affects strength properties. Although this is of little consequence when glucose is the ultimate product, it is a serious concern in the production of paper pulp. Most pulps from processes using catalysts have reported tensile strengths comparable to sulfite pulps but about 70% that of kraft pulps. Paszner has reported making papers from organosolv pulp that are nearly as strong as papers made from kraft pulp by post-treatment of the pulp with a dilute acid wash followed by soaking in dilute alkali (lixification) (14). Kleinert reported production of kraft-strength pulp using an uncatalyzed ethanol-based process (3, 4), but attempts to duplicate his results have been unsuccessful (20). However, sulfite pulps have been widely used in papermaking, and the need for kraft-strength pulps is exaggerated by an established papermaking infrastructure that is unwilling or unable to compensate for its dependence on high-strength pulps. Fortunately, high-strength pulps are not required for chemical applications such as cellulose acetate production, and, indeed, most chemical pulp production has employed the sulfite process. Since organosolv pulping conditions generally hydrolyze amorphous cellulose, the resulting pulp has a high α-cellulose content (> 90%), making chemical applications the most appropriate objective to target for organosolv processes.

Internal Solvent Production

Solvent-related considerations are foremost in determining the feasibility of organosolv fractionation. Solvent recovery must be both energy-efficient and thorough if utility and reagent costs are to be kept low. The prices of makeup solvents are considerably greater than the price of makeup Na_2SO_4 used in the kraft process. However, since organic feedstocks are employed, it is possible to devise methods for producing pulping solvents from other feedstock components. This provides an opportunity for "self-sufficient" pulp mills that need not depend on the purchase of major raw materials. Costs for producing solvents internally could be much less than the purchase price. Ethanol-based organosolv processes demonstrate the greatest potential for self-sufficiency.

One alternative to the recovery of pentose sugars from the pulping liquor following solvent recovery is to remove pentose sugars before organosolv pulping by dilute acid prehydrolysis. This involves treatment of feedstock with 0.5% sulfuric acid at 180°C for 12 sec in a plug flow reactor (21). This treatment will also hydrolyze a small portion of the cellulose that exists in a more accessible, amorphous state and will afford a pulp with a high α-cellulose content. Therefore, other advantages of prehydrolysis include a reduction in pentose degradation that occurs over the prolonged digestion period at high temperature and a reduction in the mass of material being fractionated. The reduction in pentose contaminants facilitates subsequent fermentation of the pentose sugars.

In addition to the production of ethanol from cellulose-derived glucose, species of fungi, bacteria, and yeast have been found that can convert pentose sugars to ethanol as well. Currently, the yeasts *Candida shihatae* and *Pachysolen tannophilus* appear the most promising (22). These yeasts can ferment both glucose and xylose simultaneously, although ethanol yields are somewhat low (84% of theory vs. 96% with *Saccharomyces cerevisiae* on glucose), and conversion rates are slow. Development of a viable pentose-to-ethanol fermentation would allow a self-sufficient organosolv mill to produce pulp without sacrificing valuable product for making solvent and would provide a new application for the use of pentose sugar wastes from existing pulp mills.

There are other benefits in integrating organosolv pulping with distillery operations besides having a low-cost source of pulping solvent. Since ethanol pulping uses a 50:50 ethanol water mixture, fermentation broth can be distilled to 50% ethanol in a first stage and then used directly as pulping liquor. After pulping, solvent recovery can be achieved simultaneously with distillation of the ethanol to high purity if the system can be designed for the effective recovery of lignin as it becomes insoluble. This could greatly reduce additional energy consumption by the pulping operation over the energy already required for the distillery. Figure 11.2 provides a simplified flow diagram of a self-sufficient organosolv pulp mill using ethanol. The estimated facility investment for the basic organosolv pulp mill shown in Figure 11.2 is $91.5 million (1981 $) for a 1000 dry ton/day plant. More recent estimates indicate that a grass-roots facility would cost at least twice this estimate. Cost estimates for kraft mills vary from source to source depending on the year of the analysis because of rapidly escalating equipment prices. An unbleached kraft pulp mill producing 250,000

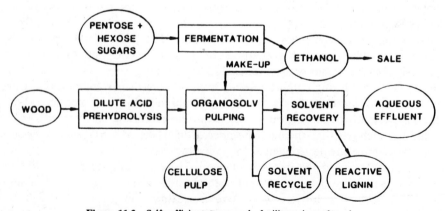

Figure 11.2 Self-sufficient organosolv facility using ethanol.

ton/yr of unbleached pulp (1083 dry ton/day wood feed) costs $225 million (1981 $). A 1000 dry ton/day (wood feed) bleached kraft pulp mill with a cogeneration plant costs $304 million (1981 $). At this time it appears that an organosolv mill might be somewhat less expensive but not by a compelling margin. The estimated sales price for the organosolv pulp is 15.9¢/lb versus 12.4¢/lb for kraft pulp.

ACETIC ANHYDRIDE PRODUCTION BY THE EASTMAN PROCESS

For every pound of cellulose acetate produced, 1.43 lb of acetic anhydride is required. The Eastman process, mentioned earlier, reduced the cost of acetic anhydride. The coal gasification was combined with the recycled acetic acid generated in producing cellulose acetate. The chemistry of Eastman and ketene processes for making acetic anhydride are shown in Figure 11.3. In the present study, wood was used as the gasification feedstock, and the Eastman process was compared to the conventional route to acetic anhydride, which employs acetic acid made from oil and gas. The comparison is shown in Table 11.2.

The Eastman route employing wood gasification is about 25% cheaper than the conventional ketene route, because raw material costs are lower for the Eastman process. This is in spite of the fact that the 29.4¢/lb sale price of methanol from small-scale wood gasification is far greater than the current market price for methanol of 10.9¢/lb. The biomass gasification system hypothesized for this analysis was based on the Battelle-Columbus dual-fluidized bed steam gasification system (9). The cost of production of the synthesis gas is actually quite low ($5.31/$10^6$ Btu). Wood gasification, however, produces

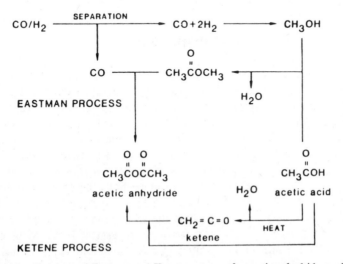

Figure 11.3 Chemistry of Eastman and Ketene processes for acetic anhydride production.

TABLE 11.2   Summary of Acetic Anhydride Economics[a]

	Process	
Economic Factor	Eastman	Ketene
Fixed investment ($10⁶)		
Battery limits	136.0	165.5
Offsites	65.3	67.5
Total	201.3	233.0
Production costs (¢/lb)		
CO		7.2
Acetic acid		26.5
Methanol	29.4	
Production costs (¢/lb)		
Raw materials	12.48	17.62
Utilities	1.54	3.50
Operating costs	1.61	1.84
Overhead expenses	5.17	5.90
Net cost of production	20.80	28.86
Sales price at 15% DCF (¢/lb)	27.1	41.6
Process energy (Btu/lb)	3600	9200

[a]Basis: 1981, 573 × 10⁶ lb/yr of product.

both methanol and CO for use in the carbonylation step of the Eastman process. A separation step is therefore required to obtain purified CO. Cryogenic separation, the method considered by Chem Systems, is capital-intensive, especially at small scales. A simplified means for obtaining CO from wood could greatly improve process economics.

CELLULOSE ACETATE PRODUCTION

The process elements for self-sufficient cellulose acetate manufacture are shown in Figure 11.4. Although acetic anhydride production from wood can lead to tremendous savings in oil and gas consumption and is much cheaper than the conventional process, the quantity of wood needed for cellulose acetate production imposes another restriction. To supply the 11.1×10^6 standard cubic feet per year of CO and the 200.6×10^6 lb/yr of methanol required to make 573.2×10^6 lb/yr of acetic anhydride, 1050 dry ton/day of aspen would have to be gasified. If one facility were to produce both the cellulose pulp and the acetic anhydride, 2050 dry ton/day of aspen wood would be required even if wood were not used to supply process energy. The facility would be greater in size than the usual pulp mill by a factor of at least 4. It would be extremely difficult to find a site in the United States that could support such a huge feedstock requirement. But if the plant were scaled down by a factor of 4, there

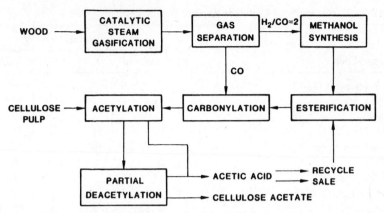

Figure 11.4 Self-sufficient production of cellulose acetate.

would be a major reduction in scale economy, and the biomass routes would be more expensive than the conventional routes. One possible way to overcome this problem would be to have the pulp mill and acetic anhydride operations in two separate facilities sufficiently far apart to not overtax the wood supply around either site but close enough that the acetic anhydride and recycled acetic acid could be transported by rail between the locations at relatively low cost. However, the capital cost for two separate facilities would be much greater than if the operations were integrated within a single facility.

The comparative costs for cellulose acetate production by conventional and self-sufficient biomass-based processes are shown in Table 11.3. The conventional and wood refinery routes are virtually identical. The only significant difference is in the price of the raw materials, cellulose, and acetic anhydride. The self-sufficient facility would produce both chemical-grade cellulose pulp and acetic anhydride at a cost below the purchase price of these raw materials. For this reason, cellulose acetate can be produced in the forest refinery at a sales price well below the current market price of 88¢/lb. However, the process based on purchased materials probably cannot compete in today's markets. It clearly benefits cellulose acetate manufacturers and users to have captive acetic anhydride production facilities; this is precisely why Eastman has gone into the acetic anhydride business. The estimated capital investment for an integrated facility producing 401 million lb/yr of cellullose acetate is $345 million. Annual revenues from the sale of cellulose acetate at 88¢/lb would be $353 million compared to $50 million for the sale of pulp alone.

Another consideration in making a product such as cellulose acetate is the size of the market. Most cellulose acetate manufactured in the United States is made by producers such as Celanese and Eastman, who use most of their production internally in such finished products as photographic film. Only about 800×10^6 lb/yr are actually sold in the United States. The volume of cellulose acetate from the proposed facility would equal 50% of the current U.S. market volume.

TABLE 11.3   Summary of Cellulose Acetate Economics[a]

	Process	
Economic Factor	Conventional	Wood Refinery
Fixed investment (10^6)		
Battery limits	35.4	39.9
Offsites	17.0	19.2
Total	52.4	59.1
Raw material prices ($¢$/lb)		
Cellulose	20.0	15.9
Acetic anhydride	38.5	27.1
Production costs ($¢$/lb)		
Raw materials	69.85	50.88
Utilities	5.47	5.98
Operating costs	0.71	0.78
Overhead expenses	2.04	2.28
Cost of production	78.07	59.84
By-product credit	−4.79	−4.79
Net cost of production	73.28	55.05
Sales price at 15% DCF ($¢$/lb)	89.6	67.1
Process energy (Btu/lb)[b]	43,260	37,440

[a]Basis: 1981 $, 401 × 10^6 lb/yr.

[b]Total energy for obtaining raw materials and for making acetate.

If one examines process energy requirements in addition to wood feedstock requirements for gasification and pulping, a self-sufficient wood-based cellulose acetate facility appears implausible. A facility using 1000 dry ton/day of pulpwood feedstock would require a comparable amount of gasification feedstock as well as a comparable amount of fuel wood to supply the required process energy. The Eastman process is energy-efficient, and the 1050 dry ton/day of gasification feedstock would supply all the Eastman process energy requirements and would generate 58 dry ton/day (2.76 × 10^9 Btu/day) of methane for sale or use elsewhere. However, the conservative design basis employed in calculating energy requirements for organosolv pulping afforded a high process energy requirement; a pound of fuelwood is needed for every pound of pulpwood. The principal contributor to the high process energy requirement for pulping is the energy required for solvent recovery. Combustion of lignin and methane coproducts would reduce the fuel wood requirement by 526 dry ton/day. Although this may be the most expedient application for these coproducts, lignin is expected to eventually prove more valuable as a commodity than as a fuel. Combustion of wood sugars obtained by prehydrolysis would not contribute significantly to meeting process energy requirements, because the dilute aqueous solution would first have to be concentrated and, even dry, the sugars would not have a very high fuel value (cellulose = 4400 Btu/lb). It is clearly impractical to consider a cellulose acetate facility requiring 2926–3450 dry ton/day of wood.

One important observation is the high cost of biomass-produced methanol; 29.4¢/lb versus a current methanol market price of 10.9¢/lb. Greatly expanded production and evolving fuel and chemical applications for methanol suggest that methanol will continue to be available in quantity at a reasonable price. It seems foolish to synthesize methanol from wood when it is available at such a low price. Another consideration is that the steam gasification of biomass produces a carbon monoxide-rich gas. In the absence of catalysts, a gas containing 46% CO (by volume), 15% CO_2, 15% H_2, and 24% hydrocarbon gases is obtained using the Battelle-Columbus system (23). When biomass CO is converted to H_2 for synthesis using the shift reaction, it is devalued by an order of magnitude on a mass basis. It therefore appears more appropriate to employ biomass gasification for CO production than for making the large quantities of H_2 needed for methanol synthesis. Furthermore, there is an efficient low-cost method for extracting CO from a mixed-gas stream in the COSORB process developed by Tenneco (24). This process, shown in Figure 11.5, uses a $CuAlCl_4$ complexing agent in toluene to selectively remove CO.

The absorber/stripper system is inexpensive and uncomplicated, operates at ambient pressure and temperature, and only requires feed gas drying. Production of CO from coal using COSORB technology costs 7.3¢/lb (100 million lb/yr plant) versus 9.4¢/lb using cryogenics. Since biomass is sulfur-free, no desulfurization is needed, and biomass CO can probably be produced for roughly 6¢/lb (vs. 7.2¢/lb for cryogenic separation). The cellulose acetate process in Figure 11.4 could be modified to eliminate catalysts from the steam gasification, to substitute the COSORB process for CO separation, and to eliminate the entire gas separation and synthesis train by purchasing methanol.

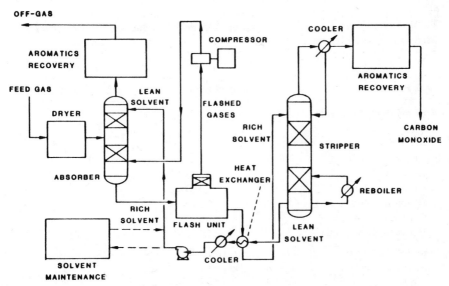

Figure 11.5 The COSORB process.

In Table 11.4 the cost of acetic anhydride by the Eastman process is recalculated using a market price of 10.9¢/lb for purchased methanol and 6¢/lb for CO produced using COSORB technology. The required acetic anhydride sales price of 19.4¢/lb is significantly less than the 27.1¢/lb price of anhydride produced in a self-sufficient facility, as is the price of cellulose acetate made with the less expensive anhydride. This analysis would indicate that biomass gasification for carbon monoxide production has great economic potential for the synthesis of chemicals such as acetic acid and acetic anhydride that can be efficiently derived from carbon monoxide. In addition to cellulose acetate production, acetic anhydride could be used to produce vinyl acetate from acetaldehyde using established Celanese technology. A Brazilian chemicals firm, Alcooquimica, is currently building a vinyl acetate plant using the Celanese process with all-biomass-derived feedstocks (26). Another attractive feature of carbon monoxide production by steam gasification is that carbon monoxide removal leaves a methane-rich high-Btu gas that can be cleaned up and pipelined or bottled for residential use. Established technology for the air gasification of biomass can also be employed, since the COSORB process is unaffected by the presence of nitrogen.

Figure 11.6 shows the modified mass and energy flows for a cellulose acetate facility using purchased methanol. The total wood feedstock requirement has been reduced by 500 dry ton/day. But 800 dry ton/day of wood needs to be gasified to supply the 281 dry ton/day of carbon monoxide required. However, instead of having both a gasification system and a separate large-scale direct combustion system, it is more logical to integrate both systems. Fuel wood would be gasified, carbon monoxide would be withdrawn, and the remaining medium-Btu gas would be used to raise steam in a gas-fired boiler. Use of a gas

TABLE 11.4 Cost Comparison of Acetic Anhydride Production Using Purchased Methanol and COSORB Technology[a]

| | Eastman Process | |
Cost Factor	Self-Sufficient	Purchased Methanol
Raw material price (¢/lb)		
CO	7.2	6.0
Methanol	29.4	10.9
Production costs (¢/lb)		
Raw materials	12.48	5.64
Utilities	1.54	1.54
Operating costs	1.61	1.84
Overhead expenses	5.17	5.90
Net cost of production	20.80	14.92
Sales price at 15% DCF (¢/lb)	27.1	19.4
Sales price of cellulose acetate	67.1	53.7

[a]Basis: 1981 $, 573 × 10⁵ lb/yr product, Eastman Process.

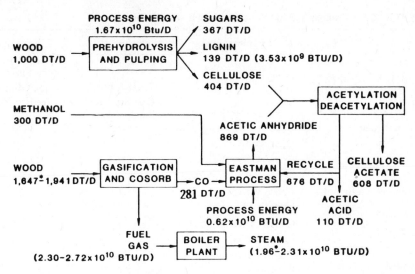

Figure 11.6 Mass and energy flows for a cellulose acetate facility using purchased methanol.

turbine for the cogeneration of electric power along with steam generation is another attractive option that could be exercised.

Although the feed requirement and heat rejected in methanol synthesis have been eliminated, the facility still has an unrealistically high feedstock requirement. It has already been noted that the process energy requirements for organosolv pulping are very high because of the conservative design basis employed (10,130 Btu/lb for pulping and 3045 Btu/lb for prehydrolysis). However, other analysts have used more liberal design bases (27), and researchers have already discovered much more energy-efficient pulping processes (7,10) and ethanol recovery methods (28). Nguyen et al. (27) employed a low liquor/feed ratio and a low ethanol recovery efficiency and calculated a process energy requirement of 1494 Btu/lb for prehydrolysis and 3045 Btu/lb for solvent pulping and recovery. Although their design basis is not economically feasible because ethanol losses would be too great, an overall process energy requirement of 5000 Btu/lb would be possible if liquor recycle and heat recovery systems were employed. Table 11.5 compares feedwood requirements using the more optimistic design basis of Nguyen et al. with those previously calculated. A reduction in the fuelwood requirement of nearly 1000 dry ton/day is achieved, and the total feedwood requirement approaches a more realistic range. This analysis clearly demonstrates the need for energy efficiency for solvent pulping to become attractive as an alternative to chemical pulping.

Methanol/water pulping systems have also attracted attention recently. If a methanol-based process were used for pulp production, then the integrated process could be further simplified by eliminating the pentose-to-ethanol fermentation component. This would enhance the near term potential of the proposed cellulose acetate facility.

TABLE 11.5   Comparison of Wood Feedstock Requirements[a]

Case	Basis	Pulping Energy Requirements[b] Total Input (10^9 Btu/day)	Total Wood Requirement (dry ton/day including feedstock)
Self-sufficient	Conservative design, lignin and CH_4 sold	16.7	3450
	Conservative design, lignin and CH_4 burned	16.7	2926
Purchased methanol	Conservative design, lignin sold	16.7	2941
	Conservative design, lignin burned	16.7	2647
	Optimistic design, lignin sold	5.0	1960
	Optimistic design, lignin burned	5.0	1664

[a] 1000 dry ton/day pulpwood scale.

[b] The energy requirement for acetic anhydride production is 6.2×10^9 Btu/day.

If a bleached kraft pulp mill were energy-self-sufficient, roughly 0.75 ton of fuelwood would be required per ton of pulp product (or per 2 tons of feedstock). Thus, the total feedwood requirement for a self-sufficient conventional pulp and paper operation is comparable to that for the optimistic case for cellulose acetate production. However, it would probably be prudent to halve the plant scale to 500 dry ton/day of pulpwood or 200 million lb/yr of cellulose acetate. Scale economy should not be excessively compromised at this smaller scale, and market saturation would be less of a problem.

There are several important conclusions from this analysis. Organosolv pulping must be energy-efficient for the technology to show promise as a viable alternative to chemical pulping. Recent developments indicate that this can be accomplished. Vertical integration of a pulp mill for cellulose acetate production is feasible provided the pulping process is energy-efficient and inexpensive methanol is available. Biomass gasification to provide carbon monoxide for synthesis offers a major new commercial opportunity to chemical producers as well as the forest products industry.

REFERENCES

1. W. H. Klausmeier, Configurations for a Forest Refinery, An Interim Report, ANL Technical Memo No. ANL/CNSV-TM-101, Argonne, Il (April 1982).
2. P. Klason, *Tekn. Tidskr. Avd. Kemi*, **23**, 53 (1983).
3. T. Kleinert and K. Tayental, U.S. Pat. 1856567 (1932).

4. T. N. Kleinert, *Tappi*, **57**(8), 99 (1974).

5. J. A. Phillips and A. E. Humphrey, *Wood and Agricultural Residues: Research on Use for Feed, Fuels and Chemicals*, E. J. Soltes, ed., Academic, New York, 1983, p. 503.

6. K. E. Pye, *Biomass Digest*, p. 3 (February 1982).

7. R. A. Young, E.-B. Wiesmann, and J. L. Davis, *Holzforschung*, and Proceedings of the 1985 International Symposium on Wood and Pulping Chemistry, Vancouver, B.C., Aug. 1985.

7a. C. E. Herdle, L. H. Pancoast, and R. H. MacClaren, *Tappi*, **47**, (10); 617 (1964).

8. M. Bucholtz and R. Jordan, "Formic Acid Wood Pulping," Gannon University, unpublished manuscript (1982).

9. E. L. Springer and L. C. Zoch, *Tappi*, **54**(12), 2059 (1971).

10. A. Johansson, J. P. Sachetto, and A. Roman, *A New Process for the Fractionation of Biomass*, Battelle Geneva Labs, Geneva, Switzerland, 1982.

11. L. P. Clemont, *Tappi*, **53**(12), 2243 (1970).

12. G. G. DeMaas and C. J. Land, *Tappi*, **57**(5), 127 (1974).

13. *Biomass Digest*, **6**(4), 1 (1984).

14. L. Paszner and P. C. Chang, International Symposium on Wood and Pulping Chemistry, Tsakuba Science City, Japan, 1983.

15. V. E. Edel, *Dtsch. Papier Wirtschaft*, **1**, 39, (1984).

16. R. Katzen, R. Fredrickson, and B. F. Brush, *Chem. Eng. Prog.*, **76**, 62 (1980).

17. R. G. Nayak and J. L. Walfhagen, presented at 9th Cellulose Conference, State University of New York, Syracuse, May 1982.

18. R. N. Busche, E. J. Shimschick, and R. A. Yates, *Biotechnol Bioeng. Symp.*, **12**, 249 (1982).

19. E. M. Lipinsky, personal communication, Battelle-Columbus Labs, Columbus, Ohio (August 1983).

20. M. Baumeister and E. Edel, *Dtsch. Papierwirtschaft*, *34*, 9 (1980).

21. H. E. Grethlein, *Biotechnol. Bioeng.*, **20**, 503 (1978).

22. T. W. Jeffries, in *Advances in Biochemical Engineering/Biotechnology*, Vol. 27, A. Fiechter, ed., Springer-Verlag, New York, 1980, p. 1.

23. H. F. Feldmann, M. A. Paisley, B. C. Kim, and H. R. Appelbaum, "Conversion of Forest Residues into a Methane-Rich Gas," Battelle-Columbus Labs, presented at the 13th Biomass Thermochemical Conversion Contractor's Meeting, Arlington, VA (Oct. 1981).

24. D. J. Hasse, "Carbon monoxide from lean gas," in *Process Technology and Flowsheets*, V. Cavaseno, ed., McGraw-Hill, New York, 1979, p. 368.

25. W. H. Klausmeier, "Biomass Chemicals Production by Thermochemical Conversion," presented at the 5th Symposium on Biotechnology for the Production of Fuels and Chemicals, Gatlinburg, TN, May 13–18, 1983.

26. W. H. Klausmeier, "Evaluation of the Technology Development Component of the First Brazilian Alcohol Project," report to the Industrial Projects Department of the World Bank by Sylvatex Corporation, April 1984.

27. X. N. Nguyen, V. Venkatesh, and J. S. Gratzl, "The Use and Processing of Renewable Resources," Chemical Engineering Challenge of the Future, *AICHE Symp. Ser. 207*, **77**, 94 (1981).

28. R. P Defilippi, "Extraction of Organics from Aqueous Solutions with Critical-Fluid Carbon Dioxide," presented at Fourth Symposium on Biotechnology in Energy Production and Conservation, Gatlinburg, TN, 1982.

PART 3

Cellulose Liquid Crystals

12

Lyotropic Liquid Crystal Solutions of Cellulose Derivatives

P. SIXOU and A. TEN BOSCH
Laboratoire de Physique de la Matière Condensée, CNRS, Université de Nice, Nice, France

Natural and renewable products such as cellulose and derivatives have great industrial importance. Synthetic polymers have been used recently to replace the older traditional cellulose industry. The discovery of the existence of cellulosic mesophases has revived interest and may reverse this tendency. Research in the field unites chemists and physicists, industrialists and scientists, and polymer and liquid crystal specialists in the search for new materials which simultaneously combine the advantages of polymers and liquid crystals. High-strength materials that conserve anisotropy in the solid phase are especially desirable. Good mechanical properties are critical for these applications, but in many other areas (e.g. optics), specific qualities due to the presence of anisotropy could also be exploited.

In this chapter we discuss the various methods to detect the presence of a mesophase and to predict phase diagrams on the basis of chain characteristics. Rheological and optical properties of polymer liquid crystals as well as the effect of external electric fields on these materials will be reviewed.

Small-molecule liquid crystals are well known to give ordered phases in the liquid either by cooling (thermotropic) or by increase in concentration (lyotropic). The description of the various types of ordered states (nematic, cholesteric, smectic) and the numerous experimental and theoretical studies have been treated in many publications (1–4).

CELLULOSE DERIVATIVES—THEORY AND EXPERIMENTS

Dilute Solutions

Intrinsic viscosity measurements have been made in many cellulose derivative/solvent systems and can be used to extract information on chain characteristics. The flexibility of semirigid polymers can be measured by the persistence length q, which correlates the direction of the first monomeric unit with that of the subsequent units. Quantities such as the radius of gyration and the end-to-end distance can be evaluated as a function of the ratio of the total contour length L divided by the persistence length (5).

Given a system of monodisperse wormlike chains, Yamakawa and Fujii (6–8) calculate the intrinsic viscosity $[\eta]$ with a method based on Kirkwood and Riseman (9). For each segment the friction is given by the local segment velocity, the local fluid velocity in absence of the segment, and the perturbation of the fluid velocity due to all other segments (method of Oseen-Burgers). These frictional forces are solutions of an integral equation that depends on the equilibrium conformation of the chain. An approximate expression of the variation of $[\eta]$ with the molecular weight M is given as a function of the diameter and persistence length of the molecular chain. Conversely from a measurement of $[\eta]$ as a function of M, a numerical fit will give both the diameter D and persistence length q. The method has been applied to cellulose derivatives (Table 12.1) and extended to the case of polydisperse samples. Because of the relatively low polydispersity of the sample used (1.4) the values of q and D for the mono- and polydisperse samples are not very different. The data also indicate that the solvent and the nature of the substituent are of importance. The applicability of the wormlike chain model of Yamakawa and Fujii can also be tested by studying the quasi-elastic light scattering from dilute solutions of cellulosic derivatives. The experiment gives the translational diffusion coefficient D and the hydrodynamic radius R_H which is substantially smaller than the radius of gyration R_G. The results from the latter experiment are in agreement with those obtained from intrinsic viscosity.

Observation of Liquid Crystalline Phases—Phase Diagrams

The simplest way to demonstrate the existence of an anisotropic phase in a polymer melt or solution is with a polarizing microscope. Some cellulose derivatives, such as hydroxypropylcellulose (HPC), are thermotropic. On heating, an optically anisotropic phase appears in the melt. At higher temperatures a transition from a liquid crystal to an isotropic phase occurs. For lyotropic liquid crystals, such as HPC in acetic acid, solutions at high polymer concentration are anisotropic. Below a certain critical concentration the solution becomes isotropic at constant temperature.

In cellulose and cellulose derivatives, the ordered phase is cholesteric (10).

TABLE 12.1 Influence of the Solvent on the Hydrodynamic Diameter D and the Persistence Length q for Cellulose Triacetate Samples at 25°C

Solvents	Monodisperse		Polydisperse	
	D (Å)	q (Å)	D (Å)	q (Å)
Dimethylacetamide (DMAC)	8	66	8	69
Acetone (A)	6	57	6	59
Trifluoroacetic acid (TFA)	10	55	11	56
Dichloromethane (DCM)	6	39	5	42
Trichloromethane (TCM)	8	35	9	36
Tetrachloromethane (TECM)	8	35	10	35

Comparison of the Values of the Hydrodynamic Diameter D and of the Persistence Length q of Some Polymer–Solvent Systems (T = 25°C)

Polymer/Solvent	D (Å)	q (Å)
CTC/Dioxane	13	154
NC/A	11	156
CTA/A	6	57
CTA/TFA	10	55

CTC = Cellulose tricarbanilate; NC = nitrocellulose; CTA = cellulose triacetate.

The type of texture observed in the polarizing microscope depends on the surface conditions of the glass plates enclosing the sample, on thermal history, and especially on the concentration of the solution.

In Figure 12.1a polygonal texture is reproduced, which was obtained by cooling a HPC/acetic acid solution from the isotropic state. Figure 12.1b shows a texture of cholesteric spherulites obtained from a mixed solvent solution (11). By registering the transmitted light intensity between crossed polarizers of a microscope, phase diagrams can be plotted (12). As for the small molecule liquid crystals, the anisotropic–isotropic transition can be also be studied by differential scanning calorimetry. The transition is of first order, and the enthalpy of the transition is small (13).

Other methods can be used to characterize the transition: the refractive index (at the transition, the refractive index splits into two parts owing to ordinary and extraordinary transmission of light), the static dielectric constant, the optical absorption or the circular dichroism, and the NMR spectra all show significant modifications which reflect the anisotropic character of the medium.

The experimental phase diagrams have shown that, for a given cellulose derivative, the type of solvent, the degree and nature of substitution, and especially the degree of polymerization are factors that determine the critical concentration. Some examples are discussed in the following.

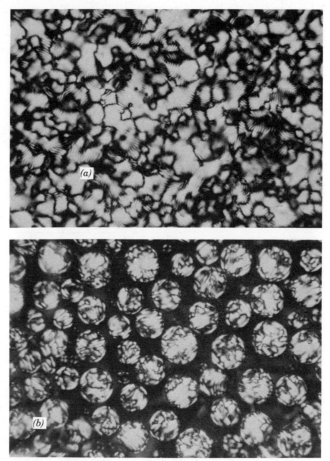

Figure 12.1   (a) Texture with fingerprint patterns in HPC solution in acetic acid. (b) Texture with spherulites obtained from HPC in a mixed solvent (water + acetic acid).

1.   The transition temperature of the thermotropic polymer increases with degree of polymerization but attains a limiting value (14) (Fig. 12.2). This type of behavior has been observed in many other mesomorphic polymers.

2.   Hydroxypropylcellulose forms anisotropic solutions in many different solvents: water, acetic acid, and so on. A linear decrease in the transition temperature (here, temperature for complete disappearance of the anisotropic phase) with decreasing polymer concentration is observed in "inert" solvents. In the case of strongly interacting solvents with specific interactions such as H-bonding, a curvature in the plotted transition temperature is found (Fig. 12.3) (15). In mixed solvents it is possible to vary the cholesteric pitch by variation of the respective concentrations (11, 16).

3.   Mixtures of flexible and mesomorphic polymers were also studied. Segregation is observed as in flexible polymer systems at high relative degree of

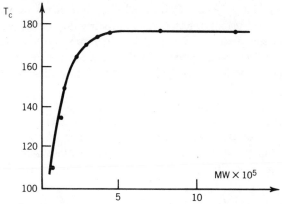

Figure 12.2 Isotropic- anisotropic transition temperature of a cellulosic thermotropic polymer as a function of molecular weight.

polymerization and can be followed more easily when one of the components is itself anisotropic. In ternary systems part of the simple solvent of a lyotropic solution can be replaced by a flexible polymer, and the anisotropic properties are strongly modified (12, 15). The formation of the inverted gel in the HPC/water can also be affected by replacing part of the water by a third component (17).

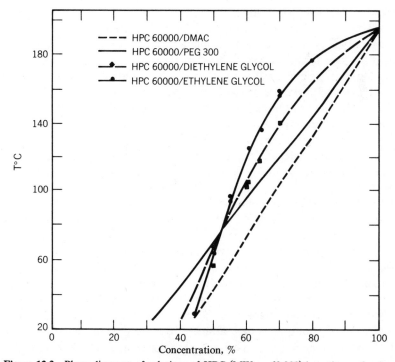

Figure 12.3 Phase diagrams of solutions of HPC (MW = 60,000) in various solvents.

Polymers with stiff linear chains such as cellulose derivatives were first predicted by Flory (18) to form mesophases because of the rigidity of the polymeric chain. Generally, sufficient chain rigidity remains an important requirement for the formation of polymeric liquid crystals. As in small-molecule liquid crystals, an additional, attractive orientation dependent interaction will assist the appearance of an ordered phase.

The earliest theories for phase separation of polymeric liquid crystals were calculated based on a model of long rigid rods. Onsager (19) showed that a tendency to order occurs owing to steric hindrance of the rods. Flory (20) calculated the entropy of a rigid rod system on a lattice. An order parameter, y, was introduced by defining the average (denoted by brackets) orientation of the rods

$$y = \left\langle \frac{x}{j} \sum_{i=1}^{j} \sin \psi_i \right\rangle$$

where j is the total number of rods, ψ_i is the angle of rod to the preferred direction, and the length to width ratio of the rod is given by x.

The free energy of mixing is minimized with respect to y, and phase diagrams in a simple solvent are calculated by equating the chemical potential in the ordered ($y \neq 0$) and disordered ($y = 0$) phases. The phase diagrams show the interaction parameter, generally assumed to vary as $1/T$, as a function of volume concentration. A large low temperature biphasic region and a narrow zone at high temperatures separate the isotropic and anisotropic phases. Later, anisotropic dispersion forces were introduced in the model (21). As a result, the high-temperature biphasic interval is tilted to obtain a finite disordered phase in the pure polymer melt. Many extensions of the original theory can be found in the literature. The rigid rod model was extended to semirigid chains by introducing a parameter representing the proportion of aligned neighboring monomers. Short, rodlike solvents were also explored as well as ternary systems (22). These theories have been applied to interpret experimental phase diagrams in cellulose-derivative systems. Problems arise owing to ill-defined biphasic separation in the experiments. Furthermore, many cellulose derivatives form cholesteric phases and are thought to have helical or ribbonlike conformations.

In some cellulose derivatives, there is a postulated change in conformation from helix to the random coil (23). Although no conclusive evidence of this effect has been presented, this transition may explain certain results in the tricarbanilate of cellulose. Pincus and deGennes (24) suggested a theory for the transition. The parameters of interest are the interaction energy between helical segments and the energy required to transform a helical into a coil segment. The rigidity of the chains is given by the fraction of monomers in rigid segments. At high concentrations, an order parameter is defined as in Maier-Saupe theory of liquid crystals, $S = \frac{1}{2} \langle (3 \cos^2 \phi - 1) \rangle$ where ϕ is the angle between the orientation of a rod segment and the mean alignment axis. A helix−coil transition is found, which can be first order and mesomorphic at high polymer concentrations—even for small coil-helix wall energies. In the mesomorphic phase, the polymer chain is stiffer than prior to the transition. This is characteristic of induced rigidity, and we return to this point later on.

Recently an extension of the Onsager theory to solutions of semi-flexible macromolecules has been proposed by Khokhlov (25), and an application to experimental phase diagrams has been discussed (14, 26). Perez and Chanzy (27) have taken a very different approach for the cellulose/morpholine systems (7) by energy conformational studies of the isolated molecule.

Interest has recently been revived in the model of stiff chains described by a string with bending elasticity (28-31) to explain nematic ordering in polymer systems. This model can also be used for helical macromolecules; the success of the Yamakawa and Fujii viscosity formula was our incentive to apply the wormlike chain to phase diagrams as well (32-35).

In the model, the stiff chain is represented by a differential space curve $\mathbf{r}$ (s). The elastic energy of bending is given by

$$\frac{1}{2} \epsilon \, (\partial \mathbf{r} / \partial s)^2$$

at contour length s. The polymer end-to-end distance is

$$\langle R^2 \rangle = 2qL \, (1 - (q/L) \, (1 - e^{-L/q}))$$

so that the persistence length is $q = \epsilon / kT$ (28).

The bend elasticity favors conformations of constant curvature. By introducing an anisotropic interaction $(uS \, (\frac{3}{2} \cos^2\phi(s) - \frac{1}{2})$ and an order parameter $S = \langle \frac{3}{2} \cos^2\phi(s) - \frac{1}{2} \rangle$, a first-order nematic isotropic transition can be found. Here $\phi(s)$ is the angle of the chain at contour length s to the director and u is an effective mean field parameter which tends to align the chains. The relevant parameters of the model are then q/L and uq/kT, L being the degree of polymerization. Two methods were used to solve for the free energy. First a numerical calculation of the distribution function for a given chain conformation and therefore the free energy was possible by use of theories developed for flexible chains. In the pure polymer, the transition temperature is shown to increase as L or q is increased (Fig. 12.4). This variation as a function of degree of polymerization was not obtained within the Flory theory. A large "induced rigidity" or extension of the polymer chain at the transition was also predicted. For comparison, a Landau–deGennes expansion of the free energy, valid near the transition temperature

$$F = F_0 + AS^2 + BS^3 + CS^4$$

was also calculated (35). The coefficients A, B, C are given in the parameters of the wormlike chain. The transition temperatures are in good agreement with

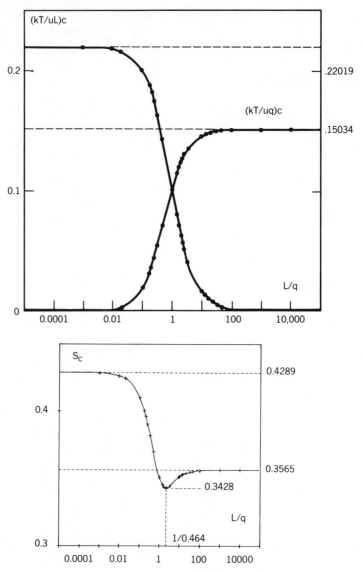

Figure 12.4 Theoretical transition temperature of a semirigid polymer; effect of flexibility and molecular weight.

the former calculations. In both calculations at low L/q, the order parameter tends to the 0.43 of rigid rods (this result must be corrected in ref. 35).

In solutions the pseudotransition temperature T_P can be plotted against concentration. This is the temperature at which the isotropic phase ceases to be thermodynamically stable and in solutions is not equivalent to the experimental temperature usually given, at which the anisotropic phase disappears. If the biphasic region is narrow, only small deviations are expected. In the simple case of an inert solvent that does not interact in any way with the polymer, a linear law for T_P as a function of polymer concentration was found when the persistence length was taken to be constant. If the persistence length varied as $1/T$ (as postulated within the bend elastic chain model) then $T_P \sim C^{1/2}$. This was applied to HPC/dimethyl acetamide solutions, and the square root law was reasonably well reproduced (36). In a solution of the benzoic acid ester of (2-hydroxy-propyl)cellulose (BzPc) an excellent fit to $C^{1/2}$ was found, perhaps owing to the chemical inactivity of the completely substituted lateral chains (14). The temperature variation of the persistence length can therefore be an important factor in phase diagrams of cellulose derivatives.

In other solvents, isotropic interactions not included in this model must be taken into account. This is especially evident in a small-molecule liquid crystal solvent. Experimentally, many different liquid crystal/cellulose derivative solutions were attempted; however, none were homogeneous. Another extreme case is the solution with a flexible incompatible polymer. These effects as well as the possibility of screening of the polymeric interactions, well known in flexible polymer systems, are currently under investigation (37).

Cellulose derivatives are often used for the spinning of fibers. The quality of the fiber could be greatly enhanced by the presence of a preordered phase. This phase may even be induced by the flow field, as was discussed in the Flory model (38, 39), and we also extended our calculations to include external fields such as elongational flow (40). The transition temperature was shown to be raised, and the order present in the melt or solution was found to increase for sufficiently high external fields.

In conclusion, although many types of phase diagrams have been investigated in cellulose derivatives, information on other quantities of interest is still lacking. No measurements of order parameters are available. Also of interest are the Frank elastic constants useful for the interpretation of textures and nonequilibrium phases in external electric and magnetic fields. On the other hand, no appropriate model at the present seems capable of interpreting the copious experimental data on the viscosity and the cholesteric pitch.

Optical Properties

At high concentrations, solutions of cellulose derivatives are irridescent, a characteristic of cholesteric liquid crystals. Selective reflection of light is caused by the helical order present in a cholesteric phase. For a planar texture, a cholesteric phase with right-handed pitch will reflect left-handed circular polarized light at a certain wavelength given by the helical pitch for normal incidence. The pitch can also be measured by apparent optical absorption or by circular dichroism. It can thus be determined as a function of concentration or temperature. The pitch can also be measured for wavelengths outside the visible spectrum.

The peak of intensity of scattered light measured as a function of scattering angle is associated with Bragg reflection in the cholesteric phase. As in small molecule cholesteric phases, the alternating dark and light lines of homeotropic textures that appear on observation under a polarizing microscope can also be used. The appearance of these lines is related to the anisotropy of polarizability due to the orientation of the molecular axis relative to the direction of propagation of the light wave. Figure 12.5 illustrates an example of the results on the pitch obtained by three different methods for HPC in acetic acid (16).

It was possible to study the role of various parameters on the pitch: concentration, temperature, molecular weight, solvent type, mixtures of solvents, mixture with a polymer. Qualitative comparison was made with the theories developed for (rigid) small-molecule cholesteric phases. Effects due to the semirigidity of the chain or to the helical conformation are thus not considered

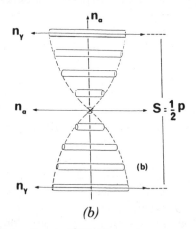

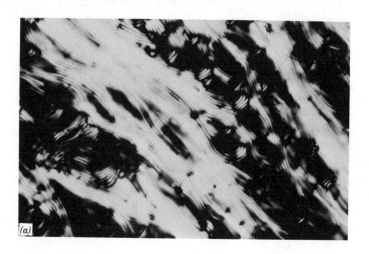

Figure 12.5 (*a*) Fingerprint pattern in HPC ($M_w = 100,000$) in acetic acid C = 31% (weight); dark and bright lines correspond to orientation of macromolecules as shown in (*b*). Opposite page: (*c*) Optical absorption of a solution of HPC in acetic acid. (*d*) Angular light scattering of solutions of HPC in acetic acid. Numbers correspond to concentrations (weight).

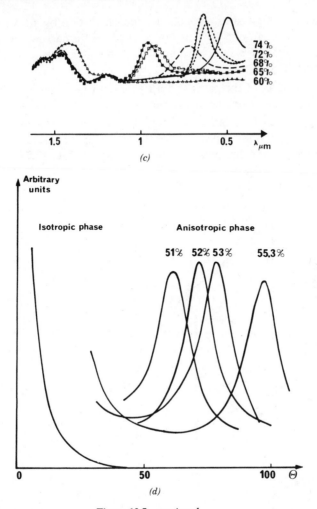

Figure 12.5 *continued.*

but rather the presence of some type of asymmetric potential in addition to the orientational interaction. From our analysis of experimental results we find (16, 41, 42) the following.

1.   On increasing the temperature, the solutions pass from blue to red. This is the opposite of the variation found in small-molecule cholesteric phases. We conclude that anharmonic effects, important in the latter case, can be neglected in the polymers we studied.

2.   The dielectric constant of the solvent does not appear to be an essential parameter. The same polymer dissolved in two solutions of different static permittivity can lead to the same value for the cholesteric pitch. Inversely, in a ternary mixture of constant static permittivity but varying proportions of the various components, the pitch varies considerably.

3.   The pitch is given as a function of concentration by a law:
$P = P_o C^{-\alpha}$ with an $\alpha \geqslant 3$. This law is valid over a narrow range of concentration. A fit with a single α is no longer possible over a larger range of concentration (Fig. 12.6).

4.   An inversion of the pitch and a transition to a nematic phase, observed in certain solutions of polypeptides, were not found in HPC-solutions.

5.   It appears that a relation between the pitch and the rigidity of the chains (or the concentration for appearance of the anisotropic phase) is possible. Often a larger pitch is found for chains of greater stiffness (Fig. 12.7).

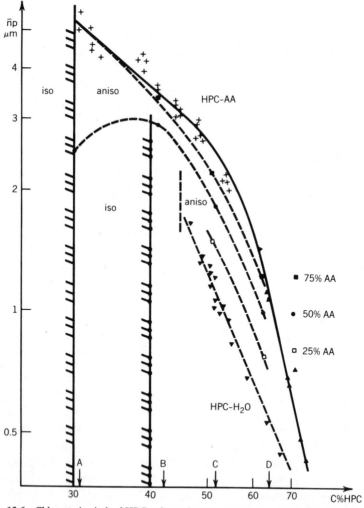

Figure 12.6   Chloresteric pitch of HPC polymer in pure acetic acid and water and in mixed solvents of various relative concentration in acetic acid.

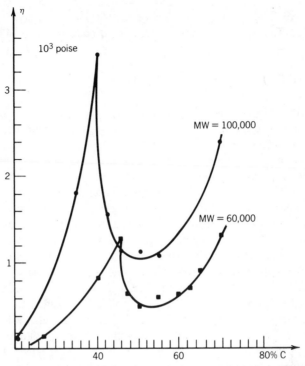

Figure 12.7 Shear viscosity of HPC in water for two molecular weights (shear gradient = 6.2 sec^{-1}).

6. The effect of degree of polymerization is generally very weak in the range investigated.

7. The effect of the molar volume of the solvent or of side groups on the chain appears to be important. If the substitution side chains are sufficiently cumbersome, a thermotropic film can be obtained with a pitch in the visible spectrum.

In conclusion, the very specific interactions between polymer–solvent (H-bonds) appear to the determining factors for the formation of the pitch.

Rheological Properties

The variation of the viscosity in lyotropic solutions of cellulose derivatives is characteristic of polymeric mesophases (Fig. 12.7): an initial increase with polymer concentration due to entanglements, then from a critical concentration on, the viscosity decreases due to appearance of the anisotropic phase and orientational ordering of the rigid polymer segments. Further increase of the concentration leads again to an increase of the viscosity after passing through a minimum. A power law of the shear as a function of the velocity gradient is observed. The behavior is Newtonian for high and low concentrations and rheofluidifing in the vicinity of the critical concentration (43).

EFFECT OF EXTERNAL FIELDS

In small-molecule liquid crystals, there is a collective organization of the molecules and the intermolecular coupling is stronger than the thermal agitation. On application of an external field (flow, magnetic, or electric fields), a modification of the ordered phase can occur. These effects are weaker in liquid crystalline polymers because of the high viscosity but are unusual for polymer systems. The most remarkable effects are found in flow fields. We have indicated (40) under which conditions an isotropic solvent can become ordered in an elongational flow. Because of the anisotropy of viscosity, instabilities can also be observed. Magnetic fields couple only weakly to cellulose derivatives owing to the small values of the anisotropy of susceptibility. This is in contrast to polymers with benzene rings in the principle chain (44).

Electric fields lead to more interesting observations. A liquid crystal can be oriented by an electric field, and electrohydrodynamic instabilities can occur at sufficiently high fields (1, 4). These phenomena are also found in liquid crystal polymers (45–48). In cellulose derivatives, the cholesteric order can also be changed by an electric field (49). Evidence for these effects has been presented, but further investigations are necessary to understand and apply these remarkable properties.

REFERENCES

1. M. J. Stephen and J. P. Straley, *Rev. Mod. Phys.*, **46**, 617 (1974).
2. P. G. deGennes, *The Physics of Liquid Crystals*, Oxford University Press, Oxford, 1976.
3. E. B. Priestley, P. J. Wojtowicz, and Sheng Ping, *Introduction to Liquid Crystals*, Plenum, New York, 1976.
4. H. Kelker and R. Hatz, *Handbook of Liquid Crystals*, Springer-Verlag, New York, 1980.
5. O. Kratky and G. Porod, *Recl. Trav. Chim.*, **68**, 1106 (1949).
6. H. Yamakawa and M. Fujii, *Macromolecules*, **6**, 407 (1973).
7. H. Yamakawa and M. Fujii, *Macromolecules*, **7**, 128 (1974).
8. H. Yamakawa and M. Fujii, *Macromolecules*, **7**, 649 (1974).
9. F. G. Kirkwood and J. Riseman, *J. Chem. Phys.*, **16**, 565 (1948).
10. R. S. Werbowyj and D. G. Gray, *Mol. Cryst. Liq. Cryst.*, **34**, 97 (1976).
11. F. Fried and P. Sixou, *J. Polym. Sci.*, **22**, 239 (1984).
12. M. J. Seurin, J. M. Gilli, F. Fried, A. Ten Bosch, and P. Sixou, in *Polymeric Liquid Crystals*, A. Blumstein, ed., Plenum, New York, 1984.
13. P. Navard, J. M. Haudin, S. Dayan, and P. Sixou, *J. Appl. Polym. Sci.*, **37**, 211 (1983).
14. S. N. Bhadani, S. L. Tseng, and D. G. Gray, *Makromol. Chem.*, **184**, 1727 (1983).

15. M. J. Seurin, A. Ten Bosch, J. M. Gilli, and P. Sixou, *Polymer*, **25**, 1073 (1984).

16. F. Fried, J. M. Gilli, and P. Sixou, *Mol. Cryst. Liq. Cryst.*, **98**, 209 (1983).

17. M. J. Seurin, J. M. Gilli, and P. Sixou, *Eur. Polym. J.* , **19**, 683 (1983).

18. P. J. Flory, *Proc. R. Soc. Lond. Ser. A.*, **234**, 60 (1956).

19. L. Onsager, *Ann. N.Y. Acad. Sci.*, **51**, 627 (1949).

20. P. J. Flory, *Proc. R. Soc. Lond. Serv. A*, **234**, 73 (1956).

21. P. J. Flory and G. Ronca, *Mol. Cryst. Liq. Cryst.*, **54**, 289 (1979).

22. M. Warner and P. J. Flory, *J. Chem. Phys.*, **73**, 6327 (1980).

23. A. K. Gupta, E. Marchal, and W. Burchard, *Macromolecules*, **6**, 843 (1975).

24. P. Pincus and P. G. deGennes, *Polym. Preprints*, **18**, 131 (1977).

25. R. Khokhlov, *Phys. Lett.*, **68A**, 135 (1978).

26. M. A. Aden, E. Bianchi, A. Cifferi, G. Conio, and A. Tealdi, *Macromolecules*, **17**, 2010 (1984).

27. S. Perez and H. Chanzy, in Conference on Polysaccharides Solutions and Gels, Torviscosa, Italy, Nov. 4–6 1981.

28. N. Saito, K. Takahashi, and Y. Yunoki, *J. Phys. Soc. Jpn.*, **2211**, 219 (1967).

29. J. Ronca and D. Y. Yoon, *J. Chem. Phys.*, 76 (1982) 3295.

30. P. G. de Gennes in *Polymer Liquid Crystals*, A. Cifferi, W. R. Krigbaum and R. B. Meyer, eds., Academic, New York, 1982.

31. F. Jahning, *J. Chem. Phys.*, **70**, 3279 (1979).

32. A. Ten Bosch, P. Maissa, and P. Sixou, *Phys. Lett.*, **94A**, 298 (1983).

33. A. Ten Bosch, P. Maissa, and P. Sixou, *J. Chem. Phys.*, **79**, 3462 (1983).

34. A. Ten Bosch, P. Maissa, and P. Sixou, *Nuovo Cimento*, **D3**, 95 (1984).

35. A. Ten Bosch, P. Maissa, and P. Sixou, *J. Phys. Lett.*, **L44**, 105 (1983); V. V. Rusakov, M. I. Shliomis, *J. Phys. Lett.* **L46**, 935 (1985).

36. M. J. Seurin, A. Ten Bosch, and P. Sixou, *Polym. Bull.*, **9**, 450 (1983).

37. A. Ten Bosch and P. Sixou, *J. Chem Phys.*, **83**(2), 899, (1985).

38. G. Marrucci and A. Cifferi, *J. Polym. Sci. Polym. Lett. Ed.*, **15**, 643 (1977).

39. P. Sixou, S. Dayan, J. M. Gilli, F. Fried, P. Maissa, M. J. Vellutini, and A. Ten Bosch, *Carbohydr. Polym.*, **2**, 238 (1982).

40. P. Maissa, A. Ten Bosch, and P. Sixou, *J. Polym. Sci.*, **21**, 757 (1983).

41. J. M. Gilli, A. Ten Bosch, J. F. Pinton, P. Sixou, A. Blumstein, and R. B. Blumstein, *Mol. Cryst. Liq. Cryst.*, **105**, 375 (1984).

42. F. Fried and P. Sixou, *Mol. Cryst. Liq. Cryst.*, **98**, 209 (1983).

43. S. Dayan, J. M. Gilli, P. Sixou, *J. Appl. Polym. Sci.*, **28**, 1527 (1983).

44. J. M. Gilli, G. Maret, A. Ten Bosch, P. Maissa, P. Sixou, and A. Blumstein, *J. Phys. Lett.*, **L46**, 329 (1985).

45. J. M. Gilli, H. W. Schmidt, J. F. Pinton, P. Sixou, O. Thomas, G. Kharas, and A. Blumstein, *Mol. Cryst. Liq. Cryst.*, **102**, 49 (1984).

46. J. M. Gilli, J. F. Pinton A. Ten Bosch, and P. Sixou, *Mol. Cryst. Liq. Cryst.* **131** (1985).

47. J. M. Gilli and P. Sixou, *Mol. Cryst. Liq. Cryst.*, **113**, 179 (1984).

48. J. M. Gilli, J. F. Pinton, P. Sixou, A. Blumstein, and O. Thomas, *Mol. Cryst. Liq. Cryst.*, **1**, 123 (1985).

49. J. M. Gilli, A. Ten Bosch, J. F. Pinton, P. Sixou, A. Blumstein, and R. B. Blumstein, *Mol. Cryst. Liq. Cryst.*, **105**, 375 (1984).

13

Structural Investigations on Some Cellulose Derivatives in the Crystalline and Liquid Crystalline State

P. ZUGENMAIER

Institute of Physical Chemistry, Technical University of Clausthal, Clausthal-Zellerfeld, Federal Republic of Germany

Liquid crystals of cellulose derivatives in highly concentrated solutions were found and extensively studied by Gray and co-workers (1). The threshold volume fractions were determined and are summarized for the biphasic equilibria for the systems hydroxypropyl cellulose, cellulose acetate, and ethyl cellulose in various solvents in ref. (2). In our laboratory the liquid crystalline systems cellulose tricarbanilate (CTC) in ethyl methyl ketone (MEK) and 2-pentanone (3,4) and the system ethyl cellulose (EC) in glacial acetic acid were investigated with respect to unusual optical properties that are closely related to the pitch of the helicoidal cholesteric structures present in these systems.

The pitch of the helicoidal structure is a function of temperature and concentration. The handedness of the helicoidal structure is able to change by temperature increase. Studies on polypeptide liquid crystals reveal that the temperature and concentration dependence of the pitch and the handedness can be described by theoretical considerations (5). Czarniecka and Samulski (6) attribute the inversion of the handedness that is sometimes observed in polypeptide liquid crystals with increasing temperature and concentration to the inversion of handedness of the molecular conformation. Such an inversion of handedness of molecular conformation is not necessary in the theoretical considerations of Kimura et al. (7).

In this report, we present experimental data on the helicoidal pitch and handedness of cellulose derivatives (CTC and EC) in cholesteric phases and discuss the molecular conformation of single helices that is determined in the crystalline state. We are able to produce changes in conformation of some single cellulose derivative helices by interaction with solvent molecules built into the crystalline lattice while other single helix conformations remain unchanged when interacting with solvent molecules. These results give hints for possible conformations and changes in conformations in the liquid crystalline state. Moreover, a solvent complexed crystalline state has to be taken into consideration besides the well-known liquid crystalline and "dry" crystalline state when experiments are described with decreasing temperature and when spherulitic growth appears in the samples.

STRUCTURAL INVESTIGATIONS IN THE CRYSTALLINE STATE

Highly concentrated solutions of CTC in MEK or in 2-pentanone exhibit isotropic phases at elevated temperatures. A cholesteric phase appears at the cooling of such a solution, and on further cooling a crystalline phase is eventually obtained with solvent molecules built into the crystalline lattice (3). Such a crystalline phase is more easily studied and important features as unit cell and chain conformation are determined when fibers of these samples are available. Fibers of cellulose derivatives are normally produced from highly concentrated solutions with a glass rod dipped into the solution and immediately pulled out with still some sample attached to the rod. These samples are drawn to a certain length without breaking the connecting fiber between rod and solution. The fibers dry in a short period of time. The crystallinity of the fibers is considerably increased by annealing at elevated temperatures. After this procedure the solvent is completely lost. When these fibers are brought back into an atmosphere of a solvent–nonsolvent mixture (the nonsolvent mostly ethanol), crystalline fibers with built-in solvent molecules are observed.

In Figure 13.1a an x-ray pattern of a CTC fiber in an atmosphere of MEK/ ethanol is shown (3). The fiber repeat c which is the identity period along the cellulose backbone was determined to $c = 15.25$ Å. A different x-ray pattern (Fig. 13.1b) is obtained with the identical fiber repeat c when the same fiber was placed in a vacuum chamber. A third pattern (Fig. 13.1c) appeared by annealing the fiber without the MEK/ethanol atmosphere with the fiber repeat of $c = 15.25$ Å again. These experiments prove that the same helix conformation of a single chain is present in all three differently treated fibers but that the packing of the chains is different. A conformational analysis reveals that left-handed threefold (3/2) helices fit the experimental x-ray data. A single CTC helix is shown in Figure 13.2 in two projections—one along the helix axis that coincides with the c direction, and the other in a plane perpendicular to c. Hydrogen bonding between adjacent residues is not possible, as has often been

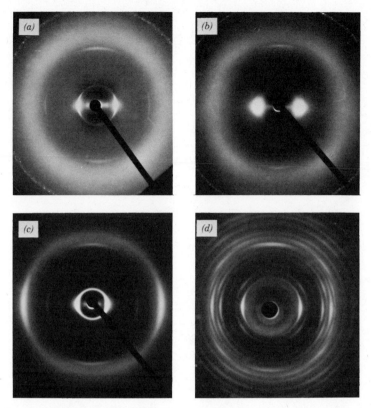

Figure 13.1 X-ray fiber diagrams of differently prepared cellulose tricarbanilate: (*a*) fiber complexed with methyl ethyl ketone: (*b*) dry fiber: (*c*) fiber annealed at 280°C: (*d*) for comparison, a fiber of tribenzoylcellulose.

proposed. The phenyl ring is in the vicinity of a perpendicular position to the chain axis.

This statement still holds for tribenzoylcellulose (TBC) with a missing NH group in the side chain as compared to CTC. The TBC derivative was quite recently investigated (8). The x-ray fiber pattern contains many discrete layer lines (Fig. 13.1*d*) in contrast to CTC. The conformation of a single chain is also a 3/2 helix (Fig. 13.3) and very closely resembles the CTC helix.

A different conformation of a single chain appears in cellulose triacetate (CTA) with a 2/1 helix and a fiber repeat of approximately 10.5 Å for both modifications, one obtained by heterogeneous (CTA I) (9) and the other by homogeneous (CTA II) (10) acetylation of native cellulose. A 2/1 helix is still present when two of the three acetate substituents are replaced by methyl groups. The conformation of 2,3-dimethyl-6-acetylcellulose (11) is represented in Figure 13.4.

A drastic change in conformation of the cellulose chain and unit cell size occurs in cellulose triacetate when nitromethane is built into the crystal lattice. A left-handed 8/5 helix (12) (Fig. 13.5) is the most probable conformation with

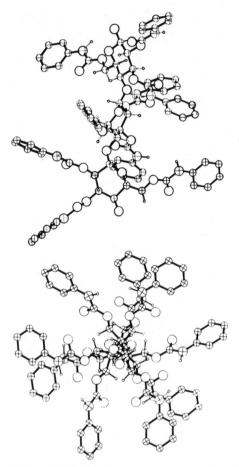

Figure 13.2   Left-handed threefold (3/2) helix of cellulose tricarbanilate in two projections. Top: along the chain axis; bottom: perpendicular to the chain axis.

nitromethane interaction in the crystalline phase. Both modifications of cellulose triacetate, CTA I and CTA II, lead to the same form of a CTA-nitromethane complex (CTA-N) with a fiber repeat of 41.2 Å. The shape of a cross section of the 8/5 helix (Fig. 13.5) is almost circular, although only small changes appear in the projection along c when the first two monomer rings of Figure 13.5 are compared with the representation of the rings of the 2/1 helix in Figure 13.4.

A left-handed threefold (3/2) helix (13) with a fiber repeat of $c = 15.0$ Å is found in triethyl cellulose (TEC) and in commercially available ethyl cellulose with a degree of substitution of $\sim$ 2.5 and is shown in Figure 13.6.

The following conclusions are derived from crystalline x-ray fiber studies and conformational analysis (13): All conformations of single chains exhibit a rise per residue of about 5 Å with 2 (cellulose, CTA) over 2.67 (CTA-N) to 3 (CTC, TBC, TEC) monomers per turn. All helices show a rod or cylinder type of conformation except the 2/1 helices. The cellulose backbone chain is in an

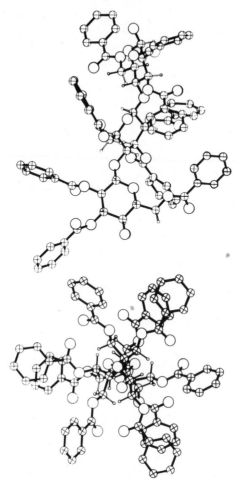

Figure 13.3   Left-handed threefold (3/2) helix of tribenzoylcellulose in two projections. Top: along the chain axis; bottom: perpendicular to the chain axis.

extremely extended form. The maximum length of a glucose residue is 5.45 Å, which compares to a 5 Å rise per residue.

STRUCTURAL INVESTIGATIONS IN THE LIQUID CRYSTALLINE STATE

Highly concentrated solutions of cellulose derivatives show optically anisotropic properties. Textures are observed in the polarization microscope that are characteristic of cholesteric phases (3). The first appearance of cholesteric phases in a sample is concentration- and temperature-dependent and is also a function of molecular mass and the solvent chosen. At low temperatures, spherulites are observed in a limited concentration range. The results are summarized in the

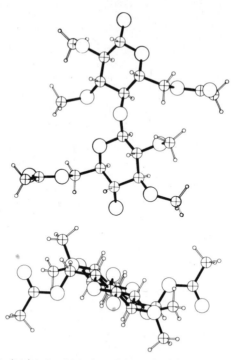

Figure 13.4 Twofold (2/1) helix of 2,3-dimetyl-6-acetylcellulose in two projections. Top: along the chain axis; bottom: perpendicular to the chain axis.

phase diagram of Figure 13.7 for CTC in MEK for two degrees of polymerization (DP) whose average values are 100 and 250. Samples of CTC with DP = 100 dissolved in 2-pentanone give the results shown in Figure 13.8. The cholesteric–isotropic transition temperature of CTC/MEK lies up to 25°C higher than the one for CTC/2-pentanone for the same concentration and DP. Conoscopic observations with the polarizing microscope along the optical axis show an unusual positive birefringence of the cholesteric phases for CTC in both solvents.

Handedness of Cholesteric Structures

Cholesteric phases consist of twisted nematic planes with a certain pitch and handedness of the so-called helicoidal structure. Therefore, cholesteric phases are often referred to twisted nematic phases. Theoretical considerations of de Vries (14) predict that right-handed circularly polarized light is totally reflected by a right-handed cholesteric helicoidal structure at a wavelength λ_0 which is correlated to the pitch p of the helicoidal structure and the mean refractive index $\bar{n}$ of a nematic plane by

$$\lambda_0 = \bar{n}p \tag{1}$$

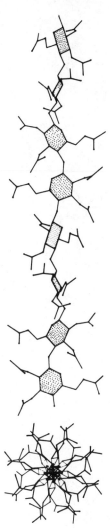

Figure 13.5 Left-handed 8/5 helix of cellulose triacetate–nitromethane complex (CTA-N) in two projections. Top: along the chain axis; bottom: perpendicular to the chain axis.

The reflected wave does not change handedness. Analogous, left-handed circularly polarized light is totally reflected by a left-handed helicoidal cholesteric structure. The optical and the helicoidal axes are assumed to be parallel in this theory.

Measurements of selective reflection or transmission of circularly polarized light may serve in determining the handedness and the pitch of the helicoidal structure of cholesteric phases. Figure 13.9 represents the transmission spectra of unpolarized, right- and left-handed circularly polarized light of 0.6 g/mL CTC (DP = 100) in 2-pentanone. Left-handed circularly polarized light passes

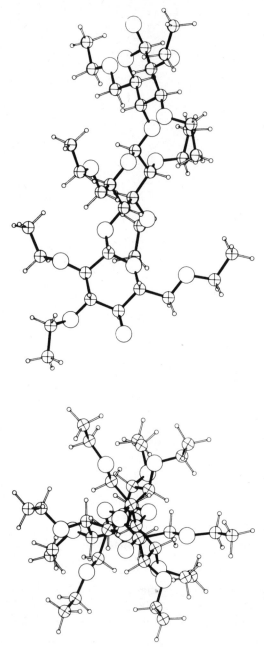

Figure 13.6 Left-handed threefold (3/2) helix of triethylcellulose in two projections. Top: along the chain axis; bottom: perpendicular to the chain axis.

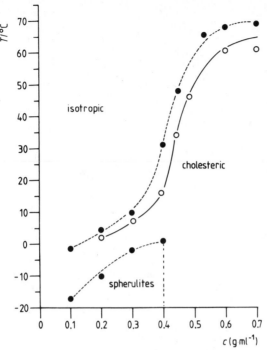

Figure 13.7 Phase diagram: transition temperatures T versus polymer concentration for cellulose tricarbanilate/methyl ethyl ketone for two degrees of polymerization. ($\bigcirc$), $DP = 100$; ($\bullet$), $DP = 250$.

through the sample without attenuation, whereas right-handed circularly polarized light is weakened. The sample was surface aligned and placed between two glass plates 50 μm apart. The optical axis of the cholesteric structure coincides with the direction of propagation of light. A right-handed helicoidal structure of the cholesteric phase has to be concluded from this experiment. The pitch of this structure can be calculated with Eq. (1) when the refractive index $\overline{n}$ is known. Ethyl cellulose (degree of substitution ~ 2.5/DP unknown) in glacial acetic acid weakens left-handed circularly polarized light (Fig. 13.10) at all measured concentrations and temperatures in the cholesteric region of the phase diagram which indicates the presence of a left-handed helicoidal cholesteric structure for this cellulose derivative/solvent system.

Refractive Indices

Cholesteric phases are optically anisotropic, and according to the models proposed for these phases, the uniaxial optical axis lies parallel to the axis of the helicoidal structure. Two refractive indices, an ordinary $n_{o,ch}$ and an extraordinary $n_{e,ch}$, characterize the cholesteric phase. They can be measured on surface-aligned

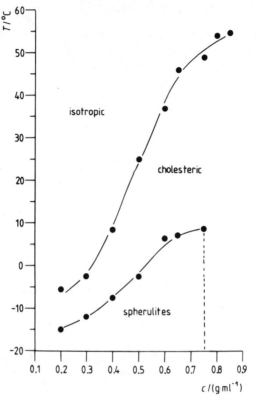

Figure 13.8 Phase diagram: transition temperatures T versus polymer concentration for cellulose tricarbanilate/2-pentanone ($DP = 100$).

samples with an Abbé refractometer. Two refractive indices, $n_{o,n}$ and $n_{e,n}$, can also be assigned to the nematic plane on which the helicoidal structure is based. A correlation is established between the two pairs of refractive indices by the simple relationship (15):

$$n_{o,n} = n_{e,ch} \tag{2a}$$

$$n_{e,n} = (2n_{o,ch}^2 - n_{e,ch}^2)^{1/2} \tag{2b}$$

and with the birefringence $\Delta n_{ch} = n_{e,ch} - n_{o,ch}$ small compared to 1 (4):

$$\Delta n_n = -2\Delta n_{ch} \tag{2c}$$

A negative sign in the birefringence of a cholesteric phase results in a positive sign of the nematic one and vice versa (see Eqs. (2)). Almost all cholesteric phases studied so far show a negative birefringence. This fact may serve with some model considerations for a characterization of a cholesteric phase.

The refractive indices are dependent on the structure of the cholesteric phases.

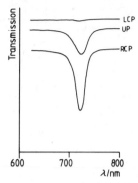

Figure 13.9 Transmission spectra of unpolarized (UP), right- (RCP) and left-handed (LCP) polarized light of 0.6 g/mL cellulose tricarbanilate in 2-pentanone at 16.7°C.

Therefore, they are functions of the pitch of a cholesteric structure and consequently of all the quantities that alter the pitch, as temperature, concentration, solvent, and so on.

The temperature and concentration dependence of the cholesteric refractive indices are summarized for selected samples in Figures 13.11–13.14. A plot of refractive indices of 0.6 g/ml CTC (DP = 100) in 2-pentanone versus reduced temperature is shown in Fugure 13.11. It is common to introduce a reduced temperature $T^* = T/T_c$, with T the actual temperature in K and T_c the clearing temperature or transition temperature in K from the cholesteric to the isotropic phase. At the clearing temperature ($T^* = 1$) in Figure 13.11, the refractive indices $n_{o,ch}$ and $n_{e,ch}$ discontinuously jump over into the isotropic refractive index n of a regular solution. This behavior signifies a first-order transition from the cholesteric to the isotropic phase, which is accompanied by a change in enthalpy measured by differential scanning calorimetry (DSC) (3). A discontinous drop in the refractive indices is seen at the low temperature range of Figure 13.11, where spherulitic growth is observed.

The birefringence of the cholesteric phase is positive and holds for all CTC samples studied. This unusual behavior is a consequence of the conformation of single helices and will be discussed later.

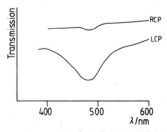

Figure 13.10 Transmission spectra of right- (RCP) and left handed (LCP) polarized light of ethyl cellulose in glacial acetic acid at −8.3°C.

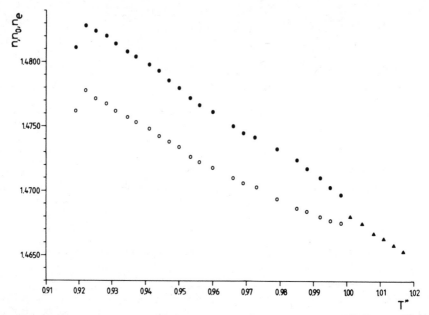

Figure 13.11 Refractive indices n (▲), n_o (○), or n_e (●) of 0.6 g/mL cellulose tricarbanilate ($DP = 100$) in 2-pentanone versus reduced temperature T^* (wavelength of light 589 nm; indices e and o signify extraordinary and ordinary, respectively).

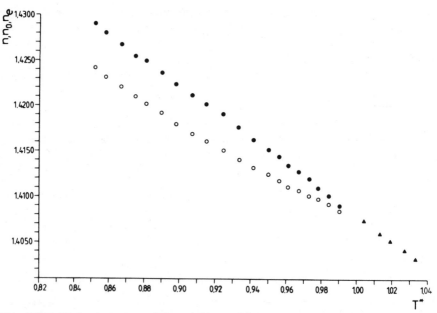

Figure 13.12 Refractive indices n (▲), n_o (●), or n_e (○) of 1 g/mL ethyl cellulose in glacial acetic acid versus reduced temperature T^*.

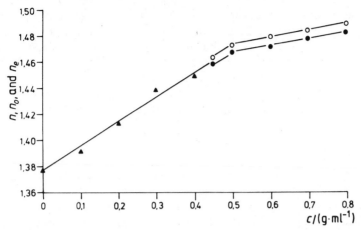

Figure 13.13 Refractive indices n (▲), n_o (●), or n_e (○) of cellulose tricarbanilate in ethyl methyl ketone versus concentration at room temperature.

The temperature dependence of refractive indices is represented for EC in glacial acetic acid in Figure 13.12. Here, a time dependence of the clearing temperature T_c was observed that might be attributed to a continuous alteration of the sample as, for example, a degradation of the chain. The time effect is compensated in the representation of the refractive indices with the introduction of the reduced temperature. The birefringence is negative as usual for a cholesteric phase and is in the same order of magnitude as for CTC in MEK or 2-pentanone.

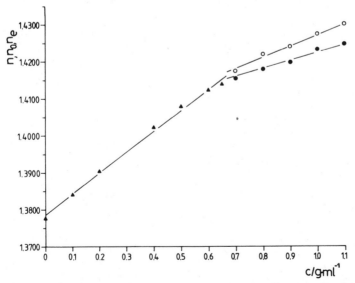

Figure 13.14 Refractive indices n (▲), n_o (○), or n_e (●) of ethyl cellulose in glacial acetic acid versus concentration at room temperature.

The concentration dependence of refractive indices of CTC in MEK (Fig. 13.13) and EC in glacial acetic acid (Fig. 13.14) exhibits some differences. The onset of birefringence occurs at different concentrations. The slope dn/dc is different for both systems. For CTC/MEK it remains the same from the isotropic into the beginning of the anisotropic region, but for EC/glacial acetic acid the slope immediately changes to a lower value with the appearance of the cholesteric phase. CTC/2-pentanone is comparable with EC in glacial acetic acid.

Transmission Spectroscopy

Unpolarized natural light always has components of circularly polarized light. Circularly polarized light of a certain wavelength will be weakened by passing through a cholesteric phase as shown in Figure 13.7 and according to Eq. (1). The influence of external parameters on the wavelength of selective reflection λ_o and consequently on the pitch can easily be determined by transmission spectroscopy. The spectra of 0.6 g/mL CTC in 2-pentanone are recorded as a function of temperature in Figure 13.15. A shift of the peak toward larger wavelengths with increasing temperature is evident. For temperatures smaller than 9.5°C the peak heights decrease, and the peak almost vanishes at 5.6°C. At this temperature spherulitic growth is observed (cf. phase diagram of Fig. 13.8).

The wavelength of selective reflection λ_o is determined from the position of the minima, and the pitch of the helicoidal structure is calculated with the knowledge of the mean refractive index of a nematic plane in the cholesteric phase, $\bar{n} = n_{o,n} + n_{e,n}$, using Eq. (1).

Optical Rotatory Dispersion

Helicoidal structures normally show a very high optical rotatory power. According to de Vries (14), the optical rotatory power θ for light that propagates along the optical axis of a cholesteric phase is expressed by the following equation:

$$\theta = \frac{\pi}{4p}\left(\frac{\Delta n_n}{\bar{n}}\right)^2 \left(\frac{1}{\lambda'^2}\right)\left(\frac{1}{1-\lambda'^2}\right) \tag{3a}$$

Where p =pitch of the helicoidal structure (the sign of the pitch is positive for right-handed helicoidal structures), $\bar{n}$ =mean refractive index of a nematic plane, Δn_n =birefringence of a nematic plane, $\lambda'=\lambda/\lambda_o$ (λ=actual wavelength of the experiment), and λ_o =wavelength of selective reflection. For $\lambda' << 1$ the approximation holds:

$$\theta = \left(\frac{\pi}{4}\right) p\, \Delta n_n^2 \left(\frac{1}{\lambda^2}\right) \tag{3b}$$

where θ is given in radian per unit length. A sufficiently thick sample is assumed in the derivation of Eq. (3a,b).

In Figure 13.16 the measured optical rotatory power for a sample of 0.6 g/

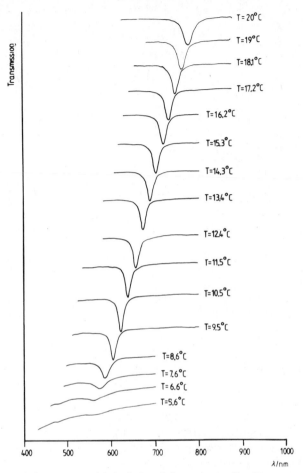

Figure 13.15 Transmission spectra of 0.6 g/mL cellulose tricarbanilate/2-pentanone at various temperatures.

mL CTC (DP = 100) in 2-pentanone is shown with the calculated curve of Eq. (3a). The birefringence was measured and assumed to remain constant over the whole wavelength range. Deviations between measurements and calculations in Figure 13.16 occur at small wavelength, which might be due to the beginning of increasing absorption of the phenyl groups. The wavelength of selective reflection is easily determined by the zero value of optical rotatory power at the singularity, and with Eq. (1) the pitch can be calculated. The positive optical rotatory power below λ_o indicates a right-handed helicoidal structure for the system CTC/2-pentanone.

The optical rotatory dispersion curves sometimes broaden in the vicinity of their extremes as shown in Figure 13.17 for 0.5 g/mL CTC/2-pentanone at different temperatures and can no longer quantitatively be described by the de Vries equation (3a). The wavelength of selective reflection can still accurately be determined and increases with increasing temperature.

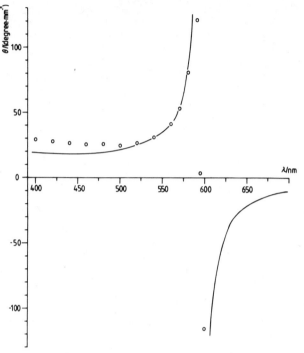

Figure 13.16 Optical rotatory power of 0.6 g/mL cellulose tricarbanilate in 2-pentanone (spacer 36 μm) at 7.1°C.

The left-handed helicoidal structures of ethyl cellulose in glacial acetic acid exhibit a negative optical rotatory power below λ_o and very broad maxima and minima (Fig. 13.18). The selective wavelength of reflection decreases with increasing temperature. Iridescent colors are observed when the selective wavelength of reflection is in the visible range as for the samples of Figures 13.16–13.18.

If the wavelength of selective reflection λ_o is outside the detectable range, Eq. (3b) can be used for the determination of the pitch of the helicoidal structure. Figure 13.19 shows a plot of the optical rotatory power θ versus λ^{-2} for the system CTC/MEK which gives a straight line with the slope $(\pi/4)p\Delta n_n^2$ according to Eq. (3b). The pitch p can be calculated with knowledge of the birefringence Δn_n. The pitch obtained is in accord with other measurements (4), although the straight line of Figure 13.19 does not pass through the origin as required by Eq. (3b).

Grandjean Lines

A direct determination of the pitch of cholesteric phases is available by a method proposed by Grandjean (16) and experimentally modified by Cano (17). If an optical lens, as shown in Figure 13.20, is placed upon a cholesteric sample spread on a glass plate, dark and bright rings are seen in a polarizing microscope. The

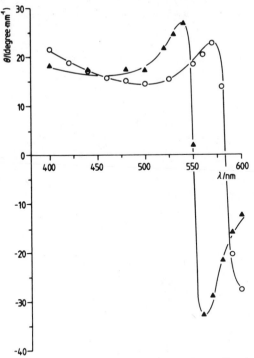

Figure 13.17 Optical rotatory power of 0.5 g/mL cellulose tricarbanilate in 2-pentanone at various temperatures (spacer 36 μm). $T = 3.8°C$ (O), $T = 1.8°C$ (▲).

cholesteric structure is oriented through surface effects at the top (lens) and at the bottom (glass plate) and can only nicely order when the distance between lens and glass plate s is a multiple of $p/2$. The distance s is correlated to the geometry of the device, and by measuring the diameter of the rings, seen in the microscope, the pitch p can be calculated (4). Figure 13.21 represents a photograph of a sample of 0.5 g/mL CTC in MEK observed through a polarizing microscope with crossed polars. The so called Grandjean lines or Cano rings are sometimes preserved in the dry state when the solvent slowly evaporates.

OVERVIEW—CELLULOSE DERIVATIVES IN THE CRYSTALLINE AND LIQUID CRYSTALLINE STATE

The pitch of the cholesteric helicoidal structure of some lyotropic liquid crystalline cellulose derivatives was determined by different methods, and the influence on the pitch of external parameters, especially temperature, concentration, and solvent effects, was ascertained in this report. The results are summarized in Figures 13.22–13.29 for selected samples of CTC in 2-pentanone (18), CTC in MEK (4), both with DP = 100, and EC in glacial acetic acid (18).

The pitch of the helicoidal structure of CTC in MEK and 2-pentanone in-

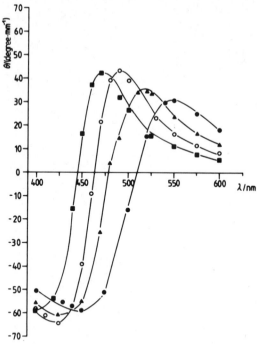

Figure 13.18 Optical rotatory power of 0.8 g/mL ethyl cellulose in glacial acetic acid at various temperatures (spacer 36 μm). $T = 21.9°C$ (■), $T = 16°C$ (○), $T = 6.0°C$ (▲), $T = -3.8°C$ (●).

creases with increasing temperature and concentration (Fig. 13.22–13.25). In contrast to a statement in a previous paper (4), the system CTC/MEK and also CTC/2-pentanone form right-handed helicoidal structures with positive birefringence. The pitch is strongly solvent-dependent and is about four times larger for CTC/MEK than for CTC/2-pentanone at the same conditions. Any size of the pitch between the two extremes of CTC/2-pentanone and CTC/MEK can be produced by mixing the two solvents in an appropriate ratio. A straight line in Figure 13.26 is obtained by plotting the pitch p of 0.6 g/mL CTC/mixture of 2-pentanone and MEK versus the volume fraction of MEK X_{MEK}.

The size of the pitch decreases with increasing temperature or concentration for EC (degree of substitution ~ 2.5) in glacial acetic acid as shown in Figures 13.27 and 13.28. The pitch of the left-handed helicoidal structure has a negative sign by definition and is determined from the wavelength of selective reflection in Figure 13.27 with the mean refractive index $\bar{n} = 1.4210$ at room temperature with Eq. (1). The error of using a temperature-independent $\bar{n}$ in the calculation of the pitch is less than 1.5% and is by far smaller than the one in the determination of the concentration. The accurate measurement of the concentration is a difficult task, because solvent evaporates during handling of the solutions. The birefringence of EC/glacial acetic acid is negative.

The influence of the DP on the temperature dependence of the pitch is shown in Figure 13.29 for 0.6 g/mL CTC in 2-pentanone with DP = 100 and

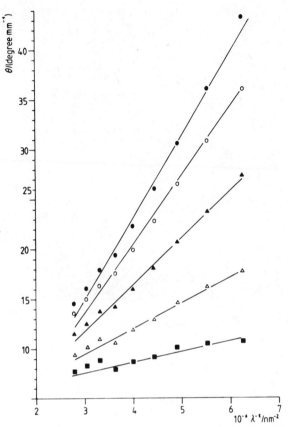

Figure 13.19 Optical rotatory power of 0.5 g/mL cellulose tricarbanilate in methyl ethyl ketone versus λ^{-2} at various temperatures (spacer 36μm). $T = 29.9°C$ ($\blacksquare$), $T = 27.9°C$ ($\triangle$), $T = 25.9°C$ ($\blacktriangle$), $T = 23.8°C$ ($\bigcirc$), $T = 21.8°C$ ($\bullet$).

DP = 250. As the clearing temperatures of the two samples are far apart, the reduced temperature was introduced for normalization purposes. The temperature gradients of the pitch are different with a smaller value for the larger DP.

The investigation of cellulose derivatives in the crystalline and the liquid

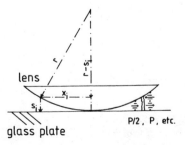

Figure 13.20 Schematic drawing of the setup for the observation of Cano rings.

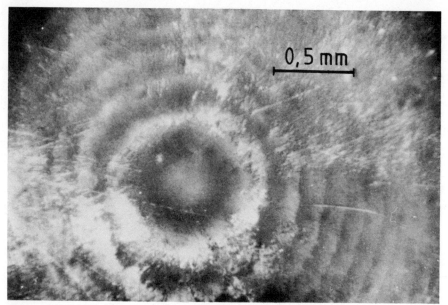

Figure 13.21 Cano rings of 0.5 g/mL cellulose tricarbanilate in ethyl methyl ketone observed through a polarizing microscope with crossed polars.

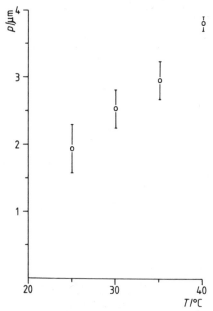

Figure 13.22 Pitch of the helicoidal cholesteric structure p of 0.5 g/mL cellulose tricarbanilate in methyl ethyl ketone as a function of temperature T. Method of measurement: Cano rings (4).

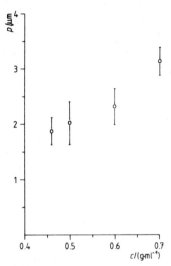

Figure 13.23 Pitch of the helicoidal cholesteric structure p of cellulose tricarbanilate in methyl ethyl ketone as a function of concentration c at room temperature. Method of measurements: Cano rings (4).

crystalline state reveals a variety of different structures. In the crystalline state 2/1, 3/2, 5/3, and 8/5 helices of single chains are found (12); all of them are left-handed except the 2/1 helix, where such a characterization fails. A change in handedness has never been observed. The single-chain conformation forms a very extended molecule and the cellulose backbone is regarded to be very stiff. Interaction with solvent in the crystalline state sometimes alters the single-chain conformation slightly. A planar form (2/1 helix) of cellulose triacetate changes

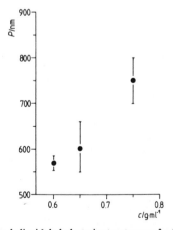

Figure 13.24 Pitch of the helicoidal cholesteric structure p of cellulose tricarbanilate in 2-pentanone as a function of concentration c at room temperature. Method of measurement: Cano rings.

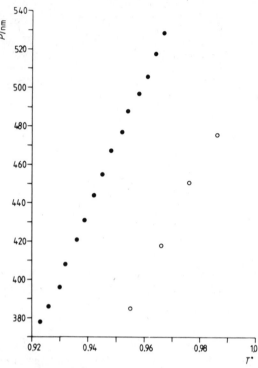

Figure 13.25 Pitch of the helicoidal cholesteric structure p of cellulose tricarbanilate in 2-pentanone as a function of reduced temperature T^* at two concentrations: $c = 0.6$ g/mL ($\bullet$); $c = 0.5$ g/mL ($\bigcirc$). Method of measurement: transmission spectroscopy.

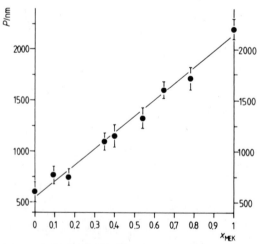

Figure 13.26 Pitch or the helicoidal cholesteric structure p of 0.6 g/mL cellulose tricarbanilate in a mixture of methyl ethyl ketone and 2-pentanone at room temperature. X_{MEK} is the volume fraction of methyl ethyl ketone. Method of measurement: Cano rings.

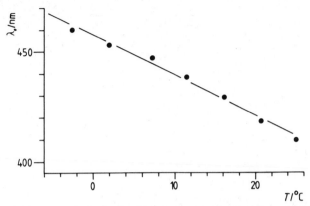

Figure 13.27   Selective wavelength of reflection λ_0 of a several-week-old sample of 0.8 g/mL ethyl cellulose in glacial acetic acid as function of temperature T. Method of measurement: transmission spectroscopy.

into a more cylinder or rod type of molecule with interaction of nitromethane. The left-handed threefold helix, a rodlike molecule, is a common conformation of cellulose derivatives in the dry and solvent-complexed crystalline state. A stiff, rodlike molecule with different polarizabilities along and perpendicular to the polymer chain axis is a prerequisite for molecules to form liquid crystalline phases.

The helicoidal structure of cellulose derivatives in the lyotropic cholesteric phase consists of twisted nematic planes, based on single-chain helices that are closely packed in a highly parallel fashion in the nematic plane (18). The pitch

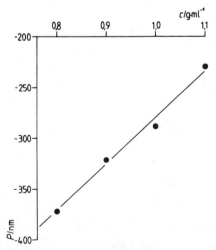

Figure 13.28   Pitch of the helicoidal cholesteric structure p of ethyl cellulose in glacial acetic acid as a function of concentration c at 0.3°C of a week-old sample. Method of measurement: transmission spectroscopy.

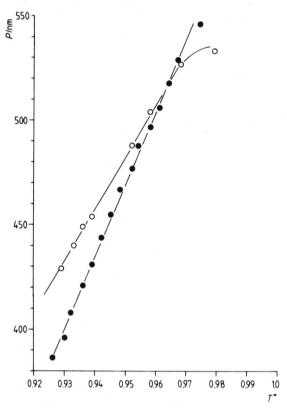

Figure 13.29 Pitch of the helicoidal cholesteric structure p of 0.6 g/mL cellulose tricarbanilate of different degree of polymerization in 2-pentanone as a function of reduced temperature T^*. ($\bullet$), DP = 100; ($\bigcirc$), DP = 250, Method of measurement: transmission spectroscopy.

and handedness of the helicoidal liquid crystalline structure depend on several parameters, such as temperature, concentration, solvent, and degree of poly-merization and substitution. The temperature dependence of the results presented here can be described by a formula given by Kimura et al. (7). The negative concentration gradient for the pitch in the system EC glacial acetic acid seems to lie outside the scope of the theory, as does a jump in the handedness of the helicoidal structure with increasing concentration of cellulose dicarbanilate in pyridine (18).

The usual negative birefringence of cholesteric phases may be attributed to a larger polarizability along the rod-type molecule compared with the one per-pendicular to the axis of the rod. Because of the placement of the phenyl rings in CTC (see Fig. 13.2), the polarizability perpendicular to the chain axis might be the larger one, which then results in a positive birefringence. All cholesteric cellulose tricarbanilate samples and also tribenzoylcellulose (see Fig. 13.3) with a similar placement of the phenyl groups (19) exhibit positive birefringence in various solvents.

REFERENCES

1. R. S. Werbowyj and D. G. Gray, *Mol. Cryst. Liq. Cryst.*, **34**, 97 (1976).
2. P. J. Flory, *Adv. Polym. Sci.*, **59**, 1 (1984).
3. P. Zugenmaier and U. Vogt, *Makromol. Chem.*, **184**, 1749 (1983).
4. U. Vogt and P. Zugenmaier, *Makromol. Chem.*, **4**, 759 (1983).
5. I. Uematsu and Y. Uematsu, *Adv. Polym. Sci.*, **59**, 37 (1984).
6. K. Czarniecka and E. T. Samulski, *Mol. Cryst. Liq. Cryst.*, **63**, 205 (1981).
7. H. Kimura, M. Hoshino, and N. Nakano, *J. Phys. (France)*, **40**, C3-174 (1979).
8. W. Gutknecht, Diplomarbeit, Institut für Physikalische Chemie der TU Clausthal, Clausthal-Zellerfeld, Jan. 1983.
9. A. J. Stipanovic and A. Sarko, *Polymer*, **19**, 3 (1978).
10. E. Roche, H. Chanzy, M. Boudeulle, R. H. Marchessault, and P. Sundararajan, *Macromolecules*, **11**, 86 (1978).
11. R. Möller, Diplomarbeit, Institut für Physikalische Chemie der TU Clausthal, Clausthal-Zellerfeld, Dec. 1982.
12. P. Zugenmaier, in *Polysaccharide*, W. Burchard, ed., Springer-Verlag, Berlin, 1985, p. 271.
13. P. Zugenmaier, *J. Appl. Polym. Sci. Appl. Polym. Symp.*, **37**, 223 (1983).
14. H. de Vries, *Acta Crystallogr.*, **4**, 219 (1951).
15. W. U. Müller and H. Stegemeyer, *Ber. Bunsenges. Phys. Chem.*, **77**, 20 (1973).
16. F. Grandjean, *C. R. Acad. Sci.*, **172**, 71 (1921).
17. R. Cano, *Bull. Soc. Fr. Miner. Crist.*, **91**, 20 (1968).
18. U. Vogt, Dissertation, Institut für Physikalische Chemie der TU Clausthal, Clausthal-Zellerfeld, 1985.
19. M. Siekmeyer, Diplomarbeit, Insitut für Physikalische Chemie der TU Clausthal, Clausthal-Zellerfeld, March 1985.

14

Rheological Behavior of Isotropic and Lyotropic Solutions of Cellulose

P. NAVARD and J. M. HAUDIN
Centre de Mise en Forme des Matériaux, Ecole Nationale Supérieure des Mines de Paris, Sophia Antipolis, Valbonne, France

Amine oxides are not new solvents for cellulose, since their discovery was reported in 1939 (1). It was only around 1970 that the first patents describing cellulose–amine oxide solutions appeared (2, 3). Since then, several studies have been published, mainly dealing with N-methylmorpholine-N-oxide (MMNO) (4–10). Because of its solvent properties, MMNO and its hydrates have also been studied (11–14), and this further clarified the MMNO–cellulose behavior. The N–O group can form two hydrogen bonds with substances like water or polysaccharides (12). The anhydrous MMNO is thus an excellent solvent for cellulose. Its melting point (184°C) (11), however, leads to a large degradation of the cellulose molecule (4). The monohydrate, with a lower melting point (76°C), is a much better candidate, and most of the studies have been conducted in this solvent (4–9). When hydrated by more than two water molecules, the MMNO is no longer a solvent for cellulose, MMNO being preferentially bonded to water. With a water content smaller than two water molecules per MMNO molecule, MMNO–cellulose forms true solutions (7), with a mesomorphic phase at high concentration and low water content (5). With a water content greater than two, the cellulose is only swollen (9). This property can be used to help dissolution (3). Degradation is a serious problem and is prevented by addition of an antioxidant during dissolution (15).

Here we report our results on the rheology of cellulose–MMNO monohydrate in the isotropic and anisotropic phases and on the spinning of a solution. Rheology investigations of cellulose solutions have mainly focused on dilute solutions (16, 17), with a few exceptions using cadoxen (18). Studies with more concentrated solutions face the problem of finding a suitable solvent (19), and no rheological data have been reported on concentrated cellulose solutions. Many studies deal with cellulose derivatives for which a large variety of solvents are available. Nevertheless, a few papers report the formation of lyotropic phases in concentrated cellulose solutions, the first one being with MMNO (5). Ordered structures are formed when there is less than a one molar water/MMNO ratio.

Cellulose–MMNO monohydrate, even at the highest concentration, does not give a mesomorphic phase. The exact reason for this is unknown, but it is probably related to the stiffening of the cellulose molecule by the MMNO due to steric hindrance and the decrease of available chain conformations (20), which may reduce the number of glucose residues in a high-energy flexible form (21). The entity undergoing the isotropic–mesomorphic transition is a complicated, rigid cellulose–MMNO complex. The nature of the lyotropic phase is unknown. Trifluoroacetic acid is a solvent known to give cholesteric phases with certain cellulose derivatives (22–25), and its mixture with chlorinated alkanes also gives cholesteric cellulose solutions (26). Optical rotation was the only experiment performed with this mesomorphic system. The last mesomorphic cellulose solution reported was in N,N-dimethylacetamide + LiCl (27), but only in metastable conditions. The mesophase was also reported to be cholesteric.

MATERIALS AND MEASUREMENTS

MMNO was obtained in hydrated form and was dried under reduced pressure to obtain the monohydrate and anhydrous forms. Two samples of cellulose were used: an alpha cellulose of $\overline{DP}$ 600 and one of $\overline{DP}$ 900. Solutions were prepared in several ways. When cellulose was dissolved in MMNO monohydrate, dissolution took place upon stirring between 80°C and 100°C, depending on concentration. The preparation of mesomorphic solutions required a water content of less than one molecule of water per MMNO, and a mixture of anhydrous and monohydrate MMNO was used. The dissolution temperature was around 130°C. Solutions used for spinning were prepared by first slowly evaporating the excess water from a MMNO–water–cellulose mixture and then extrusion at 120°C. The extruded solution had a very high degree of homogeneity. Less than 1% of propylgallate was added as antioxidant (15) in some cases, and this will be noted for each experiment. Concentrations are given in weight percent.

The capillary rheometer was equipped with four capillaries, 0.6 mm in diameter (lengths 12 and 24mm) and 1.25 mm in diameter (lengths 24.75 and 49.5 mm) and an oven (temperature range: 70–120°C). With a single charge of material, the machine was operated to give eight different plunger speeds from low to high speed. The program was set so that the melt height in the barrel was the same at a given drive speed. The data were corrected for pressure loss in the barrel and at the capillary entrance using the two capillaries of same diameter. The shear rate at the capillary wall was calculated from the speed of the plunger with the Rabinowitsh correction. The same rheometer was used for spinning when equipped with a spinning line comprising a take-up motor and two 1.5-m-long baths filled with a 5% sodium hypochlorite/water solution. The solution was poured into the reservoir of the rheometer and equilibrated at 95°C and was forced through a sand filter and a 0.37-mm capillary. After a 19-cm drawing in air, the filament plunged into the first bath and coagulated, and then into the second cleaning bath. Filament diameter was determined after the first water bath using an optical system. The force for drawing the filament in air was measured using a balance immerged in the first bath. When this balance is not in the first bath, the filament turns 90° in water with the help of glass guides. The addition of sodium hypochlorite has two functions to lubricate the contact of the filament on the glass guides and to help MMNO leave cellulose.

The fibers were then collected and washed for several hours in water. Mechanical properties were measured at 20°C and 65% relative humidity. The data are mean values of ten measurements. The following notations are used: τ, shear stress; $\dot{\gamma}$, shear rate; η, viscosity. In the case of capillary flow: τ_w, shear stress at the capillary wall; $\dot{\gamma}_{wa}$, shear rate at the capillary wall without Rabinowitsh correction; $\dot{\gamma}_w$, shear rate at the capillary wall with Rabinowitsh correction; η_a, apparent viscosity; m, strain rate sensitivity parameter $d(\ln\tau_w)/d(\dot{\gamma}_{wa})$.

ISOTROPIC SOLUTIONS

The flow curves of MMNO monohydrate were determined with a Couette rheometer at different temperatures. The flow is Newtonian up to the transition where Taylor vortices are formed (28). The viscosity η is given in Figure 14.1. The shear flow activation energy E_a is defined as

$$\eta = A\exp\left(\frac{E_a}{RT}\right)$$

where T is the absolute temperature, R the gas constant, and A a constant.

In the case of MMNO monohydrate, E_a is 45 kJ/mole. Solutions of cellulose $\overline{DP}$ 600 in MMNO monohydrate without antioxidant were studied with a cone-and-plate rheometer. Concentrations were between 1% and 6%. At low shear rates the flow curves present a rather peculiar feature, as seen in Figure 14.2. The first run between $\dot{\gamma} = 0$ sec^{-1} and $\dot{\gamma} = 12$ sec^{-1} has a very pronounced hysteresis. After shearing at higher shear rates, this phenomenon disappears, and flow curves are reproducible. For these solutions, a yield has to be overcome in order to have a stable flow behavior. It is thought that some kind of physical network is formed at rest (29), and this is destroyed upon shearing. This network is then formed again after some time. Figure 14.3 shows the flow curves, after shearing at high shear rates, of two cellulose solutions. It is difficult to find a very clear plateau region, and viscosity was arbitrarily measured at $\dot{\gamma} = 10$ sec^{-1}

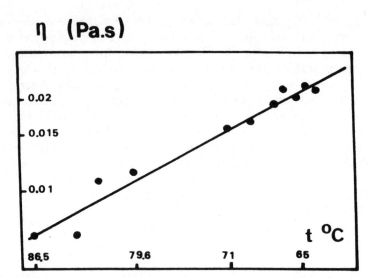

Figure 14.1 Viscosity η versus temperature for N-methylmorpholine-N-oxide monohydrate.

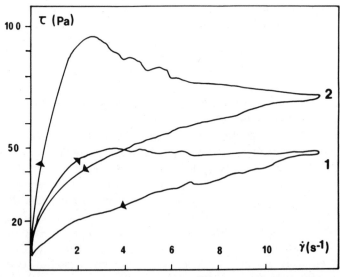

Figure 14.2 Shear stress τ versus shear rate $\dot{\gamma}$ for a $\overline{DP}$ 600–3% solution at 73°C. 1 and 2 refer to the first and second runs. A few minutes at rest are left between the two experiments.

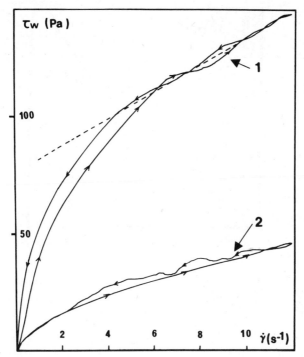

Figure 14.3 Shear stress τ versus shear rate $\dot{\gamma}$ at 73°C for a $\overline{DP}$ 600–6% (curve 1) and $\overline{DP}$ 600–4% (curve 2).

in order to compare different concentrations, as seen in Figure 14.4. Figure 14.5 shows a flow curve up to $\dot{\gamma} = 1200$ sec^{-1}. The flow behavior is pseudoplastic,

$$\tau = K\dot{\gamma}^m$$

where K is a constant.

Two more concentrated solutions were prepared, 10% and 15% $\overline{DP}$ 600–MMNO monohydrate with antioxidant. For the 10%, Figure 14.6 gives the viscosity as measured by cone-and-plate and capillary rheometers. Good agreement is found between these two methods. Viscosity is found to increase, with decreasing shear rate, at shear rates as low as 0.1 sec^{-1}. No plateau region can be seen. Cellulose–MMNO solutions exhibit a pseudo-plastic behavior, m increasing from 0.14 at 70°C to 0.26 at 95°C. Viscosity is much higher, at the same concentration, than what has been previously measured without addition of propylgallate. The influence of degradation is here greatly diminished. The dependence of viscosity on temperature, expressed by the activation energy E_a, is 29 kJ/mole. No yield phenomenon was noticed. The solution is viscoelastic, and by measuring the first difference of normal stresses $\sigma_{xx} - \sigma_{yy}$, it is possible

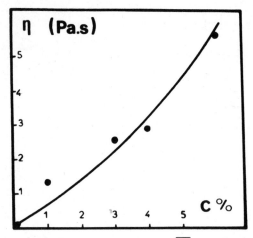

Figure 14.4 Viscosity η versus concentration ($\overline{DP}$ 600 cellulose at 73°C).

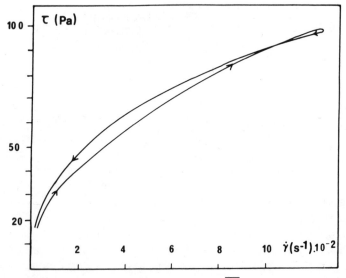

Figure 14.5 Shear stress τ versus shear rate $\dot{\gamma}$ for a $\overline{DP}$ 600–3% solution at 73°C.

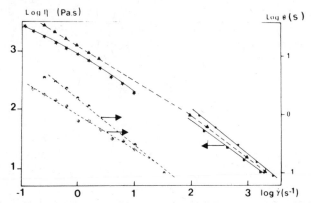

Figure 14.6 Viscosity η versus shear rate $\dot{\gamma}$ for cellulose–MMNO–10%. Data are obtained from cone-and-plate rheometry at low shear rates and from capillary rheometry at high shear rates. ■, 95°C; ★, 102°C; ▲, 85°C; ●, 70°C, Relaxation time θ versus shear rate $\dot{\gamma}$; (◓), 85°C; (⊕), 102°C. (Reprinted from ref. (43) by courtesy of Marcel Dekker, Inc.)

to calculate the viscometric relaxation time θ: $\theta = (\sigma_{xx} - \sigma_{yy})/2\eta\dot{\gamma}^2$ versus shear rate $\dot{\gamma}$ as plotted in Figure 14.6. The same remarks are valid for the 15% solution, and the results are shown in Figure 14.7.

MESOMORPHIC SOLUTIONS

The stability of the mesophase is controlled by the relative amounts of cellulose, water, and MMNO, the $\overline{DP}$ of the cellulose, and the temperature. Measurements of $\overline{DP}$ after several residence times in MMNO monohydrate show that the $\overline{DP}$ drops from 600 to 200 in 10 min at 130°C and in 50 min at 90°C (30). Dissolution took place at 130°C, and a strong degradation occurred without antioxidant.

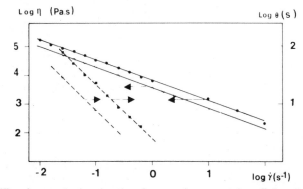

Figure 14.7 Viscosity η and relaxation time θ versus shear rate $\dot{\gamma}$ for cellulose–MMNO–15%, ● and ■, 70°C; ○ and □, 96°C.

Degradation, which is a shortening of cellulose chains, is very quick to a $\overline{DP}$ of around 150 and is slower after that. This could explain why the concentration at which the mesophase appears is insensitive to $\overline{DP}$ above 130, since these determinations were carried out without antioxidant (5). The critical concentration above which a mesophase appears is noted C^*, and the mesomorphic–isotropic transition temperature is T^*. All the experiments in this chapter have been conducted without antioxidant.

A difference between the viscosity of the two phases enabled us to characterize the transition between the anisotropic and isotropic phases. For concentrated MMNO–cellulose solutions, we found such a difference by plotting ln η_a versus inverse absolute temperature (Fig. 14.8). This difference of viscosity between the two phases decreases when the shear rate $\dot{\gamma}_w$ increases. This has usually been observed with lyotropic polymer solutions when the viscosity-versus-concentration curve is plotted at different shear rates. This viscosity difference around T^* cannot be explained by a viscous heat generation in the capillary. Calculations (31) show that the increase in temperature due to the shearing of the solution is not important ($< 0.5°C$) because of the small $\dot{\gamma}_w$ applied to these solutions. Moreover, there is a difference between the shear flow activation energies of the two phases at the transition temperature. The influence of shear rate on the transition temperature T^* could not be detected because of the lack of experimental points (the temperature variation between two experiments was 2°C).

To test the above results, we examined the solutions with an optical microscope under crossed polarizers (Fig. 14.9). Anisotropic phases can be observed because of their characteristic birefringence, and therefore it is possible to determine optically the transition temperature T^*. For example, for a $\overline{DP}$ 600–24% solution, T^* was found to lie between 90 and 92°C by microscopy and between 87 and 92°C by rheometry. So this transition, observed by both optical microscopy and rheometry, is undoubtedly the anisotropic–isotropic transition.

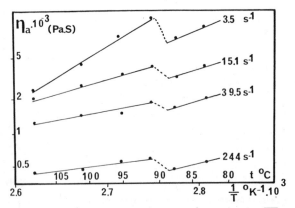

Figure 14.8   Temperature dependence of capillary apparent viscosity η_a of a $\overline{DP}$ 600–24% solution at different shear rates $\dot{\gamma}_w$.

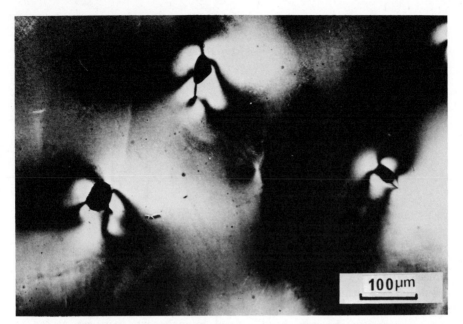

Figure 14.9 Optical micrograph of a $\overline{\text{DP}}$ 600–30% solution at room temperature.

For each solution, flow curves were plotted over a $\dot{\gamma}_w$ range of $1–500$ sec^{-1}. For example, a $\overline{\text{DP}}$ 900–25% solution has a transition temperature T^* around 85°C. Below this temperature (80–82°C) τ_w versus $\dot{\gamma}_w$ is described by a power law equation. Above 85°C (95–96°C), strain rate sensitivity parameter m is no longer independent of $\dot{\gamma}_w$. The trend of the flow curve is similar to those of molten polymers or of concentrated polymer solutions. Beyond a certain $\dot{\gamma}_w$ value, shear stress becomes constant. It may be due to a slipping of the solution at the capillary wall. The shear activation energy is a monotonic decreasing function of shear rate. A similar behavior has been observed with other polymers (32).

Plotting by capillary rheometry the viscosity-versus-concentration curve, we used several dissolutions with different concentrations. The higher the concentration, the longer the time of dissolution. Degradation is expected to be slightly different for the different concentrations, and as a result we could not get enough reproducible data to plot a viscosity-versus-concentration curve. Figure 14.10 shows $\ln \eta_a$ versus T^{-1} for three concentrations of $\overline{\text{DP}}$ 600 cellulose. When the cellulose concentration is 24 and 30%, transition occurs around 90°C. When the cellulose concentration is 20%, no transition is observable. Thus, for cellulose $\overline{\text{DP}}$ 600–MMNO solutions C^* is greater than 20% but less than 24%, with $\dot{\gamma}_w = 15$ sec^{-1}. Non-Newtonian behavior of macromolecular solutions is generally due to chain entanglements (33), assuming that for rigid or semirigid polymers, entanglements are a result of frictional interactions between adjacent

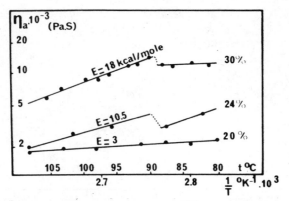

Figure 14.10 Temperature dependence of viscosity of $\overline{\text{DP}}$ 600–MMNO solutions at different concentrations.

molecules (34). In cellulose solutions, density of entanglement increases with cellulose concentration. Therefore, as the activation energy is connected with the motion of chains (35) and so with the density of entanglements, E_a increases with concentration because the density of entanglements increases, as can be seen in Figure 14.10, in the isotropic domain.

m^* is defined as the strain rate sensitivity parameter for $\dot{\gamma}_{wa}$ ranging from 5 to 20 sec^{-1}. It is possible to characterize the anisotropic–isotropic transition by plotting m^* versus temperature. The rapid variation of m^* around T^* is due to the difference of flow curves between the anisotropic and isotropic phases. The variation of m^* with temperature is much more important for the $\overline{\text{DP}}$ 900–25% solution than for $\overline{\text{DP}}$ 600–30%. This may be explained either by a different degradation between the two solutions during dissolution or by an influence of concentration.

Plots of extrudate diameters versus $\dot{\gamma}_w$ for different temperatures are shown in Figure 14.11. The $\overline{\text{DP}}$ 900–25% solution has a transition temperature around 85°C. Above 85°C, in the isotropic region, the extrudate diameter D increases

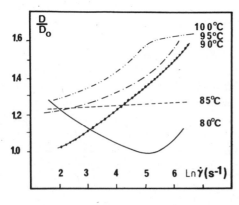

Figure 14.11 D/D_o ratio versus shear rate $\dot{\gamma}_w$. D, extrudate diameter; D_o, capillary diameter.

with $\dot{\gamma}_w$ $(1.2 < D/D_o < 1.6$, where D_o is the capillary diameter). This is indicative of a pronounced viscoelastic behavior. On the other hand, below 85°C, in the mesomorphic region, the influence of stress relaxations at the capillary exit is not important, and the extrudate diameters are lower than 1.2 D_o.

Two types of flow instabilities were observed. First, regular oscillations of the entrance pressure, as much above as below the transition temperature, occurred at shear rate lower than 1 sec^{-1}. Periods are between 30 and 200 s. These self-sustained relaxation oscillations may be due to the rheometer, where the barrel and the capillary may form a linear oscillator. These oscillations were characterized by an entrance pressure that grew slowly, then quickly decreased while a plug was ejected from the capillary. As for HDPE (36), the oscillation period is in inverse ratio to the speed of the plunger and hence to the entrance flow. Second, irregular flow instabilities were observed in the isotropic region for higher shear rate $(\dot{\gamma}_w > 100$ sec$^{-1})$. Similar instabilities have been observed with molten polymers and with polymer solutions. Several explanations may be put forward; formation of inhomogenieties (37) or temporary networks (flexible polymers) (38), breakdown of an entanglement structure at the entrance region of the capillary (39), or formation of aggregates (40). Problems of dissolution of cellulose in MMNO, leading to inhomogenieties and unmolten cellulose fibers, may be responsible for these instabilities, which increase when shear rate increases. In the anisotropic region, flow instabilities are observed whatever the shear rate.

SPINNING

Single filament spinning was performed at 95°C. Low viscosity caused trouble in the spinning line, whereas high viscosity required a large force to push the solution through the sand filter. The speed of the filament at the capillary exit is V_0, and the take-up speed is V_1. As the filament is assumed to crystallize and solidify in the first millimeters of its path in water, the take-up speed was set equal to its water entrance speed. Thus, the draw ratio D_r is equal to V_1/V_0.

With the conditions described above, the spinning stability was investigated. As usually found, the filament exhibits periodical diameter oscillations above a critical draw ratio D_r^*. D_r^* was found to be 55 and independent of V_0 for V_0 ranging from 0.14 to 3.3 m/min. The stability of spinning is greatly increased when heat transfer in the air is important and when E_a/R increases. In our case, E_a/R is in the normal range for polymers, and no measurement was performed for estimating the heat transfer between the filament and the air. There must be a rather low heat transfer, since spinning exhibits a critical draw ratio despite the rather large air gap between capillary and water. Regular and irregular amplitude oscillations were recorded. PET (41) exhibits a similar spinning behavior. The diameter variation, expressed as the difference between the maximum diameter D_{max} and the minimum D_{min} over the mean diameter $\overline{D}$ is plotted in Figure 14.12 as a function of D_r. The period T of these oscillations decreases as D_r increases, contrary to PET findings and theoretical predictions

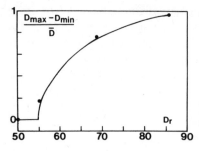

Figure 14.12. Fiber diameter versus draw ratio D_r. D_{max}: maximum diameter, D_{min}: minimum diameter, $\bar{D}$: mean diameter. (Reprinted from ref. (43) by courtesy of Marcel Dekker, Inc)

(41). Figure 14.13, a plot of TV_0/L (L is the air gap (19 cm)) versus D_r, illustrates this behavior. The Deborah number for spinning defined as $De = \theta V_0/L$ is approximatively 1.5×10^{-3}. Following Denn and Marucci (42), this would give a critical draw ratio of 20 for isothermal spinning. Obviously this requirement is far from being fulfilled, since D_r^* is 55. The athermal behavior can also be seen when the force on the filament is measured. For the isothermal spinning of a newtonian fluid, F is given by

$$F = \pi \eta_L R_c^2 \frac{V_0}{L} \ln D_r$$

where η_L is the elongation viscosity ($\eta_L = 3\eta$ for a newtonian fluid) and R_c is the capillary radius.

F was found to be about ten times greater than the calculated value owing to cooling of the filament in the air (F was found to be around 0.2 gf with $D_r = 10$). Cellulose–MMNO solutions could provide a convenient system for studying isothermal spinning, since the temperatures are low and the water can be easily maintained at the spinning temperature. By varying the spinning conditions and measuring the filament diameter and the take-up force together with the shear rheology, the elongational behavior of cellulose solutions could be more fully understood.

The filament becomes a soft, elastic swollen cellulose fiber when going through

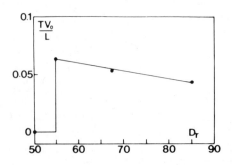

Figure 14.13 (TV_0/L versus draw ratio D_r: period of oscillation; V_0: speed of filament at the capillary exit (0.66 m/min); L: air gap (19 cm). (Reprinted from ref (43) by courtesy of Marcel Dekker, Inc).

TABLE 14.1 Mechanical Properties of Cellulose Fibers

V_o (m/min)	D_r	Dimension (deciTex)	Modulus (N/Tex[a])	Elongation (%)	Tenacity (CN/Tex)
6.6	18	12.5	13.7	6.5	20.6
3.3	18	12.0	14.1	7	20.6
1.3	18	9.3	13.7	7.7	21.6
1.3	31	5.2	12.6	5.1	16.7
1.3	46	6.6	9.8	4.2	12.8
1.3	61	3.8	12.7	5.5	20.6
1.3	92	2.8	12.6	8.6	30.4

Source: From ref. 43 by courtesy of Marcel Dekker, Inc.

[a] For cellulose, 1 N/Tex = 1.5 GPa. V_o, exit speed of filament; D_r, draw ratio.

water, and it is then squeezed between other filaments during take-up. The shape during this stage is retained when drying and is not cylindrical. All the fibers exhibit the same shape and the same skin undulations, which could result from surface irregularities on the glass guide. The dimensions and mechanical properties of the experimental fibers are summarized in Table 14.1. The mechanical properties compare well with those of rayon, even though our processing conditions (concentration, degree of polymerization, temperature, and all the spinning parameters) have yet to be optimized. V_0 has no influence on the mechanical properties of the cellulose fibers, as can be seen from the upper part of Table 14.1. There are only a small decrease of the diameter and a small increase of elongation at break as V_0 decreases. The properties follow a more complex behavior as a function of the draw ratio, as seen in the lower part of Table 14.1. The critical draw ratio seems to be the boundary between a decreasing property zone as D_r increases below D_r^* and an increasing property zone above D_r^*. Diameter irregularities do not appear to affect the mechanical properties.

SUMMARY

MMNO and amine oxides in general are very interesting solvents for cellulose. They lead to new ways of manufacturing this natural polymer. At a time when old processes are being shut down and demand for cottonlike fibers is and will be growing, such new methods are important. As shown in this study, spinning is relatively easy and produces good fibers without special care. Several additional points have now to be studied. From a fundamental point of view, the rheological behavior at low concentration and the nature of the mesophase are important. From a technological point of view, the spinning of mesomorphic cellulose solutions remains to be done, and this is probably not a trivial problem owing to the high viscosity of these solutions.

REFERENCES

1. C. Graenacher and R. Sallman, *U.S. Pat.* 2, 179, 181 (1939).

2. D. L. Johnson, *U.S. Pat.* 3,447,939 (1969); *U.S. Pat.* 3,508,941 (1970).

3. C. C. McCorsley and J. K. Varga, *Belgian Pats.* 868,735, 868,736, and 868,737 (1978).

4. H. Chanzy, M. Dube, and R. H. Marchessault, *J. Polym. Sci. Polym. Lett. Ed.*, **17**, 219 (1978).

5. H. Chanzy, A. Peguy, S. Chaunis, and P. Monzie, *J. Polym. Sci. Polym. Phys. Ed.*, **18**, 1137 (1980).

6. P. Navard and J. M. Haudin, *Br. Polym. J.*, **12**, 174 (1980).

7. D. Gagnaire, D. Mancier, and M. Vincendon, *J. Polym. Sci. Polym. Chem. Ed.*, **18**, 13 (1980).

8. H. Chanzy, S. Nawrot, A. Peguy, P. Smith, and J. Chevalier, *J. Polym. Sci. Polym. Phys. Ed.*, **20**, 1909 (1982).

9. H. Chanzy, P. Noe, M. Paillet, and P. Smith, *J. Appl. Polym. Sci. Appl. Polym. Symp.*, **37**, 239 (1983).

10. M. Dube, Y. Deslandes, and R. H. Marchessault, *J. Polym. Sci. Polym. Lett. Ed.*, **22**, 163 (1984).

11. P. Navard and J. M. Haudin, *J. Therm. Anal.*, **22**, 107 (1981).

12. E. Maia, A. Peguy, and S. Perez, *Acta Crystallogr.*, **B37**, 1858 (1981).

13. E. Maia and S. Perez, *Acta Crystallogr.*, **B38**, 849 (1982).

14. H. Chanzy, E. Maia, and S. Perez, *Acta Crystallogr.*, **B38**, 852 (1982).

15. A. Brandner and H. G. Zengel, *European Pat.* 0047 929-A2 (1981).

16. N. Gralen and T. Svedberg, *Nature*, **152**, 625 (1943).

17. A. V. Kuz'mina and L. A. Yudakhina, *Izv. Akad. Nauk. Kirg. SSR*, **5**, 57 (1981).

18. E. Riande and J. M. Perena, *Makromol. Chem.*, **175**, 2923 (1974).

19. S. P. Papkov, A. G. Kudryavtseva, S. I. Banduryan, and M. M. Iqvleva, *Khim. Volokna*, **2**, 33 (1978).

20. S. Perez, *Carbohydr. Polym.*, **2**, 303 (1982).

21. D. A. Brandt, *Carbohydr. Polym.*, **2**, 232 (1982).

22. S. M. Aharoni, *Polym. Prepr.*, **22**(1), 116 (1981).

23. M. Panar and O. B. Willcox, *French Pat.* 2,340,344 (1977).

24. G. H. Meeten and P. Navard, *Polymer*, **23**, 483 (1982).

25. G. H. Meeten and P. Navard, *Polymer*, **23**, 1727 (1982).

26. D. L. Patel and R. D. Gilbert, *J. Polym. Sci. Polym. Phys. Ed.*, **19**, 1231 (1981).

27. G. Conio, P. Corazza, E. Bianchi, A. Tealdi, and A. Ciferri, *J. Polym. Sci. Polym. Lett. Ed.*, **22**, 273 (1984).

28. G. I. Taylor, *Phil. R. Soc. Lond.*, **223**, 289 (1923).

29. J. Shurz, *Cell. Chem. Technol.*, **11**, 3 (1977).

30. D. Loubinoux and S. Chaunis, paper presented at the International Man Made Fibers Congress, Dornbirn, Austria, Sept. 26–28, 1984.

31. S. Middleman, *The Flow of High Polymers*, Wiley, New York 1968, p. 32.

32. E. C. Berhardt, *Processing of Thermoplastic Materials*, Rheinhold, New York, 1969, p. 40.

33. W. W. Graessley, *Adv. Polym. Sci.*, **16**, 1 (1974).

34. D. G. Baird, in *Liquid Crystalline Order in Polymers*, A. Blumstein, ed., Academic, New York, 1978, p. 237.

35. S. Glasstone, K. Laideler, and H. Eyring, *The Theory of Rate Processes*, McGraw-Hill, New York, 1941.

36. A. Weill, *Rheol. Acta*, **19**, 623 (1980).

37. A. S. Lodge, *Polymer*, **2**, 165 (1961).

38. A. Peterlin and D. T. Turner, *J. Polym. Sci. Polym. Lett. Ed*, **3**, 517 (1965).

39. P. D. Criswold and J. A. Cuculo, *J. Polym. Sci. Polym. Phys. Ed*, **15**, 1291 (1977).

40. A. Peterlin, C. Quan, and D. T. Turner, *J. Polym. Sci. Polym. Lett. Ed.*, **3**, 521 (1965).

41. Y. Demay, Thèse Doctorat d'Etat, Université de Nice, 1983.

42. M. M. Denn and G. Marucci, AIChE J., **17**, 101 (1971).

43. P. Navard and J. M. Haudin. *Polym. Process Eng.*, **3**, 291 (1985).

PART 4

Cellulose Hydrolysis and Degradation

15

Comparative Effectiveness of Various Acids for Hydrolysis of Cellulosics

MORRIS WAYMAN
Department of Chemical Engineering and Applied Chemistry,
University of Toronto, Toronto, Ontario, Canada

The acid hydrolysis of cellulose and of cellulosics such as wood for the production of fermentable sugars has long been studied in the laboratory. Industrial facilities for wood hydrolysis have been operating for the past 40 years. Even earlier, under the stimulus of wartime conditions, European and North American facilities were built for acid hydrolysis of wood. The sugars produced were fermented to ethanol, which is an acceptable motor fuel. With the rapid expansion of the petrochemical industry in the years following about 1935, most such fermentation facilities became uneconomical, and most ethanol production switched to petroleum-derived feedstock. Some fermentation ethanol continued to be produced with molasses as the substrate. The rapid increase in the price of oil in the past 10 years has altered that position substantially, and now once again large amounts of ethanol are made by fermentation. This reversion to fermentation as the preferred process has created a need for a cheap and widely available source of fermentable sugars. Cellulosics such as wood and agricultural crop residues appear to meet these requirements.

Brazil's numerous fermentation fuel alcohol plants are fed by the by-products of their large cane sugar industry. Their experience demonstrates that fermentation ethanol is a perfectly satisfactory motor fuel at 95% alcohol 5% water. At least 500,000 Brazilian automobiles operate on undried alcohol continuously, and most of the rest of their fleet operate on this fuel some of the time—for example, on weekends, when only alcohol is available at the gas station. Brazil also sells a "gasohol" with 20% anhydrous ethanol in gasoline, and this is considered a transition until their fermentation capacity can be increased to meet the full demand.

In the United States there were, at the end of 1982, 65 fermentation alcohol plants in operation, mostly based on corn. Of these, half made 4×10^6 L/yr or more, and five each made 200×10^6 L/yr or more. The economics of ethanol production is highly sensitive to scale (1), and these very large plants make ethanol by fermentation more cheaply than it can be made from petrochemicals. The total output of these U.S. plants is about 2×10^9 L/yr—a large volume but even so, less than 1% of U.S. motor fuel demand.

All of these fermentation plants are based on grain or sugar, which are human food or animal feed. There is widespread intuitive concern about driving automobiles on food- or feed-based fuel. This conflict is exacerbated by the need for relatively good land for these crops, resulting in competition for land. Inevitably, it is feared, such competition will raise food costs, with harmful social consequences. As a result, much attention has been devoted to the much more difficult problem of converting cellulosics—wood and agricultural crop residues—to power alcohols (2).

In Canada, waste wood and waste crop residual—surplus to needs of soils and animals—are in relatively concentrated locations so that they can be delivered to a conversion facility at acceptable costs (3, 4). There is enough to make 6×10^9 L/yr of fermentation ethanol—that is, about 20% of current Canadian gasoline consumption. In addition, there is available a great deal of abandoned farmland and other poor land on which to plant fast-growing wood hybrids and other energy crops, enough to supply all of the rest of the motor fuel Canada requires. It would require only 5×10^6 ha of relatively poor land to provide the full amount of cellulosic raw material for the full amount of motor fuel.

In the United States, logging and wood processing residue are estimated (5) at 53×10^6 tonnes/yr, having a potential ethanol yield of 7×10^9 L/yr. To supply the rest of the motor fuel for the United States, by planting fast-growing wood hybrids and other energy crops, would require $20-25 \times 10^6$ ha of relatively poor land. These calculations suggest that the provision of motor fuel in North America entirely from replenishable resources is technically feasible.

Of the many factors that delay implementation of such a fuel supply program, failure to agree on a satisfactory technology is significant. All commercial plants for producing fermentable sugars from wood have been based on hydrolysis by dilute sulfuric acid in percolation towers. There are about 40 such plants in operation in the Soviet Union. Such processes have poor yields, many operating problems (6), and unsolved metallurgy, and they give rise to humic substances which inhibit fermentation (7). Furthermore, the economics of such plants is questionable (8).

A hopeful alternative appeared with the introduction of autohydrolysis as a pretreatment for cellulose (9). Autohydrolysis—that is, steam cracking followed by rapid quenching—is highly effective for hydrolysis of hemicelluloses of deciduous woods such as poplar and of agricultural crop residues. In addition, much of the lignin becomes soluble in dilute alkali, or in organic solvents, and extraction of the lignin leaves a relatively pure cellulose for hydrolysis or for other uses. The autohydrolysis–extraction–hydrolysis process (10) results in good yields of fermentable sugars and ethanol (11). It is schematized in Figure 15.1.

Until now this process has not proved useful for coniferous woods such as pine or spruce, which are the most plentiful residues of the forest industries. In an effort to improve the process by extending its raw material base and to find a satisfactory alternative to sulfuric acid as hydrolysis catalyst, we have studied catalyzed autohydrolysis, which should more properly be called prehydrolysis, and have compared several acids as hydrolysis catalysts for production of fermentable sugars.

BACKGROUND

Cellulosics pose special problems as a feedstock for hydrolysis because of their physical and chemical properties. Wood and agricultural crop residues are bulky solids, which for reasonably rapid and uniform chemical treatment require subdivision into relatively fine particles. Physical methods of subdivision such as grinding are expensive, expecially for wood. One of the virtues of the "steam explosion" processes is that wood chips can be converted to finely divided fibrous products suitable for further processing as fluid slurries while at the same time effecting useful chemical changes. Chemically, cellulose is homogeneous, but biomass has several components, namely, cellulose, hemicellulose, lignin, and extractives. Hemicelluloses, like cellulose, are polymeric carbohydrates (12), but cellulose is a highly crystalline polymer of glucose exclusively, and hemicelluloses are amorphous polymers made up of many monomers including D-glucose, D-xylose, D-mannose, L-arabinose and D-galactose. In addition, hemicelluloses also contain acetyl and uronic acid groups.

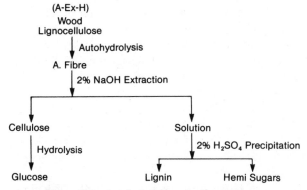

Figure 15.1 Autohydrolysis-extraction process.

The proportions of these sugar residues vary widely in the hemicelluloses of different plant species, with coniferous woods having a higher proportion of glucomannans and deciduous woods more xylans and arabinoxylans. This is an important distinction, since mannose is readily fermentable by ordinary yeasts, whereas the pentoses, xylose and arabinose, are fermentable by only a few unusual yeasts. The carbohydrate composition of hemicelluloses of straws, bagasse, and other agricultural crop residues is more complex, but in general they tend to be very high in pentosans.

In plants, the hemicelluloses are combined with lignin although by relatively weak bonds. Lignin is an amorphous phenolic polymer made up of three monomers—coniferyl alcohol, sinapyl alcohol, and p-coumaryl alcohol (13). The proportions of these monomeric residues vary with species. The lignin of conifers is richer in residues derived from coniferyl alcohol, whereas that from deciduous species has relatively more sinapyl-derived residues. In consequence of this difference, the lignin in conifers is more condensed and less tractable to chemical attack than the lignin in hardwoods. Grasses and agricultural crop residues have relatively little lignin, and they are relatively high in p-coumaryl derivatives.

Extractives may vary in quantity from 1 to 7%—a few species have more—and these are highly species-specific. Generally, they include fats and fatty acids, resin acids such as abietic acid, sterols ("phytosterols"), and a great number of other aliphatic and aromatic organic chemicals. These may affect processing conditions (14).

In addition to these compounds, pectins may be present in crop residues, and, finally, all cellulosics contain inorganic ash, which may be less than 0.5% in wood but may be a few percent of silica in some crop residues.

Thus cellulosics are highly heterogeneous both physically and chemically. The rates of reaction of the various components with hydrolyzing agents are quite different. In particular, hemicelluloses are hydrolyzed by dilute acids at a much more rapid rate and at much lower temperature than cellulose. The use of high temperature for dilute acid hydrolysis of cellulose has resulted in decomposition of the sugars formed in hydrolysis, and the resultant poor yields have led to a continuing search for better processes for hydrolysis of cellulosics.

Among the first cellulose hydrolyzing agents are the three strong acids—72% H_2SO_4, 41% HCl, and 85% H_3PO_4. The action of all three acids is to dissolve the carbohydrates completely at low temperatures (4–20°C). With biomass, the residue is mainly lignin (Klason lignin in the case of H_2SO_4; Willstatter lignin with HCl). The dissolved carbohydrates are mainly oligomers which may be hydrolyzed almost completely to monomers by careful dilution to 3–6% acid and heating at 100–120°C for 30–360 min. These fundamental observations have been used as a basis for industrial installations, using fortified HCl and concentrated H_2SO_4. The strong acid processes that reached an industrial level have been described by Wenzl (15). A more recent survey by Kosaric et al. (5) is based on Ladisch (16) and brings the literature on strong acid hydrolysis to about 1979.

Since then, interest in HCl hydrolysis has continued, particularly as a result of a project undertaken by Battelle-Geneva (17). In the United States, Goldstein at North Carolina State University and Gaddy at the University of Arkansas have continued studies of HCl (18). In addition, considerable research effort has been directed to HF (19). Of these various processes, those based on 41% HCl have industrial potential, but the cost of recovery of the rather expensive acid is a deterrent to large scale implementation (see also Chap. 17).

All of today's industrial installations and many pilot plants use dilute sulfuric acid as the hydrolysis catalyst. The existing industrial facilities are nearly all in the Soviet Union, with one in Brazil and another in New Zealand, both in the start-up phase. Russian technology has been described by Tokarev (20). The hydrolysis reactors in all of these plants are percolation towers in which batches of wood chips are exposed to dilute (about 0.5–1%) H_2SO_4 at high temperature (to about 190°C) for a long time (2–4 hr). Sugar solutions are withdrawn from the bottom of the towers.

The Brazilian plant employs the latest Russian titanium-based metallurgy and is generally based on their technology but modified to produce both fermentation alcohol and metallurgical coke. The New Zealand technology is a version of the New Zealand Forest Research Institute pilot plant (6, 7), probably modified along the lines suggested by Wayman (8). The Brazilian plant will use eucalyptus as the feedstock, and the New Zealand plant will use pine sawmill residues. Of the several pilot plants in operation, those at New York University (21) and at the Toronto pilot plant of Biohol use extruders as the reactors. The latter has a capacity of about 1 ton of wood feed per day, including processed crop residues. The sugars produced are fermented to alcohol by *Zymomonas mobilis* (22). The extruders have very high capacity considering their small size and have not experienced the serious metallurgical problems that have characterized the percolation tower installations. However, they have not been operated continuously over a long enough time for a definitive statement on corrosion. The other problems associated with dilute sulfuric acid, namely low yields (about 50% of theory) and the formation of fermentation inhibitors, remain.

Grethlein (23) has proposed a different approach to dilute H_2SO_4 hydrolysis of cellulosics by the use of a continuous plug flow reactor at very high temperatures, up to 280°C, for very short times. Such a reactor would take advantage of the kinetics described by Saeman (24), who demonstrated that the rate of cellulose hydrolysis increases faster with increasing temperature than does the rate of glucose destruction in the acid medium. At the high temperature used by Grethlein, yields of 70–80% were obtained. A pilot plant scale reactor capable of operation at high temperatures for short times is under construction (25).

Because of the heterogeneous nature of cellulosic biomass, laboratory and pilot plant studies have generally adopted a two-stage approach to hydrolysis (5, 10, 16). The first stage, prehydrolysis or autohydrolysis depending on conditions, is designed to hydrolyze hemicelluloses, with the resulting sugars removed before the second hydrolysis stage (cf. Chap. 16). In the industrial percolation towers a similar effect is achieved by beginning the hydrolysis reaction at a relatively lower temperature, 150°C, and gradually raising the temperature to 190°C. The sugars are removed as formed, from the bottom of the reactor. We have generally adopted a two-stage approach. Because of the limitations of our equipment, we have generally carried out the hydrolysis in several steps. The sugars are removed as formed to give good yields (11).

To overcome the problems encountered in industrial practice of dilute acid hydrolysis of wood, paticularly low yields, corrosion, and the formation of fermentation inhibitors, we have compared several acids as prehydrolysis and hydrolysis catalysts. Another object of our work has been to broaden the range of feedstocks that can be treated by autohydrolysis, including conifers.

AUTOHYDROLYSIS OR PREHYDROLYSIS

The raw materials used in this research were red pine residues and aspen chips. Their analyses are given in Table 15.1. Samples of these woods were heated in small autoclaves in the presence of a small amount of water, with a liquid/solid ratio of 4:1 for autohydrolysis and with very dilute SO_2 solutions for prehydrolysis. The time at temperature for autohydrolysis was 15 min at 190°C and for prehydrolysis was 20 min at 150°C. The results are shown for pine in Table 15.2 and for aspen in Table 15.3.

In pine autohydrolysis, about 59% of the hemicellulose nonglucose sugars were hydrolyzed to soluble carbohydrates, and of these about 22% occurred as oligomers rather than monomers. In the presence of a small amount of SO_2, hemicellulose was hydrolyzed completely with a somewhat smaller amount, 17%, as soluble oligomers. In this prehydrolysis, the fibrous residue was less than on autohydrolysis, 69% compared to 74%. The alkali-soluble lignin decreased from about 9% to about 5%.

Aspen hemicelluloses are much more readily hydrolyzable than those of pine under conditions of autohydrolysis, producing 88.4% of the theoretical yield of nonglucose sugars. In the presence of SO_2 this was raised to the full theoretical level. There is a large difference in the completeness of hemicellulose hydrolysis.

TABLE 15.1 Summative Analysis

	% of Dry Weight	
	Red Pine Sawmill Residues	Aspen Poplar Chips
Extractives	5.0	5.4
Polysaccharides	68.0	67.7
(of which glucan)	(50.7)	(42.2)
Lignin:		
Klason	25.8	21.5
Acid-soluble	1.2	5.4
Total	100.0	100.0

TABLE 15.2 Autohydrolysis or Prehydrolysis of Pine

	Glucose %	TRS %	Fiber Residue	Alkali Extract
	Autohydrolysis (Steam), 190°C, 15 min			
Monomer	1.52	11.78	74.33	9.22
Oligomer	2.32	3.32		
Total	3.84	15.10		
% of Theory				
Glucose	6.9			
Nonglucose	59.2			
	Prehydrolysis, 2% SO₂/wood, 150°C, 20 min			
Monomer	5.29	20.55	69.06	4.91
Oligomer	—	4.32		
Total	5.29	24.87		
% of Theory				
Glucose	9.5			
Nonglucose	102.9			

In autohydrolysis, 41.7% of the soluble sugars are oligomers, whereas in the presence of SO_2 catalyst, only 5.3% are oligomers. While the fiber residue is similar in the two procedures (69.1% and 67.1%), the alkali-soluble lignin is much greater in autohydrolysis than in prehydrolysis—14.4% compared to 7.9%. The reduced solubility of the lignin is due to acid-catalyzed polymerization (26).

One can conclude that the use of a small amount of SO_2 as catalyst in

TABLE 15.3 Autohydrolysis or Prehydrolysis of Aspen

	Glucose %	TRS %	Fiber Residue	Alkali Extract
	Autohydrolysis (Steam), 190°C, 15 min			
Monomer	1.41	15.36	69.1	14.40
Oligomer	1.47	11.00		
Total	2.88	26.36		
% of Theory				
Glucose	6.2			
Nonglucose	88.4			
	Prehydrolysis, 3% SO₂/wood, 150°C, 20 min			
Monomer	2.56	29.07	67.1	7.86
Oligomer	0.81	1.62		
Total	3.37	30.69		
% of Theory				
Glucose	7.8			
Nonglucose	102.0			

pretreatment of these woods permits the use of lower temperatures and results in complete hydrolysis of the hemicellulose. The solubility of the lignin in alkali is decreased in the presence of SO_2 under these conditions, presumably because of acid-catalyzed polymerization.

HYDROLYSIS OF PINE WOOD

The above results led us to adopt the general procedure of a first-stage acid-catalyzed prehydrolysis, followed by an alkaline extraction and a second-stage acid hydrolysis, with the same acid used in both prehydrolysis and hydrolysis.

The results of SO_2 acid-catalyzed hydrolysis are reported in Table 15.4. Very high yields were obtained, 84.2% of theory for glucose and 92.2% of theory for total reducing sugars, using 2% SO_2 on fiber with a liquid/solid ratio of 4. The conditions were 0.5% SO_2 solution, at 150°C for 20 min at temperature in prehydrolysis and 1% SO_2 on residual fiber for a total time-temperature equivalent of 24 min at 190°C for hydrolysis. The residue, mainly lignin, was 34.1% of the starting material. The higher total reducing sugar yield is, of course, a consequence of complete hydrolysis of hemicellulose in prehydrolysis. Of the total sugars, 12% occurred as soluble dimers and oligomers, and the remaining 88% was monomeric sugars. It is believed that these are the highest yields reported for dilute acid hydrolysis of cellulose.

In Table 15.5 are given the corresponding results of sulfuric acid hydrolysis. The yields are also good, 86.5% of the glucose in pine as sugars, although almost one-third of these are dimers and oligomers (cf. 11). The yield of total reducing sugars, 83.1%, is high, although significantly lower than that obtained with SO_2, 92.2%. The ligneous residue was almost the same, 34.3% compared to 34.1%.

TABLE 15.4 SO_2 Hydrolysis of Pine Sawmill Residues

	Glucose		Total Reducing Sugars	
	Monomer	Oligomer	Monomer	Oligomer
Prehydrolysis	5.53	1.37	25.28	4.56
Hydrolysis	34.60	5.50	35.31	3.85
	40.13 +	6.87	60.59 +	8.41
Total		= 47.00		= 69.00
% of Theory	71.9 +	12.3	81.0 +	11.2
Total		= 84.2		= 92.2

Prehydrolysis: 2.00% SO_2 on fiber: liquid/fiber = 4, SO_2 solution = 0.5%; 150°C, 20 min at temperature.

Hydrolysis: 1.0% SO_2 on residual fiber, 190°C, 4 × 4 min at temperature, total time–temperature equivalent 24 min at 190°C.

Ligneous residue: 34.09%.

TABLE 15.5 H_2SO_4 Hydrolysis of Pine Sawmill Residues

	Glucose		Total Reducing Sugars	
	Monomer	Oligomer	Monomer	Oligomer
Prehydrolysis	4.11	3.38	23.74	1.75
Hydrolysis	29.19	11.56	36.08	0.58
	33.30 +	14.94	59.82 +	2.33
Total	=	48.24	=	62.15
% of Theory	59.7 +	26.8	81.0 +	3.1
Total	=	86.5	=	83.1

Prehydrolysis: 3.00% SO_2 on fiber: liquid/fiber = 4, SO_2 solution = 0.75%; 150°C, 20 min at temperature.

Hydrolysis: 1.5% SO_2 on residual fiber, 190°C, 4 × 4 min at temperature, total time–temperature equivalent 24 min at 190°C.

Ligneous residue: 34.33%.

In Table 15.6 the results of hydrolysis with these and other acids—nitric, hydrochloric, and formic—are presented. It is clear that SO_2 is the most effective catalyst. The results with HCl and formic acid were quite disappointing, since HCl has been proved to be a highly effective hydrolysis catalyst for starch (27) and the autohydrolysis effect has been attributed to formic acid. Formic and acetic acids are split from the hemicellulose in the wood during steaming.

While different optimal concentrations are used in the experiments reported in Table 15.6, the ratio of [H$^+$] to carbohydrate is remarkably constant. In one series of experiments, an attempt was made to augment the effectiveness of SO_2 by adding just enough formaldehyde to form the addition compound, which has a higher dissociation constant than H_2SO_3. However, the effects observed were the same as those found for SO_2 alone.

There are important benefits of SO_2 as a hydrolysis catalyst, particularly as

TABLE 15.6 Hydrolysis of Pine Sawmill Residue: Comparison of Acids

	% on Fiber		Yield: % of Theory			
			Glucose		Total Reducing Sugars	
	Prehydrolysis	Hydrolysis	Monomer	Total	Monomer	Total
SO_2	2	1	71.9	84.2	81.0	92.2
H_2SO_4	3	1.5	59.7	86.5	80.0	83.1
HNO_3	2	1	53.2	72.4	68.8	77.4
HCl	1.3	0.65	47.0	57.6	70.5	66.7
Formic	3	—	0.8	7.0	6.5	14.8

compared to sulfuric acid, which has been the industrial standard to date. Since SO_2 is a gas and can be added either ahead of or with the steam, it distributes better through the wood substance than H_2SO_4, to give a more uniform reaction. Some of the difficulties with H_2SO_4 are attributable to uneven distribution in the wood. In addition, SO_2 costs less than H_2SO_4, less is used, and the cost of neutralization following hydrolysis is reduced.

There is much experience in the pulp and paper industry in generation and use of SO_2 in pressure vessels. Therefore we can expect that stainless steels are available that are resistant to attack by SO_2, although metallurgy associated with H_2SO_4 at these temperatures remains a serious problem. While it is apparent that the lignin is somewhat polymerized during prehydrolysis, thereby reducing the solubility in dilute NaOH, it does not display the humic characteristics of H_2SO_4-treated lignin. Lignosulfonates made by the reaction of SO_2 with lignin are the basis of a substantial chemical industry such as vanillin production. Preliminary work suggests that these lignins will be quite suitable for such chemical conversion. The amout of SO_2 used, a total of 3% on fiber, is much less than the amount used in sulfite pulping of wood, which is about ten times as much.

The use of SO_2 as a hydrolysis catalyst for cellulosics has attracted attention for quite some time. Although there is little on this subject in the ordinary technical literature, the patent literature has many examples, one of the earliest is by Ewen and Tomlinson (28). These men built a commercial plant in Georgetown, North Carolina, during World War I. Saeman (29) comments that this plant, and another like it at Fullerton, Louisiana,

> produced alcohol profitable into the 20's. Both plants benefited from very low cost sawdust and a lack of environmental controls. The operators coped successfully with corrosion problems presented by dilute *sulphuric acid* at over 170°C, and with fermentation of impure and variable substrates. The hydrolysis was carried out by adding a small amount of acid to green wood and heating with steam. Essentially these same conditions could be used today in a batchwise or continuous configuration for prehydrolysis or for pretreatment of lignocellulose.

Saeman's statement about sulfuric acid raises some questions, since the Ewen and Tomlinson patent described only the use of gaseous SO_2, "an amount of SO_2, say, equal to about three percent by weight of the (dry weight of the) sawdust." Later Tomlinson patents (30) mention dilute sulfuric acid or sulfurous acids. There is no suggestion in these patents that sulfuric acid is superior to sulfurous acid or SO_2 gas.

TWO-STAGE SO_2 HYDROLYSIS OF OTHER CELLULOSICS

The procedure adopted for hydrolysis of cellulosics was SO_2 prehydrolysis for 20 min at 150°C in 0.5 or 0.75% SO_2 solution (2–3% on fiber). The hemicellulose sugars were then washed from the fibrous residue with water, and the fibrous residue was hydrolyzed with 1% SO_2 on fiber (0.25% solution) at 190°C, using the procedure described in reference 11.

The results are shown in Table 15.7. As expected, hemicellulose hydrolysis was essentially complete in all cases. However, the amount of sugar recovered from the residual cellulose varied widely among these five species of biomass. The procedure gave good results for pine, 84.3% of the theoretical yield of glucose. All other glucose yields were less than 70%, varying from 68.7% for aspen to only 54.5% for wheat staw. Since in all cases the amount of fibrous residue at the end of the process was about what would be expected from the lignin content, the difference is due not to differences in rate of hydrolysis but rather to differences in the rates of destruction of the glucose as formed; or perhaps a better formulation might be in our failure to preserve the glucose as formed.

FUEL ALCOHOL YIELD FROM CELLULOSICS

Based on the yields of hexose and pentose sugars, yields of ethanol and butanol–acetone–ethanol have been estimated and are listed in Table 15.8. Ethanol yields are based on fermentation with *Candida Shehatae* (31).

Fermentation by *Clostridium acetobutylicum* to butanol–acetone–ethanol (32) would make use of both hexose and pentose sugars. Corn cobs, wheat straw, and pine hydrolyzates would give the highest yields of these fuels; yields from hydrolyzates of poplar and bagasse would be about 20–29% less.

TABLE 15.7 Sugar Yields on Two-Stage SO_2 Hydrolysis of Cellulosics[a]

	Red Pine Sawmill Residue	Aspen Poplar Chips	Bagasse Louisiana	Corn Cobs	Wheat Straw
Prehydrolysis					
Glucose	6.9	3.3	3.0	3.0	2.5
Nonglucose sugars	21.1	23.3	24.8	42.9	43.8
Hydrolysis					
Glucose	40.1	28.9	20.4	20.5	20.2
Nonglucose sugars	(1.0)	3.3	4.0	7.1	2.0
Total					
Glucose	47.0	32.2	23.4	23.5	22.7
Nonglucose sugars	20.1	26.6	28.8	50.0	45.8
% of Theoretical Yield					
Glucose	84.3	68.7	60.4	66.2	54.5
Nonglucose	103.7	91.5	101.0	99.0	113.2

[a] Percentage of dry raw material weight.

TABLE 15.8   Fuel Alcohol Yields on Two-Stage SO$_2$ Hydrolysis of Cellulosics[a]

	Red Pine Sawmill Residue	Aspen Chips	Bagasse Louisiana	Corn Cobs	Wheat Straw
Ethanol	403	342	297	406	379
Butanol–acetone–ethanol	252	220	196	276	257

[a] Estimated from sugar yields; L/tonne dry raw material.

EQUIPMENT CONFIGURATIONS

Some equipment previously discussed (10) as suitable for autohydrolysis would, if made of 316 ELC stainless steel or better, prove suitable for SO$_2$ hydrolysis. The Masonite gun configuration shown in Figure 15.2 and the Stake reactor in Figure 15.3 are capable of accepting wood chips and processing them through SO$_2$ prehydrolysis. Since prehydrolysis is about 150°C, saturated steam pressure at this temperature is about 4 atm (380 kPa), and the commercial equipment of these two designs can easily withstand this or higher pressure. Alternatives may include defibrators and extruders (21). Extraction of lignin following prehydrolysis can be carried out with conventional washers, diffusers, or presses. It is an optional step, which would only be taken if markets for the lignin are available.

There are several configurations of equipment suitable for SO$_2$ hydrolysis. A continuous plug flow reactor has proved very successful in starch hydrolysis and can be modified for cellulose hydrolysis. A schematic of the St. Lawrence reactor is shown in Figure 15.4, and a model of a full-scale reactor, now in commercial operation on starch, is shown in Figure 15.5. Modifications of the

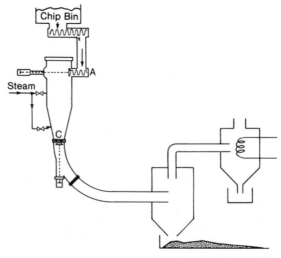

Figure 15.2   Masonite gun configuration.

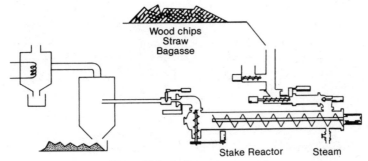

Figure 15.3 Stake reactor.

St. Lawrence reactor for cellulose hydrolysis will include provision for operation at higher temperature and possibly for removal of sugar as formed.

Existing commercial cellulose hydrolysis plants are based on dilute sulfuric acid percolating through a bed of wood. These would profit by conversion from H_2SO_4 to SO_2 catalyst, especially in the modified configuration suggested in references (1) and (8). This would lower costs, protect lignin, and substantially reduce problems of metallurgy.

SUMMARY AND CONCLUSIONS

Of several acids tested as catalysts in hydrolysis of pine sawmill residues, SO_2 proved to give the highest yield of total reducing sugars—92% of theoretical. In the same tests, the yield with a sulfuric acid catalyst was 83%. These yields were obtained in a three-stage process, prehydrolysis, followed by alkaline extraction of lignin followed by hydrolysis. For some purposes, the alkaline extraction is optional.

Prehydrolysis with SO_2 has advantages over autohydrolysis (steam cracking). In the absence of SO_2, pine hemicellulose was resistant to hydrolysis, with only 59% hydrolyzed to sugars, whereas in the presence of SO_2, hemicellulose hydrolysis was complete. Aspen hemicellulose is much more readily hydrolyzed by steam; 88% yields of hemicellulose sugars are realized. However, in the

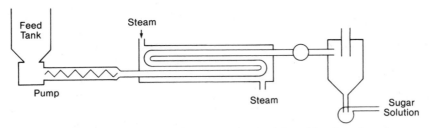

Figure 15.4 Plug flow reactor schematic. (Courtesy of National Organizing Committee, Symp. on Alcohol Fuels, Ottawa).

Figure 15.5 Photograph of plug flow reactor model which is now in full-scale operation. (Courtesy of National Organizing Committee, Symp. on Alcohol Fuels, Ottawa).

presence of SO_2 the hemicellulose was completely hydrolyzed. Furthermore, much lower temperatures could be used with SO_2 catalyst—150°C compared to 190°C with steam alone. Also, much more of the hemicellulose sugars occurred as monomers with SO_2 catalyst, about 95%, whereas in autohydrolysis only 58% of the hemicellulose sugars were monomeric.

Compared to the more usual H_2SO_4 catalyst, SO_2 as a gas distributes better in the wood; it is of lower cost and less is required, thereby lowering the cost of neutralization before fermentation. Stainless steels that fail with H_2SO_4 are resistant to SO_2. Lignin quality is better preserved.

Well-proven equipment is available for the various stages of this process. A continuous plug flow reactor can be used for hydrolysis, and the present St. Lawrence starch reactor is undergoing modification for this use.

REFERENCES

1. M. Wayman and A. Dzenis, *Can. J. Chem. Eng*, **62**, 699, (1984).

2. Proceedings, International Symposium on Ethanol from Biomass, Royal Society of Canada, Winnipeg, Oct. 1982.

3. R. D. Hayes, "Availability of Crop Cellulosics for Ethanol Production," in ref. 2, 1982, pp. 118–141.

4. M. R. Clarke and C. B. R. Sastry, "Availability of Wood Cellulosics," in ref. 2, 1982, pp. 142–160.

5. N. Kosaric, A. Wieczorek, G. P. Cosentino, R. J. Magee, and J. E. Prenosil, in *Biotechnology*, H. J. Rehm and G. Reed, eds., Vol. 3, Verlag Chemie, Weinheim, 1983.

6. R. J. Burton, "The New Zealand Wood Hydrolysis Process," in ref. 2, 1982, pp. 247–270.

7. K. Mackie, K. Deverell, and I. Callander, "Aspects of Wood Hydrolysis via the Dilute Sulphuric Acid Process", in ref. 2, 1982, pp. 271–311.

8. *Assessment of the New Zealand Forest Research Institute Process for Pine Wood Hydrolysis and Fermentation to Ethanol*, Dept of Energy, Mines and Resources of Canada, Ottawa, 1983.

9. J. H. Lora and M. Wayman, *Tappi*, **61** (6), 47–50 (1978); M. Wayman and J. H. Lora, *Can.*

Pat. 1,147,105; K. Okamura and H. Morikawa, in *Recent Technologies and Prospects of Wood Chemicals*, J. Nakano and T. Higuchi, eds., CMC Report No. 40, CMC Publishers, Kyoto, Japan, 1983.

10. M. Wayman, "Alcohol from Cellulosics: The Autohydrolysis–Extraction Process," *Proceedings IV International Alcohol Fuels Technology Symposium*, Guaruja, Brazil, Vol. I, 1980, pp. 79–89.

11. M. Wayman, J. H. Lora and E. Gulbinas, *"Chemistry for Energy,"* ACS Symp. Ser. No. 90, American Chemical Society, Washington, D.C., 1978, pp. 183–201.

12. S. A. Rydholm, *Pulping Processes*, Interscience, New York, 1965.

13. K. Freudenberg, in *Lignin Structure and Reactions*, J. Marton, ed., Advances in Chemistry Series No. 59, American Chemical Society, Washington, D.C., 1966.

14. W. E. Hillis, *Wood Extractives*, Academic, New York, 1962.

15. H. F. J. Wenzl, *The Chemical Technology of Wood*, Academic, New York, (1970).

16. M. R. Ladisch, *Process Biochem.*, **14**(1), 21, (1979).

17. J. P. Sachetto and A. Johansson, ' The Battelle–Geneva Fractionation Process," presented at the 186th ACS National Meeting, Washington, Aug. 28, 1983.

18. T. J. Laughlin, D. R. Coleman, M. V. Kilgore, C. L. Lishawa, and M. H. Eley, "The Production of Fuel-Grade Ethanol from the Cellulose in Municipal Solid Waste," presented at the Solar and Biomass Energy Workshop, Atlanta, GA, April 17–19, 1984.

19. M. C. Hawley, Chapter 17 in this volume; C. M. Ostrovski, H. E. Duckworth, and J. C. Aitken, "Canertech's Ethanol-from-Cellulose Program," *Proc. VI Int. Symp. Alcohol Fuels Technol.*, **II**, 187–191 (1984).

20. B. I. Tokarev, in *The Chemistry of Wood*, N. I. Nikitin, ed., English translation, 1966.

21. B. Rugg, Report SER1/TR-1-9386-1, 105 pp., 1982.

22. G. R. Lawford, B. H. Lavers, D. Good, R. Charley, J. Fein, and H. G. Lawford, "Zymomonas Ethanol Fermentation: Biochemistry and Bioengineering," in ref. 2, 1982, pp. 482–507.

23. H. E. Grethlein and A. O. Converse, "Continuous Acid Hydrolysis for Glucose and Xylose Production", in ref. 2, 1982, pp. 312–336; D. R. Thompson and H. E. Grethlein, *Ind. Eng. Chem. Prod. Res. Dev.*, **18**(3), 166–169 (1979).

24. J. F. Saeman, *Ind. Eng. Chem.*, **37**, 43–52 (1945).

25. Anonymous, "Renewable Fuels Report," Canadian Renewable Fuels Association, Ottawa, May 1, 1984, p. 5.

26. M. Wayman and M. G. S. Chua, *Can. J. Chem.*, **57**, 2603–2611 (1979).

27. C. B. Keim, J. Nagasuye, and P. G. Assarsson, "Preparation of Acid Hydrolysed Starch Substrates and Their Fermentability," XI Int. Carbohydr. Symp., Vancouver, Abstract V-8, 1982.

28. M. F. Ewen and G. H. Tomlinson, *U.S. Pat.* 763,473, June 28, 1904.

29. J. F. Saeman, "Acid Hydrolysis of Cellulosics," in ref. 2, 1982, pp 231–246.

30. G. H. Tomlinson, *U.S. Pats.* 1,032,448 and 1,032,450, July 16, 1912.

31. M. Wayman and S. T. Tsuyuki, "Fermentation of Xylose to Ethanol by *Candida shehatae*", *Biotech. and Bioeng.*, in press; J. C. du Preez and J. P. van der Walt, *Biotech. Lett.* **5**, 357–362 (1983).

32. M. Wayman, G. R. Husted, J. D. Santangelo, and S. Levy, "Butanol Production by Extractive Fermentation Using Fluorocarbons," Proc. VI Int. Symp. Alcohol Fuels Technol, Vol. II, 1984, pp. 234–238.

16

Kinetic Modeling of the Saccharification of Prehydrolyzed Southern Red Oak

ANTHONY H. CONNER, BARRY F. WOOD, CHARLES G. HILL, JR., and JOHN F. HARRIS

USDA, Forest Service, Forest Products Laboratory and Department of Chemical Engineering, University of Wisconsin, Madison, Wisconsin

A renewed interest in the use of biomass as a chemical raw material or as an energy source has developed over the past decade due, in large part, to fuel shortages in the early 1970s, to present political unrest in the Middle East, and to dwindling petroleum supplies predicted in the near future. Pyrolysis, gasification, liquefaction, and saccharification (hydrolysis) of cellulosics have all been proposed as methods for utilizing renewable biomass as an energy or chemical source. For the near term, saccharification of cellulosics to glucose, which can be fermented to fuel alcohol or a variety of other chemicals, is the method closest to implementation.

Hydrolysis of cellulosics can be carried out either with cellulolytic enzymes or with mineral acids. Enzymatic hydrolysis occurs under mild conditions and gives good sugar yields, but the cost of the enzyme is high and the conversion rate of cellulose to glucose is slow (1).

Acid hydrolysis of cellulosics can be carried out either in concentrated [e.g., the Bergius process (2)] or dilute [e.g., the American process (3), the Scholler process (4), or the Madison process (5)] acid solution. Hydrolysis of cellulosics in concentrated mineral acids at low temperature can give near quantitative yields of glucose (6). However, large quantities of concentrated acid are needed, a fact compounded by incomplete hydrolysis at low liquid-to-solid ratios (7). Therefore, for economic reasons acid recycling is imperative. In addition, plant costs will probably be excessive owing to the need for expensive corrosion resistant equipment.

Dilute acid hydrolysis of cellulosics occurs under rather severe reaction conditions and gives poorer sugar yields. However, the conversion of cellulose to glucose is rapid, and the acid catalyst is cheaper than enzymes and does not have to be recycled. Dilute acid hydrolysis of cellulosics is the method closest to commercial implementation and is preferable from a process point of view (8).

Current efforts to hydrolyze cellulosics with dilute acid are diverse and sometimes highly innovative. Of these processes, Rugg and Brenner (9) described equipment for a screw-fed, continuous, single-stage hydrolyzer, and Thompson and Grethlein (10) investigated a plug-flow reactor. Both processes take advantage of the improved kinetics of cellulose hydrolysis relative to glucose decomposition at higher temperatures and shorter retention times to improve sugar yields. Both also have the advantage of being simple one-stage processes.

The U.S. Forest Products Laboratory (FPL) in cooperation with the Tennessee Valley Authority (TVA) has been studying a two-stage dilute acid hydrolysis process (11) based in part on studies of Cederquist (12) in Sweden during the 1950's. The first stage (prehydrolysis) selectively removes the hemicellulosic sugars with dilute sulfuric acid at about 170°C (13) prior to hydrolysis of the lignocellulosic residue to glucose in the higher temperature (230°C) second stage. Saeman (14) pointed out that a two-stage dilute acid process has a number of important advantages: (a) The carbohydrates are fractionated into hemicellulosic sugars and glucose, which facilitates the separate utilization of each fraction; (b) glucose is isolated in moderately good yield ($\sim$ 50%); (c) the glucose solution from the second stage is moderately concentrated ($\sim$ 10–12%); and (d) the consumption of acid and steam is relatively low.

Despite recent efforts such as these to produce glucose from agricultural residues and wood, little attention has been given to improving the kinetic modeling of cellulose saccharification with dilute acid. Kinetic modeling plays an important role in the design, development, and operation of many chemical processes. Kinetic data are equally important in the design, evaluation, and operation of processes to hydrolyze cellulosics to fermentable sugar.

In this chapter we describe a new model for the dilute acid hydrolysis of cellulose that was developed at FPL in connection with our studies on the two-stage dilute sulfuric acid hydrolysis process. The model incorporates the effect of the neutralizing capacity of the substrate, the presence of readily hydrolyzable cellulose, and the reversion reactions of glucose in acid solution. Although general in nature, the model was developed specifically for application to the dilute sulfuric acid hydrolysis of prehydrolyzed wood. A computer program to simulate the new model under various hydrolysis conditions is presented. This program can be used to predict yields of free glucose, reducing sugars, reversion material, remaining cellulose, and glucose loss due to dehydration as a function of acid concentration, temperature, and reaction time.

CELLULOSE HYDROLYSIS MODEL

Luers (15) first recognized that the rate of glucose production from cellulose depends on the reaction rates for its formation and destruction and that these reactions proceed independently. However, he concluded from limited data that the two reactions were similarly affected by changes in acidity and temperature, and thus maximum glucose yields were unaffected by such changes. Saeman (16) showed that the rate of the hydrolysis reaction is increased more rapidly than that of the sugar dehydration reaction by both increased temperature and increased acid concentration. He modeled the system as two consecutive pseudo-first-order reactions in which the rate constants were functions of the applied acid concentration and temperature:

$$\text{cellulose} \xrightarrow{k_1} \text{glucose} \xrightarrow{k_2} \text{dehydration products}$$

Saeman's model, based on two consecutive reactions, is currently the only model used to simulate cellulose saccharification. This model is, however, an oversimplification of the reactions that actually occur in dilute acid solution. He recognized that glucose yields would be affected by the presence of neutralizing material (i.e., ash-forming constituents) in the cellulose substrate and by the formation of reversion materials from glucose in dilute acid solution. Since these effects were less pronounced at the reaction conditions of the percolation process he was investigating, he did not incorporate them into his model. However, both factors are especially important in predicting glucose yields at the lower acidities and liquid-to-solid ratios encountered in some of the current processes under study. In addition, Saeman's model does not include the fact that all celluloses contain material that is rapidly hydrolyzed. This factor can, however, be readily introduced (10) into his model.

Our new model, which includes elements to correct the shortcomings of the model limited to two consecutive pseudo-first-order reactions, is outlined in Figure 16.1. This model incorporates the fact that cellulose contains both an easily hydrolyzable portion and a resistant portion, that reversion products (primarily disaccharides and levoglucosan-type materials) are formed from glucose in dilute acid solution, and that the disaccharides formed during reversion can degrade at the reducing end, giving a glycoside that on hydrolysis gives glucose and degradation products.

The model depicted in Figure 16.1 can be described in terms of the following differential equations:

$$\frac{d\,[\text{Y1}]}{dt} = -k_1[\text{Y1}] \tag{1}$$

$$\frac{d[\text{Y2}]}{dt} = -k_c[\text{Y2}] \tag{2}$$

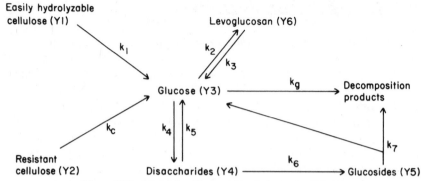

Figure 16.1 Model for dilute acid hydrolysis of cellulosics.

$$\frac{d[Y3]}{dt} = k_1[Y1] + k_c[Y2] - k_2[Y3] + k_3[Y6] - 2k_4[Y3]^2$$

$$+ 2k_5[Y4] + k_7[Y5] - k_g[Y3] \tag{3}$$

$$\frac{d[Y4]}{dt} = k_4[Y3]^2 - k_5[Y4] - k_6[Y4] \tag{4}$$

$$\frac{d[Y5]}{dt} = k_6[Y4] - k_7[Y5] \tag{5}$$

$$\frac{d[Y6]}{dt} = k_2[Y3] - k_3[Y6] \tag{6}$$

If the values of the various rate constants for a particular saccharification condition are known or can be derived, one can numerically integrate this set of equations to determine concentrations of intermediates at those conditions as a function of time.

Cellulose Hydrolysis Rate Constant

The rate constant for the hydrolysis of the resistant portion of cellulose in prehydrolyzed Douglas fir was determined by Saeman (16, 17) and Kirby (18). A recorrelation (11) of their data, accounting for the acid-neutralizing capacity of the cellulosic substrate, showed that the pseudo-first-order rate constant for the resistant portion, k_c, could be described as a function of acid concentration, $[H^+]$, and temperature as follows:[‡]

$$k_c = 2.47 \times 10^{20} [H^+]^{1.218} \exp\left(\frac{-42900}{RT}\right) \tag{7}$$

[‡]Throughout this chapter all rate constants are expressed in units of min^{-1}.

where $[H^+]$ = molal hydrogen-ion concentration, R = universal gas constant (calories/gmole–°K), and T = temperature (°K). This equation applies specifically to the sulfuric acid hydrolysis of cellulose contained in Douglas-fir lignocellulose. Saeman (16) measured the rates of hydrolysis of the resistant cellulose in five different wood species using dilute sulfuric acid. His data showed that rates differing by as much as 20% accompany changes in the substrate. Thus while the functional form of an equation to describe the cellulose hydrolysis rate constant for other species will have the same form as that of Eq. (7), the values of the parameters will change with substrate as well as catalyst.

Assuming the hydrolysis rate for southern red oak (*Quercus falcata* Michx.) is affected by temperature and acidity in a manner similar to that for Douglas fir, the following equation to describe the rate constant for the hydrolysis of southern red oak lignocellulose is obtained from experimental data (Table 16.1) by a least-squares procedure:

$$k_c = 2.87 \times 10^{20}\,[H^+]^{1.218}\,\exp\!\left(\frac{-42900}{RT}\right) \tag{8}$$

This equation differs from Eq. (7) only in the pre-exponential factor. Comparison of Eqs. (7) and (8) shows that at the same temperature and acidity southern red oak will hydrolyze slightly faster than Douglas-fir. This difference was observed previously by Saeman (16).

In addition to the resistant portion of cellulose, all celluloses contain an easily hydrolyzable portion that reacts at a much greater rate. The quantity of easily hydrolyzable material depends on both the origin and prior treatment of the cellulosic substrate (19). The data of Millett et al. (20) show that approximately 10% of the cellulose contained in wood pulps and native celluloses hydrolyzes at a rate 20–100 times as fast as the remaining 90%. For purposes of the model it was assumed that 10% of wood lignocellulose hydrolyzes at 50 times the rate of the resistant portion—that is, $k_1 = 50 \times k_c$ (Fig. 16.1).

Glucose Dehydration Rate Constant

Of the information available on glucose dehydration (16, 17, 21, 22), the data of McKibbins et al. (21, 22) include the complete range of temperature and acidity of interest in cellulose saccharification. They measured the rate constant for the disappearance of glucose as reducing sugars, k_r, in dilute sulfuric acid solution. The formation of reversion products in acid solution (see following discussion) contributes to the reducing power of the sugar solutions. Hence, k_g (Fig. 16.1), the true rate constant for glucose dehydration, is related to but cannot be equated with k_r. Values for k_g were calculated from McKibbins' experimental values of k_r using the submodel (see following discussion) developed for the reversion reactions (23).

The calculated k_g values were then correlated by a least-squares method (23). The final correlation between the glucose dehydration rate constant and acid

TABLE 16.1 Experimental Data for Saccharification of Prehydrolyzed Southern Red Oak Lignocellulose[a]

Time (Min)	Weight Loss ($\%^b$)	Glucan Remaining ($\%^c$)	Monomeric Glucose Yield ($\%^d$)	Reducing Sugars Yield ($\%^e$)
230°C Low Ash Lignocellulose (27.2 meq/kg)				
0.16	30.1 ± 2.2		30.7 ± 2.9	33.5 ± 2.3
0.67	48.1 ± 0.9		48.1 ± 2.8	49.8 ± 0.6
0.93	58.4 ± 0.8		51.6 ± 1.6	53.4 ± 0.4
1.25	62.1 ± 0.5		45.1 ± 1.9	46.6 ± 0.9
1.58	61.0 ± 1.1		34.7 ± 0.9	36.3 ± 0.7
2.01	55.8 ± 1.2		23.3 ± 0.1	25.4 ± 0.9
230°C High Ash Lignocellulose (107.2 meq/kg)				
0.58	32.5 ± 0.9	64.4	29.2 ± 1.2	34.5 ± 1.3
0.75	41.9 ± 0.4	45.2	39.0 ± 0.8	43.6 ± 0.5
0.92	48.9 ± 1.5	32.0	45.2 ± 1.2	49.4 ± 0.4
1.08	51.7 ± 1.2		44.2 ± 1.1	49.2 ± 1.0
1.25	58.1 ± 1.8	17.6	44.9 ± 2.4	50.0 ± 1.5
1.42	56.9 ± 1.6		42.8 ± 1.4	47.9 ± 1.7
1.58	58.1 ± 0.7	15.6	43.6 ± 0.4	46.7 ± 0.9
2.00	59.0 ± 1.3	9.1	36.5 ± 3.2	40.6 ± 1.4
210°C Low Ash Lignocellulose (27.2 meq/kg)				
1.0	23.1 ± 0.3	74.8	21.0 ± 0.8	24.5 ± 0.6
2.0	38.9 ± 1.1	38.8	37.8 ± 0.3	39.7 ± 0.5
3.0	48.3 ± 1.7	29.8	44.1 ± 2.2	45.7 ± 0.7
3.5	50.7 ± 0.3	24.8	43.2 ± 1.7	46.9 ± 0.5
4.0	53.1 ± 0.5	18.9	44.0 ± 0.5	46.5 ± 0.6
6.0	56.4 ± 0.1	5.3	35.5 ± 0.6	38.4 ± 0.4
8.0	56.1 ± 0.5	1.6	27.1 ± 2.2	29.8 ± 1.4
210°C High Ash Lignocellulose (107.2 meq/kg)				
2.0	28.7 ± 0.6	57.0	27.0 ± 1.1	31.4 ± 0.8
3.0	36.8 ± 1.0	48.6	37.6 ± 2.0	38.7 ± 0.8
4.5	45.3 ± 1.4	30.4	40.9 ± 0.5	43.9 ± 0.5
6.0	49.5 ± 1.0	23.7	41.7 ± 1.0	44.5 ± 0.4
7.5	52.9 ± 0.6	13.6	40.0 ± 2.3	42.0 ± 0.9
9.5	53.1 ± 0.9	9.9	34.7 ± 1.1	36.6 ± 0.7
12.5	53.3 ± 0.2	3.8	27.0 ± 1.6	29.4 ± 0.9

[a] Applied acid solution—0.8% H_2SO_4; liquid-to-solid ratio = 3.

[b] Percent of original ovendry prehydrolysis residue ± SD.

[c] Percent of original glucan.

[d] Percent of potential glucose available from residue ± SD.

[e] Percent of potential reducing sugars available from residue ± SD.

concentration and temperature was found to be:

$$k_g = 5.580 \times 10^{15} \, [\, H^+] \, \exp(0.183[G]) \, \exp\!\left(\frac{-33800}{RT}\right) \tag{9}$$

where $[G]$ = molal glucose concentration

The rate constant k_g relates to the acid-catalyzed degradation reaction of glucose, but glucose and other sugars are also destroyed by various base catalyzed and uncatalyzed reactions (24). Above 0.05 molal $[H^+]$ the disappearance of glucose is correctly predicted by Eq. (9); however, at lower acid concentrations the non-acid-catalyzed reactions become significant. Therefore, Eq. (9) should not be used below an acid concentration of 0.05 molal. Since only limited data (21, 22) are available on the rates of glucose destruction at hydrogen-ion concentrations below 0.05 molal, the application of the saccharification model is also limited by this restriction.

Reversion Reaction Rate Constants

In aqueous acid solutions, glucose not only reacts to form small quantities of its isomers, mannose and fructose, but also undergoes reversible reactions that result in disaccharides, oligosaccharides, and anhydrosugars. These reactions are referred to as reversion reactions, and the reaction products as reversion products. The reversion reactions result in components bonded only by glucosidic bonds; no evidence of ether bond formation exists. Although most of the isomeric disaccharides of glucose have been isolated under reversion-type reaction conditions (25–27), it has been found that the preferential mode of condensation is that between the anomeric hydroxyl and the C6-hydroxyl (i.e., the most prominent constituents of the reversion mixture are [1,6]-linked glucosides).

Minor (28) showed that at glucose concentrations below 20%, the reversion products formed from glucose in dilute sulfuric acid solution consist almost exclusively of the disaccharides gentiobiose and isomaltose and the internally [1,6]-linked glucosans levoglucosan and 1,6-anhydro-β-D-glucofuranose. At 180°C the disaccharides accounted for most of the reversion products, the glucosans constituting less than 30%. However, at 230°C the glucosans were the major reversion products. Smith et al. (29) also observed large amounts of levoglucosan in reversion products obtained at 220°C.

These facts led to the development of a submodel for the formation of reversion products from glucose. The reaction scheme for this submodel was incorporated into the cellulose saccharification model (Fig. 16.1). In this scheme, the reversion products are assumed to consist only of monomeric, nonreducing glucosans (exemplified by levoglucosan) and of reducing disaccharides such as gentiobiose and isomaltose. It is further assumed that cellulose hydrolysis produces only monomeric glucose and that the disaccharides result exclusively from the reversion reactions. Also incorporated into this part of the model was the fact that in acid solution the reducing end of the disaccharide can be dehydrated to give a glucoside. Indeed, 2-furfuryl-5-methylglucoside, the expected glucoside from a degradation reaction of this type, has been observed among reversion products (30). Because this reaction is identical to the reaction by which glucose is destroyed, dehydration of the disaccharides would proceed at approximately the same rate as glucose dehydration (i.e., $k_6 = k_g$). This is not a critical assumption. Large variations in the assumed value of k_6 may be tolerated, since the quantity of disaccharide seldom exceeds 5% of the total potential glucose.

Equilibrium constants for the two reversion reactions shown in Figure 16.1 can be defined as:

$$KL = \frac{[\text{levoglucosan}]}{[\text{glucose}]}$$

$$= \frac{k_2}{k_3} \tag{10}$$

$$KD = \frac{[\text{disaccharide}]}{[\text{glucose}]^2}$$

$$= \frac{k_4}{k_5} \tag{11}$$

where KL and KD are the equilibrium constants for the formation of levoglucosan and disaccharides, respectively. Values for KL and KD at 180 and 230°C were determined by Minor (28). These can be expressed analytically as:

$$KL = 9.91 \times 10^2 \exp\left(\frac{-4520}{T}\right) \tag{12}$$

and

$$KD = 7.90 \times 10^{-7} \exp\left(\frac{5444}{T}\right) \tag{13}$$

Minor's data were taken at a single acid concentration; however, the principle of thermodynamic consistency requires that the forward and reverse reactions be accelerated to the same extent by changes in acidity. Thus, the expressions obtained above for KL and KD are independent of acidity.

If one assumes that the rate constants for the hydrolysis of 1,6-glucosidic bonds are similar (i.e., $k_3 = k_5 = k_7$), and if one knows this rate constant relative to the glucose dehydration rate constant (k_g), then Eqs. (10) and (11) can be used to obtain estimates of the rate constants k_2 and k_4, expressed in terms of k_g. The rates for the hydrolysis of all glucosides (31) are much greater than that for dehydration of glucose. The rate of hydrolysis of isomaltose is approximately 35 times the glucose dehydration rate; the ratio is higher for gentiobiose. It is assumed that $k_3 = k_5 = k_7$ and that these rate constants are all equal to $35 \, k_g$. Hence as a conservative estimate, $k_2 = KL \times 35 \times k_g$ and $k_4 = KD \times 35 \times k_g$.

Effective Acid Concentration

The acidity of the reacting system, $[H^+]$, is a function of the amount and concentration of the applied acid solution, the neutralizing capacity of the substrate, and, for sulfuric acid, the extent of the secondary ionization. This latter factor may be ignored, since measurements of the bisulfate dissociation constant (32) indicate that under the usual conditions of cellulose saccharification less than 1% of the bisulfate ion dissociates. Thus, to evaluate the hydrogen-ion concentration the sulfuric acid is assumed to release only one hydrogen ion. It is also assumed that all the cations in the cellulosic substrate are readily accessible and are effective immediately. For prehydrolyzed lignocelluloses this assumption is quite valid, since most of the ash constituents of these materials result from ion exchange during the water wash that follows prehydrolysis. Thus, with the further assumption that the hydroxyl-ion concentration is negligible, the hydrogen-ion concentration is the difference between the molality of the added acid solution and the molality of the cations (expressed as eq/kg of water). At low liquid-to-solid ratios and low acid concentrations, the neutralizing capacity of the lignocellulose can significantly lower the effective acid concentration.

COMPUTER SIMULATION OF THE SACCHARIFICATION MODEL

From the preceding discussion, it is readily apparent that Eqs. (1) to (6), which describe the saccharification model, can be rewritten in terms of the known constants k_c (the hydrolysis rate constant for resistant cellulose), k_g (the glucose dehydration rate constant), KL (the equilibrium constant for the formation of levoglucosan from glucose), and KD (the equilibrium constant for the formation of disaccharides from glucose). A BASIC computer program (33) which numerically integrates these transformed differential equations by the Runge–Kutta fourth-order method (34) is listed in Appendix I. This program allows one to calculate yields of monomeric glucose, total anhydroglucose, reducing sugar, and the remaining cellulose as a function of acid concentration, temperature, and reaction time.

In the program it is assumed that the temperature varies in a manner approximating that occurring in the heat-up of the experimental reaction vessel. The temperature of the contents is assumed to have no spatial variation but to change with time according to Newton's law:

$$\frac{dT}{dX} = \alpha(TB - T) \tag{14}$$

where T = temperature of reactants, TB = temperature of bath, X = time (min), and α = characteristic heat transfer coefficient for reaction = 13.8 min^{-1} for 5-mm glass ampuls.

Calculation is simplified by assuming the water content of the reactants to be constant throughout the reaction. Although water is involved in the reactions, being produced by glucose dehydration and consumed by cellulose hydrolysis, the net variation was less than 1% over the range of experimental data. This assumption is equivalent to assuming a constant acid molality throughout the reaction. The same assumption was used in correlating the rate constants.

The variables used in the program are defined in the program listing. The program is also annotated at various points to document how the program was constructed from the parameters discussed above. Table 16.2 contains an example of the output generated by this program; the program can be easily modified to output yield data for any reaction product.

DILUTE SULFURIC ACID HYDROLYSIS OF PREHYDROLYZED SOUTHERN RED OAK

Chips of southern red oak, impregnated with dilute sulfuric acid, were prehydrolyzed by direct steam heating for 6 min at 170°C as previously described (13). One-half of the lignocellulosic residue was slurried, fiberized, and washed using distilled water; the other half was treated similarly using tap water. After being washed, the wet residues were pressed to remove as much water as possible and then air-dried. The ash contents were determined using ASTM method No. D 1102 (35). Ash samples were titrated with dilute acid to determine their neutralizing capacity (36) as meq/kg oven-dry lignocellulose. The distilled water–washed residue contained 0.18% ash and had a neutralizing capacity of 27.2 meq/kg. Corresponding values for the tap water–washed material were 0.47% and 107.2 meq/kg.

Samples of the two residues were reacted in glass ampuls (5 mm OD) at a liquid-to-solid ratio of 3, using 0.8% H_2SO_4 at 210 and 230°C. Approximately 200 mg of residue was placed in a tared 20-cm-long glass ampul sealed at one end. The sample was then dried (60°C *in vacuo*). The glass ampul was fitted with a rubber septum and evacuated. To obtain the desired liquid-to-solid ratio, acid solution was injected into the ampul through the septum with a syringe. The contents were brought to atmospheric pressure with N_2 and sealed.

The saccharification reaction was carried out by placing the ampul in a molten salt bath controlled to ±0.1°C. The reaction time was taken as the interval from immersion in the salt bath to quenching in an adjacent water bath. After reaction, the ampul was opened and its contents were washed onto a tared sintered glass funnel with hot water. The solid residue was washed with approximately 100 ml of boiling water, dried overnight (60°C *in vacuo*), and weighed. The collected filtrate was diluted to a known volume for analysis.

TABLE 16.2 Cellulose Hydrolysis Yields Calculated from Model[a]

Time (min)	Glucose Yields Total (%)	Monomer (%)	Combined (%)	Glucose Conc. (%)	Cellulose Remaining (%)	Reducing Yield (%)	Cellulose in Residue (%)
0.00	0.00	0.00	0.00	0.00	100.00	0.00	57.00
0.04	0.00	0.00	0.00	0.00	100.00	0.00	57.00
0.09	0.04	0.04	0.00	0.00	99.96	0.04	56.99
0.13	1.19	1.18	0.00	0.25	98.81	1.18	56.71
0.17	6.65	6.58	0.08	1.39	93.32	6.58	55.30
0.22	13.51	13.00	0.51	2.77	86.31	13.03	53.36
0.26	19.02	17.70	1.32	3.86	80.45	17.79	51.61
0.30	24.88	22.61	2.27	4.99	74.00	22.78	49.52
0.35	30.93	27.68	3.25	6.13	67.11	27.94	47.08
0.39	36.63	32.42	4.22	7.18	60.31	32.80	44.43
0.44	41.71	36.60	5.11	8.09	53.91	37.11	41.68
0.48	46.05	40.15	5.90	8.86	48.05	40.80	38.91
0.52	49.68	43.09	6.59	9.49	42.75	43.85	36.17
0.57	52.62	45.46	7.16	10.00	38.00	46.33	33.50
0.61	54.94	47.32	7.62	10.39	33.76	48.28	30.91
0.65	56.73	48.74	7.99	10.69	29.98	49.76	28.44
0.70	58.03	49.76	8.27	10.91	26.62	50.85	26.08
0.74	58.92	50.45	8.47	11.06	23.64	51.57	23.86
0.78	59.45	50.85	8.60	11.15	20.99	52.00	21.76
0.83	59.66	50.99	8.67	11.19	18.63	52.16	19.81
0.83	59.67	51.00	8.68	11.19	18.32	52.16	19.54*
0.87	59.62	50.93	8.69	11.18	16.54	52.10	17.98
0.91	59.34	50.68	8.66	11.13	14.69	51.84	16.30
0.96	58.87	50.27	8.60	11.05	13.04	51.42	14.74
1.00	58.24	49.74	8.51	10.95	11.58	50.87	13.30
1.04	57.48	49.09	8.38	10.82	10.28	50.20	11.99
1.09	56.60	48.36	8.24	10.67	9.12	49.43	10.79
1.13	55.62	47.54	8.08	10.51	8.10	48.59	9.70
1.17	54.58	46.67	7.91	10.33	7.19	47.68	8.70
1.22	53.47	45.75	7.72	10.14	6.39	46.72	7.80
1.26	52.31	44.78	7.53	9.95	5.67	45.72	6.99
1.31	51.12	43.79	7.33	9.74	5.03	44.69	6.25
1.35	49.90	42.78	7.12	9.53	4.47	43.63	5.59
1.39	48.67	41.75	6.92	9.32	3.97	42.57	5.00
1.44	47.43	40.72	6.71	9.10	3.52	41.49	4.46
1.48	46.18	39.68	6.51	8.88	3.13	40.41	3.98
1.52	44.94	38.64	6.30	8.67	2.78	39.34	3.55
1.57	43.70	37.60	6.10	8.45	2.46	38.27	3.16
1.61	42.48	36.58	5.90	8.23	2.19	37.20	2.82
1.65	41.27	35.56	5.70	8.01	1.94	36.16	2.51
1.70	40.07	34.56	5.51	7.80	1.72	35.12	2.24
1.74	38.90	33.57	5.33	7.59	1.53	34.10	1.99
1.78	37.74	32.60	5.14	7.38	1.36	33.10	1.77
1.83	36.61	31.65	4.97	7.17	1.21	32.12	1.57
1.87	35.50	30.71	4.79	6.97	1.07	31.15	1.40

[a]Yields were generated by the BASIC program listed in Appendix I. The program screens all calculated values for the maximum glucose yield and indicates this point with an asterisk in the rightmost column of the table. In this example two lines were printed at 0.83 min. This occurred because the time predicted for maximum glucose yield is (after rounding) the same as for a normal print interval.

Temperature = 230°C; L/S = 3; added acid = 0.8%; Neutralizing capacity of ash = 0.0272 meq/g; Cellulose content of starting material = 57%; resistant cellulose = 90%; All yields are anhydroglucose based as % of the original cellulose. Concentration reported as % glucose.

The filtrates were analyzed for both total reducing sugars and monomeric glucose. Reducing sugars were measured by Nelson's colorimetric modification of the Somogyi method (37). Glucose was determined by high-performance liquid chromotography (HPLC) (38) using a Bio-Rad HPX-85 carbohydrate column maintained at 85°C. The column was eluted with distilled water, and the effluent was monitored with a refractive index detector.

Carbohydrate analyses on the solid residues were performed using ASTM method No. D 1915 (39), except that the sugars were analyzed by HPLC as described above rather than by paper chromatography. The original prehydro-lyzed southern red oak contained 57% glucan, 3% xylan, and 33% Klason lignin (40), reported on an ash-free oven-dried basis. The unaccounted 7% is probably soluble lignin.

COMPARISON OF MODEL WITH EXPERIMENTAL DATA

The experimental data obtained on the saccharification of two prehydrolyzed southern red oak residues by the methods presented above are shown in Table 16.1. The two substrates differed only in their neutralizing capacity. This resulted in a difference in the effective hydrogen-ion concentration during the sacchari-fication of the two substrates. Thus four groups of data, constituting a 2^2 factorial experiment with two temperatures and two acidities, were obtained.

At each set of reaction conditions, measurements of monomeric glucose yield, reducing sugar yield, and remaining cellulose were made at time intervals that covered a wide extent of reaction. Comparisons of the experimental data with values of monomeric glucose and reducing sugars calculated with the model are shown graphically in Figures 16.2 and 16.3. These comparisons illustrate that the model fits the experimental data quite well.

This model has several advantages for use in process calculations. The various forms of glucose-containing species in solution can be estimated and may be treated as entities in process schemes. The estimated values can be used to interrelate values obtained from the various analytical methods used to assay carbohydrate content—for example, reducing sugars, HPLC, and enzymatic methods. The model also estimates the amount of material lost through dehy-dration. From this, one can then estimate the quantities of the various sugar degradation products (i.e., hydroxymethylfurfural and levulinic acid) in solution.

USING THE MODEL WITH OTHER SUBSTRATES AND CATALYSTS

Although the model is general in nature, it was developed specifically for ap-plication to the dilute sulfuric acid saccharifiation of prehydrolyzed wood lig-nocellulose. In principle the model can be used if other substrates or catalysts are used, but the exact values of parameters in the various equations of the

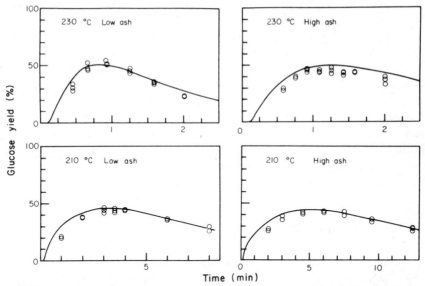

Figure 16.2 Comparison of glucose yields calculated from the model (solid line) with experimentally determined glucose yields (open symbols) for hydrolysis of prehydrolyzed southern red oak wood. The temperature for the hydrolysis reaction and the neutralizing capacities of the substrates are indicated in the figure. The liquid-to-solid ratio was 3:1. The added acid was 0.8% sulfuric.

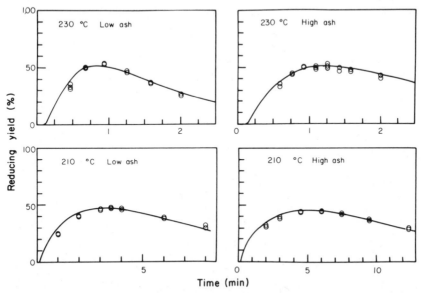

Figure 16.3 Comparison of reducing sugar yields calculated from the model (solid line) with experimentally determined reducing sugar yields (open symbols) for hydrolysis of prehydrolyzed southern red oak wood. The temperatures for the hydrolysis reaction and the neutralizing capacities of the substrates are indicated in the figure. The liquid-to-solid ratio was 3:1. The added acid was 0.8% sulfuric.

model will probably be different. Millett et al. (20), for example, have shown large differences in the rates of cellulose hydrolysis depending on the cellulosic substrate used (Fig. 16.4). It has also been shown (41) that hydrochloric acid, for example, is a much more effective catalyst than sulfuric acid (compared at equal hydrogen-ion concentrations) for the degradation of glucose, but only limited data are available on its effect on the cellulose hydrolysis rate at similar conditions.

Further experimental data are needed that will allow this model to be used with a wide variety of substrates and catalysts. Kinetic studies on the saccharification of other cellulosics are under way. Work is also planned that will permit a more accurate description of the reversion reactions. These data will extend the applicability of this model.

SUMMARY AND CONCLUSIONS

The model presented successfully simulates the dilute sulfuric acid saccharification of prehydrolyzed southern red oak lignocellulose as determined by comparison of the experimental data with the values predicted for a given set of reaction conditions. The model is a considerable improvement over the simple model based on two consecutive reactions. It can be used to advantage for process calculations, since the various forms of glucose-containing species in solution can be estimated separately and may thus be treated as entities in process schemes. It takes into consideration the facts that the neutralizing capacity of the cellulosic substrate lowers the effective concentration of the added acid, that glucose gives reversion products in dilute acid solution, and that all cellulosics contain material that is readily hydrolyzable. Although the model

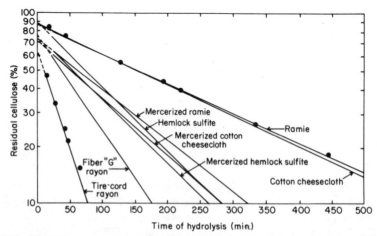

Figure 16.4 Weight loss occurring during the heterogeneous hydrolysis of cellulose (20). The semilogarithmic relation to time is illustrated for a range of natural and regenerated celluloses.

was developed specifically for application to the dilute sulfuric acid saccharification of prehydrolyzed wood lignocellulose, it can in principle be used if other substrates or catalysts are used. However, the values of parameters in the various equations of the model will be different, and they must be evaluated for the specific system of interest.

REFERENCES

1. M. R. Ladisch, *Proc. Biochem.*, **14**, 21 (1979).

2. F. Bergius, "Saccharification of Ground Wood," *German Pat.* 710216 (1941).

3. M. F. Ewen and G. H. Tomlinson, "Process of Producing Fermentable Sugar from Lignocellulose", *U.S. Pat.* 1,032,392 (1912).

4. H. Scholler, "Saccharifying Cellulosics," *French Pat.* 706678 (1930).

5. E. E. Harris and E. Beglinger, "The Madison Wood-Sugar Process," Rep. R1617, USDA, Forest Serv., Forest Prod. Lab., Madison, WI (1946).

6. I. S. Goldstein, *Tappi* **63**, 141 (1980).

7. I. S. Goldstein, H. Pereira, J. L. Pittman, B. A. Strouse, and F. P. Scaringelli, *Fifth Symposium on Biotechnology for Fuels and Chemicals*, Gatlinburg, TN, 1983.

8. H. E. Grethlein, *Biotechnol. Bioeng.*, **20**, 503 (1978).

9. B. Rugg and W. Brenner, *Proc. 7th Energy Technol. Conf. and Exposition,* Washington, D.C., 1980.

10. D. R. Thompson and H. E. Grethlein, *Ind. Eng. Chem. Prod. Res. Dev.*, **18**, 166 (1979).

11. J. F. Harris, A. J. Baker, A. H. Conner, T. W. Jeffries, J. L. Minor, R. C. Pettersen, R. W. Scott, E. L. Springer, and T. H. Wegner, "Two-Stage, Dilute Sulfuric Acid Hydrolysis of Wood," Gen. Tech. Rep. FPL-45. Madison, WI: U.S. Department of Agriculture, Forest Service, Forest Products Laboratory; 1985. 73p.

12. K. N. Cederquist, "Some Remarks on Wood Hydrolyzation. Report of a Seminar on the Production and Use of Power Alcohol in Asia and the Far East," Lucknow, India, United Nations Rep. ST/TAA/Ser. C/10 (1954).

13. R. W. Scott, T. W. Wegner, and J. F. Harris, *J. Wood Chem. Technol.*, **3**, 245 (1983).

14. J. F. Saeman, Bioenergy '80 World Congress and Exposition, Atlanta, GA, 1980.

15. H. Luers, *Z. Angew. Chem.*, **43**, 455 (1930).

16. J. F. Saeman, *Ind. Eng. Chem.*, **37**, 43 (1945).

17. J. F. Saeman, A. M. Kirby, E. P. Young, and M. A. Millett, "Rapid High Temperature Hydrolysis of Cellulose," Res. Mark. Act Stud. Rep. 30, USDA, Forest Serv., Forest Prod. Lab., Madison, WI (1950).

18. A. M. Kirby, Jr., "Kinetics of Consecutive Reactions Involved in Wood Saccharification," Masters Thesis (Chem. Eng.), University of Wisconsin–Madison, Madison, WI (1948).

19. E. L. Springer, *Cellulose Chem. Technol.*, **17**, 525 (1983).

20. M. A. Millett, W. E. Moore, and J. F. Saeman, *Ind. Eng. Chem.*, **46**, 1493 (1954).

21. S. W. McKibbins, "Kinetics of the Acid Catalyzed Conversion of Glucose to 5-Hydroxymethyl-2-Furaldehyde and Levulinic Acid," Ph.D. Thesis (Chem. Eng.), University of Wisconsin–Madison, Madison, WI (1958).

22. S. W. McKibbins, J. F. Harris, J. F. Saeman, and W. K. Neill, *Forest Prod. J.*, **12**, 17 (1962).

23. J. F. Harris, to be published.

24. R. P. Bell, *Acid–base Catalysis*, Oxford University Press, New York, 1941.

25. A. S. Spriggs, "Studies on the Composition of Starch Hydrolyzates and Glucose Reversion Mixtures," Ph.D. Thesis, Washington University, St. Louis, MO (1954).

26. A. Thompson, K. Anno, M. L. Wolfrom, and M. Inatoma, *J. Am. Chem. Soc.*, **76**, 1309 (1954).

27. S. Peat, *J. Chem. Soc.*, 586 (1958).

28. J. L. Minor, *J. Appl. Polym. Sci. Symp.*, **37**, 617 (1983).

29. P. C. Smith, H. E. Grethlein, and A. O. Converse, *Solar Energy*, **28**, 41 (1982).

30. Von A. Sroczynski and M. Boruch, *Die Starke*, **16**, 215 (1964).

31. J. Szejtli, *Sauerhydrolyse Glykosidischer Bindungen*, Akademiai Kiado, Budapest, 1976.

32. M. H. Lietzke, R. W. Stoughton, and T. F. Young, *J. Phys. Chem.*, **65**, 2245 (1961).

33. A FORTRAN version is available from the authors.

34. A. C. Norris, *Computational Chemistry: An Introduction to Numerical Methods*, Wiley, New York, 1981.

35. American Society for Testing and Materials, "Standard Test Method for Ash in Wood," ASTM Designation D 1102-56.

36. E. L. Springer and J. F. Harris, *Ind. Eng. Chem., Prod. Res. & Dev.* (accepted).

37. N. Nelson, *J. Biol. Chem.*, **153**, 375 (1944).

38. R. C. Pettersen, V. H. Schwandt, and M. J. Effland, *J. Chrom. Sci.*, **22**, 478 (1984).

39. American Society for Testing and Materials, "Standard Method for Chromatographic Analysis of Chemically Refined Cellulose," ASTM Designation D 1915-63.

40. M. J. Effland, *Tappi*, **60**, 143 (1977).

41. M. Losin and D. Sumnicht, "An Experimental Study of the Kinetics of Glucose Decomposition in Dilute Acid," unpublished student report (Chem. Eng.), University of Wisconsin–Madison, Madison, WI (1983).

17

Hydrogen Fluoride Saccharification of Lignocellulosic Materials

MARTIN C. HAWLEY, KEVIN W. DOWNEY, and SUSAN M. SELKE
Department of Chemical Engineering, Michigan State University, East Lansing, Michigan

DEREK T. A. LAMPORT
MSU-DOE Plant Research Laboratory, Michigan State University, East Lansing, Michigan

The current oil glut notwithstanding, the need for liquid fuels and chemical production schemes using renewable raw materials is evident. A potentially important source is the hydrolysis of wood and other lignocellulosics to produce sugars, which can then be converted to liquid fuels and chemicals. Processes for the conversion of lignocellulosics to sugars can be divided into concentrated acid, dilute acid, and enzymatic categories.

In 1819 Braconnot (1) discovered that concentrated acid could produce sugars from cellulose. In 1869 Gore (2) noted the ability of hydrogen fluoride to dissolve cellulose. Later, in 1894, Simonsen (1) became the first to saccharify wood via dilute acid hydrolysis. In 1899 Classen (1) used sulfurous acid in a pilot plant. In the United States, Ewen and Tomlinson (1) commercially used sulfuric acid saccharification until shortly after World War I.

With the advent of World War II, the scarcity of petroleum initiated new interest in wood hydrolysis technologies as a means of producing liquid fuels. Numerous European plants resulted, but the availability of cheap ethanol, produced from petroleum-derived ethylene, rendered these plants uneconomical following World War II. Recent oil shortages have again resurrected interest in saccharification technologies.

Hydrogen fluoride (HF) saccharification was extensively investigated in Germany (3) in the 1930s, and a pilot plant using HF technology was constructed in 1937, but little of this early work survived the war. Except for a brief investigation by Russian scientists in the 1950s (4, 5), HF saccharification was virtually forgotten until 1979, when several researchers began essentially simultaneous and independent investigations, prompted by their belief that HF saccharification has several basic advantages over competing technologies. These advantages have been demonstrated, but much more work is needed to bring the technological level of HF hydrolysis up to the level of other hydrolysis methods. Toward this end, several groups worldwide are currently conducting research on HF saccharification. Figures 17.1–17.4 show tentative process flow diagrams put forth by these researchers. The purpose of this paper is to provide a comprehensive summary of the work related to HF saccharification performed thus far, including the latest findings from all researchers.

The discovery by Gore in 1869 that HF could transform cellulose into "glutinous substances" opened the door for research into HF saccharification (2). Helferich and Bottger (6) investigated this reaction in 1929. Continuing this research in 1933, Fredenhagen and Cadenbach discovered the formation of glucosyl fluorides as an intermediate in the reaction of HF with filter paper (3). The HF and glucose were freed upon the addition of even small amounts of water, although the glucose could form reversion products of polyglucans. HF

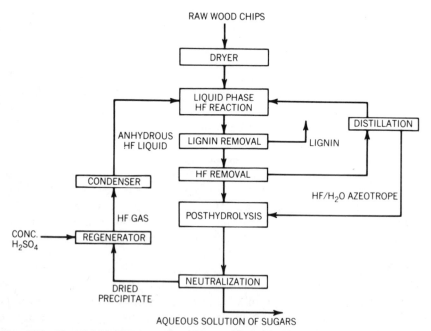

Figure 17.1   Liquid-phase HF process flow diagram. (Copyright 1982 American Chemical Society. Reprinted with permision from S. M. Selke et al., *I. & EC Prod. Res. Dev.*, **21**, 14 (1982)).

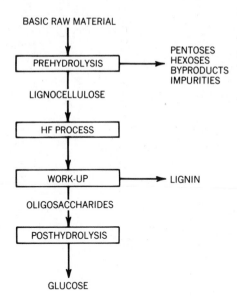

Figure 17.2 Hoechst's general HF saccharification process. (R. Franz et al., in A. Strub, P. Chartier, and G. Schleser (eds.), *Energy from Biomass*, Applied Science Publisher, London, 1982, p. 874.)

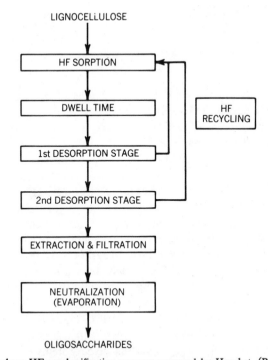

Figure 17.3 Practical gas-phase HF saccharification process proposed by Hoechst, (R. Franz et al., in A. Strub, P. Chartier, and G. Schleser (eds.), *Energy From Biomass*, Applied Science Publisher, London, 1982, p. 875.)

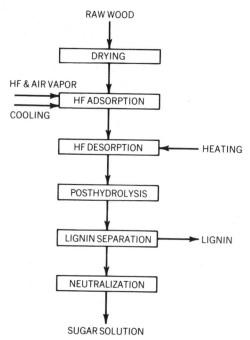

Figure 17.4 Gas-phase HF saccharification scheme proposed by Canertech. (C. M. Ostrovski, H. E. Duckworth, and J. C. Aitken, "Canertech's Ethanol-from-Cellulose Program," presented at the VI International Symposium on Alcohol Fuels Technology, Ottawa, Canada, May 21–25, 1984, in preparation.) (Courtesy of Canertech Inc., Winnepeg, Canada).

saccharification was suggested as a means of determining the lignin and carbohydrate content of the lignocellulosics. In addition to liquid-phase HF, Fredenhagen and Cadenbach studied the gas-phase HF reaction, viable with a temperature low enough to allow the formation of a liquid adsorption layer. They obtained sugar yields of up to 95% while using a 1:1 wood-to-gas phase HF ratio, by weight. The HF was removed at a temperature of 100°C (3).

In 1937, Hoch and Bohunek used gas-phase HF under low pressure which they believed made the formation of a liquid adsorption phase unnecessary, and a pilot plant in Germany ran for 6 months using their technology. The reaction parameters were temperature 35–40°C, pressure 30 mm Hg, reaction time 30 min, and an HF (95% pure) to-wood ratio (2–3% water content) of 0.4:1. Sugar yields of 90% were achieved, of which 80% were recovered for further processing. After removal of contaminating acetic acid, the HF was recycled (7, 8).

The HF saccharification work at Michigan State University began as an outgrowth of research done at the MSU Plant Research Laboratory. HF was found to be a good reagent for removing cell wall polysaccharides from cell wall protein and for deglycosylating glycoproteins (9). The idea of using HF to hydrolyze cellulose was then investigated—not a new idea, as it turned out.

Potentially, a vast array of lignocellulosic materials can be used as feedstocks for hydrogen fluoride saccharification. Because the composition of each of these materials differs, some will be preferred over others, depending on the desired end product. For example, if glucose was the desired product, then a material high in cellulose would be preferred, whereas if phenolics were desired, a material high in lignin would be selected. Table 17.1 shows the composition of several common lignocellulosics in terms of cellulose, hemicellulose, and lignin content (10).

Once a suitable feedstock has been chosen, there is still a major choice to be made between using HF in the liquid or the vapor phase. Reaction in the gas phase may allow easier HF recycling, quicker reaction rates (reducing equipment size), and smaller HF requirements. Liquid-phase reaction may reduce sugar degradation and increase hydrolysis of glucosyl fluorides. At present the gas-phase reaction seems to be the likely choice for a commercial application, but only further research can determine the optimal reaction phase.

An evaluation of the economics of an HF saccharification system for production of ethanol from wood was recently presented by Wright et al. at the Solar Energy Research Institute. They concluded that HF saccharification is the reaction system most likely to be economically competitive with the long-established dilute sulfuric acid percolation method for biomass hydrolysis (11).

TABLE 17.1 Biomass Content

Biomass	Cellulose Content (%)		Hemicellulose Content (%)		Lignin Content (%)	
Barley	Str	41.9	E	44.27–5.66	16–22	
			Hu	27.0–30.1		
Corn	Sta	34.4	Sta	23.7	Sta	10.5
	C	34.9	C	37.3	C	7.4
Sorghum	Sto	39.6	Str	25	H	13.6
Hay (loose)	A	35	Str	25	A	7.3
Oats	Str	49.3	Str	25	Str	14–22
Rye	Str	41.7	Str	24		
Soybeans	Sto	46.3	Str	25	H	11.4
					Hu	6.5
Wheat	Str	47.0	Str	25	Str	13.9
Wood						
Hardwood		45.8		30.7		20.3
Softwood		43.8		24.5		29.5

A, All hay types combined; E, endosperm; Hu, hull; Sto, stover; C, cob; H, hay; Sta, stalk; Str, straw.

Source: J. F. Bartholic et al., in *Wood and Agricultural Residues, Research on Use for Feed, Fuels, and Chemicals,* E. Soltes (ed.), Academic, New York, 1983, pp. 543–544.

HF REACTION PROCESSES

The reaction of HF with a lignocellulosic substrate is extremely complex. A wood feedstock contains not only cellulose and lignin but also hemicellulose, extractives, and ash, which complicates the reacting system. Furthermore, as the reaction progresses, the porosity and geometry of the solid substrate change both inside and out. The variation of porosity and composition during reaction when coupled with the nonuniformity of the raw material is very difficult to model. However, a great deal is already known about the reacting system, and certain simplifications yield a model that, at least with the liquid phase HF reaction, works quite well.

The reaction of HF on cellulose yields sugar fluorides according to the following reaction:

$$H(C_6H_{10}O_5)_n{-}OH + HF \rightarrow H{-}(C_6H_{10}O_5)_j{-}F$$

$$+ H{-}(C_6H_{10}O_5)_k{-}OH$$

where $j + k = n$. In the presence of even small amounts of water, the HF is liberated, yielding smaller oligosaccharides:

$$H{-}(C_6H_{10}O_5)_j{-}F + H_2O \rightarrow H{-} (C_6H_{10}O_5)_j{-}OH + HF$$

Upon removal of the HF, the sugar fluorides undergo reversion to yield a mixture of soluble oligosaccharides varying in size.

$$H{-}(C_6H_{10}O_5)_j{-}F + H{-}(C_6H_{10}O_5)_k{-}F \rightarrow H{-}(C_6H_{10}O_5)_n{-}F + HF$$

There is an equilibrium between the sugar fluoride and reversion products that is temperature- and concentration-dependent (12).

To simplify the system somewhat, consider the wood substrate to be composed of only cellulose, hemicellulose, and lignin. At the liquid–solid interface (or gas–solid, depending on the HF phase), the HF will diffuse toward the surface, then into the substrate itself. The size of the HF molecule is sufficiently small (on the order of 1 Å) to penetrate the wood pores, which account for approximately 55% of the wood volume. In the case of liquid-phase HF reaction, the cellulose and sugar fluorides will diffuse away from the solid surface. If water is present, its diffusion direction will depend on its concentration in the wood and HF solution. Reactions occur both on the surface and inside the wood particle, but outward diffusion of the large sugar fluorides from inside the particle is difficult. For gas-phase HF reaction, the sugar fluorides will not diffuse. The reaction at the surface will tend to change the substrate such that the diffusion of HF through it may be different from through unreacted wood. For further simplification at high temperature the gas-phase reaction can be viewed as being mass-transfer-limited, so there will be diffusion of HF only through previously reacted material. Similarly, the liquid-phase reaction can be viewed as having no mass transfer limitations, as long as the particle size is small enough (13), with the reaction rate as the system limiting factor. Different views of the reacting system are therefore taken depending on whether reaction takes place with HF in the liquid or the gas phase. In reality under certain conditions both mass transfer and reaction resistances are important. These conditions can be determined from well-designed rate experiments.

The above considerations do not include the effect of the lignin matrix hampering the transport of HF through the lignocellulose. Therefore, apparent or global reaction rates for cellulose are higher than for lignocellulose because of unhindered HF transport.

Selke has extensively investigated liquid phase HF saccharification (13–15). The reaction system was viewed as homogeneous, and all intermediates were neglected. A pseudo-first-order reaction model was proposed yielding the following (13–15):

$$\underset{\text{Cellulosic glucan}}{\text{C}} \quad + \text{HF} \longrightarrow \quad \underset{\text{Soluble glucan}}{\text{G}}$$

$$\frac{dC_c}{dt} = -k\,C_c, \qquad C_c = C_{c_0}\,e^{-kt}$$

and

$$\underset{\text{Hemicellulosic xylan}}{\text{H}} \quad + \text{HF} \longrightarrow \quad \underset{\text{Soluble xylan}}{\text{X}}$$

$$\frac{dC_H}{dt} = -k\,C_H, \qquad C_H = C_{H_0}\,e^{-kt}$$

In general, the first-order model appears to fit the data fairly well if an "initial delay" parameter is added to account for time required for solubilization of cellulose. A detailed discussion of the results of this analysis appears in the rate data and kinetics portion of this paper.

The gas-phase HF system is viewed as being diffusion-limited, as previously discussed. One method of modeling this system would be to view the wood chip as a layer of air where the diffusion rate of the HF through air would be the limiting step. Fick's laws of diffusion are applied. Kinetic studies for the gas-phase reaction have not been reported in the literature, but studies are currently under way at Michigan State University.

The heat of reaction of liquid HF with cellulose is on the order of −94 cal/g wood (3). In addition, there are heats of sorption involved, at least for gas-phase reaction. Therefore, coupled to the mass balance problems (reactant and product inventories), there are also energy balance problems to be investigated.

With the goal of a viable ethanol plant using HF saccharification technology, several investigators worldwide are conducting research toward solution of these mass and energy balance problems.

PRELIMINARY HF SACCHARIFICATION ECONOMICS

To help determine the feasibility of the HF saccharification process, an economic study was done for both the liquid- and gas-phase reaction systems. The results of this study are summarized in Table 17.2 (16). Because of the lack of actual data concerning the HF system, several assumptions were made. The preliminary

TABLE 17.2 Manufacturing Cost and Profitability Buildups for Base Case in 1982 Dollars[a]

	Liquid Phase	Gas Phase
Fixed capital investment (no control equipment)	15,585,000	8,330,00
Control fixed capital (5% of above)	779,000	417,000
Total Fixed Capital Investment	16,364,000	8,747,000
Manufacturing costs (annual)		
HF	363,000	363,000
Other chemicals (H_2SO_4)	670,000	670,000
Utilities	2,763,000	2,842,000
Wood	4,922,000	4,922,000
Labor	1,050,000	1,050,000
Supervision, overhead, engineering, maintenance (15% of total fixed capital)	2,455,000	1,312,000
Depreciation	1,091,000	583,000
Total Manufacturing Cost	13,314,000	11,742,000
Revenues (annual)		
Ethanol	18,200,000	18,200,000
Xylose syrup	4,282,000	4,282,000
Lignin	2,706,000	2,706,000
Total Revenues	25,188,000	25,188,000
Net income before tax	11,874,000	13,446,000
Net income after tax (48% tax rate)	6,174,000	6,992,000
Depreciation	1,091,000	583,000
Cash flow/year	7,265,000	7,575,000
Return on investment (after tax)	37.7%	79.9%

[a] Base case assumes 15-year plant lifetime, zero salvage value, straight-line depreciation. For other assumptions, see text.

Source: K. Downey, S. Selke, and M. Hawley, in Proceedings of the Seventh International FPRS Industrial Wood Energy Forum, Nashville, TN, Sept. 19–21, 1983.

economics appear favorable for both the liquid- and gas-phase systems, with the gas-phase significantly more profitable. The gas-phase reaction is favored owing to quicker reaction times (lower reactor capital costs), lower HF requirements, and greater ease in HF recycling. However, the higher temperatures associated with gas-phase HF operation may result in an increase in sugar product degradation, thus favoring reaction in the liquid phase. Only further research will determine if the gas phase is indeed the optimum phase.

COMPARISON OF HYDROLYSIS TECHNOLOGIES

Currently, various hydrolysis technologies competing with hydrogen fluoride saccharification are being investigated worldwide. These processes fall into three general categories: concentrated acid processes, dilute acid processes, and enzymatic processes.

Concentrated acid processes using sulfuric or hydrochloric acid have higher sugar yields than dilute acid (H_2SO_4) or enzymatic processes, short reaction times, and ambient operating requirements. They do have several disadvantages such as severe acid corrosion and consumption problems, expensive recycling, a posthydrolysis requirement, and a condensed lignin byproduct, although HCl lignin is of significantly better quality than H_2SO_4 lignin (17).

Dilute acid processes are cheap, have shorter reaction times than comparable enzymatic processes, and do not require a dry feedstock. They require no posthydrolysis but do require a prehydrolysis (Chap. 16). Their drawbacks are that they require high temperature and pressure and have considerable sugar product degradation, resulting in yields on the order of 50% (18).

Enzymatic hydrolyses produce relatively pure products and require only ambient operating conditions. However, they require long reaction times and extensive pretreatments and give yields on the order of only 50%. Also, the cost of the enzymes is quite high (18). Either they can be coupled to a fermentation scheme or simultaneous saccharification and fermentation can be performed. Compatability of microorganisms for this second scheme may be difficult to achieve. Major areas of research are the development of better pretreatments and more productive enzymes (cf. Chap. 18).

HF saccharification has the advantages of high sugar yield, short reaction times, ambient temperatures and pressures, and no prehydrolysis requirement, although some researchers recommend it (20). In addition, without the prehydrolysis, it does not require a massive drying of a wet material, although a relatively dry substrate is required. The near ambient boiling point of HF will facilitate less expensive recycling than sulfuric or hydrochloric acid processes (11). In fact, 98–99% HF recycle efficiencies have been reported (20–22). Corrosion problems will be less than other concentrated acid processes, and the lignin may be quite valuable. Chemical and feedstock costs for HF saccharification were estimated to be 11.9¢/lb glucose compared with 13–20¢/lb glucose for dilute acid and enzymatic processes (14).

At the present time, HF saccharification appears to be the most efficient method for the production of chemicals and liquid fuels from lignocellulosics. It suffers but one major drawback: the status of its technology. Its technological development lags well behind that of other competing processes because of limited research. Its advantages warrant a great deal more research, and only with this research can a complete and reliable economic analysis be performed.

HF LIGNIN

Because lignin constitutes approximately 30% of a lignocellulosic feedstock, it represents an important hydrolysis by-product. In fact, the choice of what to do with the lignin fraction may well determine the economic feasibility of an ethanol plant utilizing HF saccharification technology. The current consensus is that the lignin product should be burned to supply the necessary process heat for the ethanol plant. However, a preliminary energy analysis of a 10 MM gal/ yr ethanol plant using HF hydrolysis concludes that even if lignin is used to supply the required process energy, a significant fraction of this lignin remains unused. Therefore some use must be found for this extraneous lignin supply. The choices range from selling the lignin for fuel to be burned elsewhere to using it as a feedstock for chemical production. Table 17.3 lists a variety of uses for lignin (23, 24).

Potentially, a large variety of products can be produced from lignin, as seen from Table 17.3. For any of these conversion pathways other than simple combustion to be economically viable, however, a great deal of research is needed. Toward this end, research on HF lignin conversion continues worldwide.

Probably the single most important goal of HF lignin research is to deduce the structural makeup of the lignin. This is an extremely difficult task because of the complex polymeric nature of lignin and has yet to be accomplished. However, many details concerning HF lignin that will be useful in its eventual characterization have been found. These include information regarding solubility, condensation, and general physical properties.

Although a great deal more needs to be discovered about the solubility of HF lignin, investigations show a marked difference between HF lignin and kraft lignin. Kraft lignin is the sulphonated and highly condensed lignin by-product from the kraft (sulfate) pulping process. For example, at room temperature only 40% of HF lignin (recovered from a 20°C reaction temperature) was soluble in 0.1 N NaOH, whereas 100% of kraft lignin was soluble (25, 26). This is probably due to molecular weight and cross-linking differences. Since molecular weight and degree of cross-linking may be reaction-dependent, solubility studies must be performed on lignin generated at a variety of HF hydrolysis conditions as well as for both liquid- and gas-phase HF systems.

Defaye et al. have investigated the ether linkages in HF lignin obtained from birch (*Betula verucosa*) (12). Their wood samples were previously solvent-extracted and dried, and liquid-phase HF was used. They discovered that the ester

TABLE 17.3 Current and Potential Uses for Lignin

Chemical Intermediates
  Hydrogenation
    Phenols
    Substituted Mononuclear Phenols
    Cresols
    Benzene
  Hydrolysis
    Phenols
    Substituted Mononuclear Phenols
    Catechols
  Alkali Fusion
    Phenols
    Phenolic Acids
    Catechols
  Pyrolysis
    Acetic Acid
    Phenols
    Substituted Mononuclear Phenols
    Methane
    Carbon Dioxide
    Carbon Monoxide
    Carbon
    Hydrogen
  Fast Thermolysis
    Acetylene
    Ethylene
  Oxidation and Miscellaneous Processes
    Vanillin
    Vanillic Acid
    Syringaldehyde
    Syringic Acid
    Dimethyl Sulfide
    Dimethyl Sulfoxide
    Methyl Mercaptan

Polymeric Lignins
  Adhesive
  Dispersant
  Complexing Agent
  Stabilizer
  Antioxidant
  Precipitant
  Deflocculant
  Coagulant
  Pesticide Carrier
  Emulsion Stabilizer
  Soil Conditioner
  Fertilizer
  Rubber Reinforcement
  Phenol–Formaldehyde
    Resin and Extender
  Urea–Formaldehyde Resin
    and Extender
  Polyurethane Foam

and ether linkages were probably not cleaved, even under continued exposure to HF. However, a significant degree of condensation, as determined via alkaline nitrobenzene oxidation, had occurred. In fact, the condensation was to such an extent that HF lignin appeared to be more highly condensed than other acid hydrolysis lignins, as shown in Table 17.4, where each lignin was generated under the conditions for optimum carbohydrate recovery (12). Table 17.5 shows the reduction in vanillin and syringaldehyde yields with time as a function of temperature (12). Smith et al. similarly reported a decrease in yields of vanillin, vanillic acid, syringaldehyde, and syringic acid after HF hydrolysis of poplar

TABLE 17.4 Comparison between Vanillin and Syringaldehyde Yields from HF–Birch Wood Lignins and Various Acidic Birch Extraction Lignins[a]

	Vanillin (mole %)	Syringaldehyde (mole %)	Total Aldehydes (mole %)	SYR/VAN
Total birch Wood lignin	20.3	22.3	42.6	1.1
HF–lignin	10.4	9.4	19.8	0.9
HCl–lignin	9.1	16.4	25.5	1.8
H$_2$SO$_4$–lignin	8.4	10.9	19.3	1.3

[a] HCl–lignin is prepared by successive treatments of birch wood with 30% HCl at 40°C for 90 min, then 43% HCl at 50°C for 2 hr. H$_2$SO$_4$–lignin is prepared by successive treatments with 0.5% H$_2$SO$_4$ at 140°C for 1 hr, then 180°C for 2 hr.

Source: J. Defaye, A. Gadelle, J. Papadpoulos, and C. Pederson, *J. Appl. Polym. Sci.*, **37**, 668 (1983).

(27). The condensation can take place by two possible pathways: autocondensation via a Friedel–Krafts type reaction, or condensation between lignin and carbohydrate leading to C—C bond formation, thus reducing the aldehyde yield that results from nitrobenzene oxidation. To avoid excessive condensation, reaction time should be kept under 30 min (12).

Because liquid-phase HF was used in the Defaye condensation studies, fairly long reaction times were necessary because of the relatively low temperatures.

TABLE 17.5 Effect of Time and Temperature of Exposure to HF on the Vanillin and Syringaldehyde Yields Derivable from Lignin after HF Hydrolysis of Birch Wood

Time (min)	Vanillin (mole %)	Syringaldehyde (mole %)	Total Aldehydes (mole %)	SYR/VAN
Lignin extracted at 23°C				
0	20.3	22.3	42.6	1.1
10	14.3	14.3	28.6	1.0
20	12.0	12.0	24.0	1.0
30	10.4	9.4	19.8	0.9
60	8.7	6.9	15.6	0.8
120	7.2	5.7	12.9	0.8
Lignin extracted at 0°C				
0	20.3	22.3	42.6	1.1
60	12.3	12.3	24.6	1.0
120	9.0	10.0	19.0	0.9
180	8.9	7.1	16.0	0.8
300	6.8	5.4	12.2	0.8

Source: J. Defaye, A. Gadelle, J. Papadopoulos, and C. Pederson, *J. Appl. Polym. Sci.*, **37**, 668 (1983).

Long reaction times resulted in severe condensation. A gas phase where shorter reaction times and a lower HF concentration would be required could reduce condensation. However, the higher reaction temperature corresponding to gas-phase HF reaction may itself increase condensation.

HF lignin has a higher degree of thermal stability than kraft lignin. Degradation of kraft lignin occurred at 100°C, but HF lignin degradation did not occur even at 250°C (25). Also, HF lignin did not agglomerate as kraft lignin did, possibly important for use as a resin or extender. For the feasibility of proposed HF lignin conversions to be fully evaluated, the molecular weight distribution, surface area and porosity, bulk density, degree of cross-linking, and many other characteristics need to be determined as a function of reaction conditions.

POSSIBLE FLUORIDE LOSSES

Because of the relatively high cost of HF, acid losses must be kept low for any HF saccharification plant to be economically feasible. Therefore, high HF recycle efficiencies are necessary, and losses due to side reactions with impurities should be avoided.

HF losses can result from reaction with the ash, reaction with extractives, fluoridation of the lignin component, and general recycle inefficiencies. The first two can be eliminated by a prehydrolysis with dilute mineral acid, a potentially costly step. However, the HF losses due to reaction with the calcium, sodium, and potassium oxides of the ash represent only a small fraction of the overall operating cost of a 10 MM gal/yr ethanol plant (16). This loss can be reduced through the regeneration of HF via reaction of CaF_2 (the major ash salt) with concentrated sulfuric acid.

Lignin fluoridation studies by Hardt and Lamport showed that all HF losses could be accounted for by reaction with the ash content of the wood (28). Fluoride retention in the lignin and sugar fraction after HF evacuation was approximately 2 mg/g wood. With subsequent washings and grindings, this value was reduced to 0.1 mg/g wood (28). HF losses resulting from lignin–HF reaction are negligible relative to those from the HF–ash reaction.

Because of the near ambient boiling point of HF, efficient recycling should be relatively easy. Estimates of losses as a function of HF:wood ratio were made assuming a 99.5% recovery of HF and no recovery of ash fluoride losses for a 10 MM gal/yr ethanol plant, as shown in Tables 17.6 and 17.7 (16). As shown by these tables, it is advantageous to keep the HF:wood ratio as low as possible, thus favoring reaction in the gas phase. Losses for relatively high HF:wood ratios for liquid-phase reaction could be considerably higher than for the gas-phase reaction because of the larger quantity of HF contained in the process. The design ratio for gas phase was 0.35 parts HF to 1 part wood by weight (29), quite close to Hoch and Bohunek's value of 0.4:1 (7, 8). The assumed recycle efficiency of 99.5% appears attainable, since several researchers have already reported recycle efficiencies of 98–99% and above (20–22).

TABLE 17.6   HF Losses as a Function of HF:Wood Ratio for Liquid
Phase, Assuming an Ash Content of 0.3% and a 99.5% Recovery of
HF Other Than Ash Salts

HF:Wood Ratio	HF Losses ($/yr)
0.5 : 1	807,000
1.0 : 1	1,253,000
1.5 : 1	1,699,000
2.0 : 1	2,145,000
3.0 : 1	3,037,000
4.0 : 1	3,930,000
5.0 : 1	4,822,000
6.0 : 1	5,714,000
7.0 : 1	6,606,000
8.0 : 1	7,498,000
9.0 : 1	8,390,000
10.0 : 1	9,283,000

Source: K. Downey, S. Selke, and M. Hawley, Proceedings of the Seventh International FPRS Industrial Wood Energy Forum, Nashville, TN, Sept. 19–21, 1983.

PRETREATMENT AND POSTHYDROLYSIS

Research continues toward determining possible pretreatments and/or posthydrolyses that will enhance the HF saccharification of lignocellulose. Possible pretreatments include chipping, drying, solvent extraction, steam explosion, and prehydrolysis. Though moderate drying (down to 3–10% moisture) and chipping are required, solvent extraction and steam explosion are not necessary for the HF process. A prehydrolysis may have advantages. Franz et al., proponents of a dilute sulfuric acid prehydrolysis aimed at removing the hemicellulose component, report the following benefits: removal of impurities, increased porosity of the substrate due to opening of the fibrillar structure, lower HF requirements, less shrinkage of the lignocellulose during reaction, and macroscopic properties of the reacted product unchanged from those of the unreacted product (20). However, prehydrolysis requires the drying of a wet material prior to HF treatment. Thus, as Lohaus and Schlingmann state, the desirability of a prehydrolysis step is a matter of economics (30).

Optimally, the HF hydrolysis product can be directly fermented to ethanol. However, the typical fermenting organisms require the sugar substrate to be predominantly in monomeric form. Since the product resulting from HF saccharification undergoes reversion and yields an oligomeric mixture, a posthydrolysis is required to obtain monomeric glucose. A simple dilute sulfuric acid hydrolysis is sufficient to convert the reversion products (mainly α-1-6 linkages) to glucose (28).

An important area of research is to develop an organism that can directly ferment the oligomeric sugar mixture, thus eliminating the entire posthydrolysis

TABLE 17.7 HF Losses as a Function of HF:Wood Ratio for Gas Phase, Assuming an Ash Content of 0.3% and a 99.5% Recovery of HF Other Than Ash Salts

HF:Wood Ratio	HF Losses ($/yr)
0.05	406,000
0.10	450,000
0.15	495,000
0.20	539,000
0.25	584,000
0.30	629,000
0.35	673,000
0.40	718,000
0.45	762,000
0.50	807,000

Source: K. Downey, S. Selke, and M. Hawley, Proceedings of the Seventh International FPRS Industrial Wood Energy Forum 83, Nashville, TN, Sept. 19–21, 1983.

step. If posthydrolysis is necessary, there are several options, such as dilute sulfuric acid, less concentrated HF, or enzymes.

CARBOHYDRATE ANALYSIS METHODS AND YIELDS

A variety of analytical methods including NMR, mass spectroscopy, chromatography, and derivatization are used in the analysis of the carbohydrate product resulting from HF saccharification. After initial separation of the HF from the reaction mixture, residual F— can be measured via a fluoride electrode. The intermediate glucosyl fluoride was isolated via rapid termination of the reaction and neutralization with calcium carbonate (28). Its structure was identified after separation from the reaction mixture by paper chromatography, through its specific optical rotation, and by GC–MS analysis of its TMS (trimethylsilyl) derivative. The intermediate was identified as α-D-glucopyranosyl fluoride (28). ^{13}C NMR studies have also aided structural analysis of the reaction products (12).

The sugar products can be either directly analyzed via HPLC or derivatized and analyzed by GLC. The derivatives generally prepared are either trimethylsilyl derivatives of the methyl glycosides or alditol acetates. HPLC offers several advantages such as shorter analysis time and no degradation from derivatization, and it provides oligomer distribution data.

The sugar yields resulting from HF saccharification have consistently been higher than those for enzymatic or dilute acid hydrolysis technologies. These high yields have been corroborated by various researchers worldwide using several different substrates. Selke et al. at Michigan State University have reported sugar yields of up to 95% of theoretical from Bigtooth aspen (*Populus grandidentata*) and virtually quantitative yields from cellulose (filter paper) in liquid-phase HF containing up to 10% water by weight (13–15). Using birch wood, Defaye et al. in France obtained yields of 90% of both theoretical glucose and xylose (both in monomer and oligomer form) in a 4:1 anhydrous liquid HF:wood ratio at 23°C for 30 min (12). Using barley straw in a gas-phase HF reaction system, Bentsen in Denmark obtained yields of fermentable sugars in excess of 90% (22). Ostrovski et al. at Canertech, using gas-phase HF reaction, obtained yields of over 90% of theoretical glucose and over 80% of theoretical xylose from Canadian poplar (21). Finally, Franz et al. at Hoechst in Germany observed a conversion of over 90% of the carbohydrate present in the original raw material to soluble form using dilute acid prehydrolysis on a variety of woods and straw followed by reaction with gas-phase HF (20).

RATE DATA AND KINETICS FOR LIQUID-PHASE HF STUDIES

Rate data for liquid-phase HF reaction have been generated for filter paper (pure cellulose) and wood chips (*Populus grandidentata*) as a function of temperature, water concentration, and wood particle size in a specially designed Kel–F reaction system (13, 15). Temperatures ranged from −43°C to 5°C with most work at 0°C, −13°C, and 5°C. Water concentrations of 3.6% and 6.4% H_2O (measured as a percentage of HF present), 0% (anhydrous HF), and wood particle sizes of 60-mesh ground wood, small wood chips (1–3 cm × 0.2—0.8 cm × 0.05—0.2 cm), and large wood chips (2–3 cm × 2–3 cm × 0.5 cm) were used (13, 15).

Reaction rate constants and activation energies from these studies are summarized in Tables 17.8 and 17.9. Some representative data for conversion versus time together with the calculated yield curves are shown in Figures 17.5 and 17.6. A pseudo-first-order reaction model fit the kinetic data only if an initial time delay was incorporated. This is presumed to represent time required for solubilization of the substrate.

The exothermic nature of the HF–cellulose reaction coupled with the poor thermal conductivity of the experimental Kel–F reaction vessel led to deviations from isothermal behavior during the course of the reaction. To reduce possible error from the assumption of a constant average temperature, the temperature dependence of the rate, in an Arrhenius form, was incorporated into a second analysis. The temperatures between two successive sample times were averaged and taken to be a constant value, eliminating much of the error because the averaging takes place over a much smaller range. A nonlinear curve-fitting program developed at Michigan State University was then used to analyze the data. Tables 17.10 and 17.11 list the rates and activation energies calculated from this method for glucose and xylose, respectively (13, 15).

TABLE 17.8 Reaction Rates for Glucose—Average T Model

Substrate and Conditions	Average Temperature (°C)	Rate (min⁻¹)	Initial Delay (min)	Correlation Coefficient
Filter paper, anhydrous	1.3	0.235	0.6	0.896
	−1.9	0.225	3.2	0.978
	−9.5	0.029	9.8	0.991
	−11.9	0.011	16.1	0.924
Filter paper, 3.6% water	1.4	0.109	2.7	0.994
	0.3	0.102	2.9	0.973
	−7.4	0.042	14.9	0.983
	−9.0	0.023	13.0	0.952
Filter paper, 6.4% water	1.3	0.066	3.0	0.925
	−6.0	0.013	2.8	0.948
	−7.2	0.025	5.3	0.945
Wood, anhydrous	3.9	0.179	1.4	0.865
	−3.6	0.079	5.0	0.970
	−8.2	0.063	12.6	0.933
	−8.5	0.061	11.2	0.903
Wood, 3.6% water	3.1	0.091	2.2	0.921
	−9.1	0.024	5.4	0.951
	−9.6	0.032	8.5	0.953
	−10.6	0.036	9.5	0.969
Wood, 6.4% water	2.5	0.086	1.8	0.959
	−6.2	0.012	4.3	0.692
	−6.4	0.015	2.9	0.950

Source: S. M. Selke, Ph.D. Thesis, Michigan State University, 1983, p. 90.

From the kinetics studies, various trends appear. The rate constant is higher for pure cellulose than for the cellulose in wood, probably because of the hindrance of the lignin matrix in wood. The initial delay parameter for glucose appearance decreases with increasing temperature. If the delay parameter represents the time required for solubilization of the substrate, then this trend is as expected because of the increase in solubilization rate with increased temperature (13, 15).

Another trend is that the reaction rate decreases as the water concentration increases. However, the activation energies for glucose are approximately equal for 3.6% and 6.4% water content and about 10 kcal higher for anhydrous HF with filter paper. For wood, the glucose activation energies are about equal for anhydrous and 3.6% water conditions, whereas 6.4% water gave an unrealistically high value (13, 15).

For xylose, the delay parameter was very small in all cases. Therefore, the results may be improved by forcing the curve through the origin. Although few

TABLE 17.9   Reaction Rates for Xylose—Average T Model

Substrate and Conditions	Average Temperature (°C)	K (min^{-1})	Initial Delay (min)
Wood, anhydrous	3.9	0.184	0.20
	−3.6	—[a]	
	−8.2	—[a]	
	−8.5	0.084	0.29
Wood, 3.6% water	3.1	0.157	0.29
	−9.1	0.158	1.32
	−9.6	0.151	0.62
	−10.6	—[a]	
Wood, 6.4% water	2.5	0.085	0
	−6.2	—[a]	
	−6.4	0.097	(<0)

[a] No meaningful results.

Source: S. M. Selke, Ph.D. Thesis, Michigan State University, 1983, p. 129.

data are available for xylose, it appears that an increase in water content affects xylan saccharification to a much smaller degree than glucan saccharification and that xylan saccharification has a much smaller activation energy. A more realistic model postulating successive first-order reactions for solubilization and reaction was developed to eliminate the artificial "initial delay" parameter. Further data are required to develop kinetic parameters for this model (13).

LIQUID-PHASE HF SOLUTION STUDIES

Cellulose is quite soluble in anhydrous HF, and concentrated solutions can be prepared in just a few minutes. Defaye et al. have studied solutions of approximately 40–50 wt% cellulose in anhydrous HF at temperatures of −10 to 20°C (12). There was no detectable solubility at −78°C. ^{13}C NMR, GLC, and HPLC were used to analyze the compositions of the HF solutions. Solutions of cellulose in HF held either at −20°C for 30 min or at −5°C for 40 min yielded partially hydrolyzed cellulose (water-soluble but β-1-4 linked) on precipitation with ether. Reaction at 20°C for 45 min produced a completely reacted product consisting primarily of α-D-glucopyranosyl fluorides and α-(D)-linked D-glucopyranosyl residues (reversion products) in a concentration-dependent equilibrium. α-D-Glucopyranosyl fluoride is predominant at higher dilutions, whereas the amount of reversion oligomers increases as the initial cellulose concentration increases. A small quantity of β-(1→4) linked glucopyranosyl oligomers are also present. The nature of the products obtained depends also on the recovery method. When

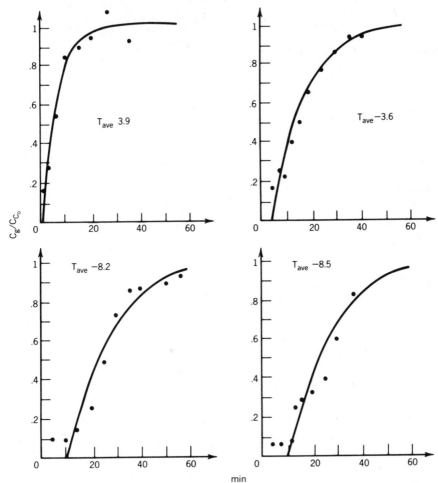

Figure 17.5 Wood treated with anhydrous HF: concentration-soluble glucan/initial concentration glucan versus time–average temperature model. (S. M. Selke, Ph.D. Thesis, Michigan State University, 1983, p. 96.)

evaporation of HF is used in place of precipitation of the product with diethyl ether, no noticeable α-D-glucopyranosyl fluoride is found, and the product is a complex mixture of oligosaccharides, with no more than 3% monomeric glucose (12).

Studies on D-xylans in HF solution yield only traces of α-D-xylopyranosyl fluoride with large amounts of xylo-oligosacharides regardless of temperature, concentration, and recovery conditions (12). Hardt and Lamport isolated α-D-xylopyranosyl fluoride from HF solvolysis of purified xylan (31).

Defaye et al. have proposed a mechanism for the HF reaction (Fig. 17.7),

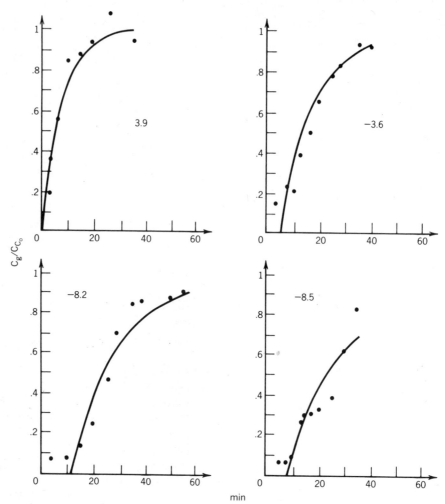

Figure 17.6 Wood treated with anhydrous HF: concentration-soluble glucan/initial concentration glucan versus time–varying temperature model. (S. M. Selke, Ph.D. Thesis, Michigan State University, 1983, p. 107.)

consisting of two general steps: (a) The cellulose undergoes an easy dissolution in the HF due to the strong tendency of HF to form hydrogen bonds, and the formation of hydrogen bonds with HF disrupts the internal hydrogen bonding between the polysaccharides; (b) the polysaccharides undergo degradation to produce sugar fluorides via protonation. The oxocarbonium ion is then formed via conjugate acid formation and is in equilibrium with the sugar fluoride. Though the oxocarbonium ion was not detected by their ^{13}C NMR, Defaye et al. postulate that it should have an appreciable lifetime (12).

The assumption that α-D-glucopyranosyl fluoride was the main intermediate was supported by the fact that reaction of D-glucose and cellulose with anhydrous

TABLE 17.10 Reaction Rates For Glucose—Varying T Model[a]

Substrate and Conditions	K (min^{-1})	E_a (kcal)	Initial Delay (min)
Filter paper, anhydrous	0.447 (0.094)	17.8 (2.4)	
1.3°C			0
-1.9°C			0
-9.5°C			24.3 (3.0)
-11.9°C			13.5 (1.9)
Filter paper, 3.6% water	0.553 (0.170)	23.5 (3.6)	
1.4°C			3.9 (0.7)
0.3°C			3.9 (0.5)
-7.4°C			14.0 (2.7)
-9.0°C			17.4 (1.9)
Filter paper, 6.4% water	0.130 (0.029)	12.6 (2.7)	
1.3°C			3.8 (0.6)
-6.0°C			22.1 (1.9)
-7.2°C			6.6 (1.6)
Wood, anhydrous	0.278 (0.076)	14.2 (2.7)	
3.9°C			1.0 (0.6)
-3.6°C			6.0 (0.8)
-8.2°C			12.2 (1.8)
-8.5°C			9.7 (1.1)
Wood, 3.6% water	0.289 (0.097)	15.8 (2.9)	
3.1°C			3.6 (0.8)
-9.1°C			5.3 (2.0)
-9.6°C			6.1 (2.0)
-10.6°C			8.9 (2.7)
Wood, 6.4% water	0.376 (0.128)	28.1 (4.2)	
2.5°C			1.4 (1.1)
-6.2°C			20.7 (3.9)
-6.4°C			6.9 (2.0)

[a] Numbers in parentheses are marginal standard deviations.
Source: S.M. Selke, Ph.D. Thesis, Michigan State University 1983, pp, 102–103.

TABLE 17.11 Reaction Rates for Xylose—Varying T Model

Substrate	K (10°C, min^{-1})	SD	E_a (kcal)	SD
Wood, anhydrous	0.184	0.060	6.5	3.1
Wood, 3.6% water	0.162	0.042	2.7	2.2
Wood, 6.4% water	0.295	0.091	16.5	3.5

Source: S. M. Selke, Ph.D. Thesis, Michigan State University, 1983, p. 134.

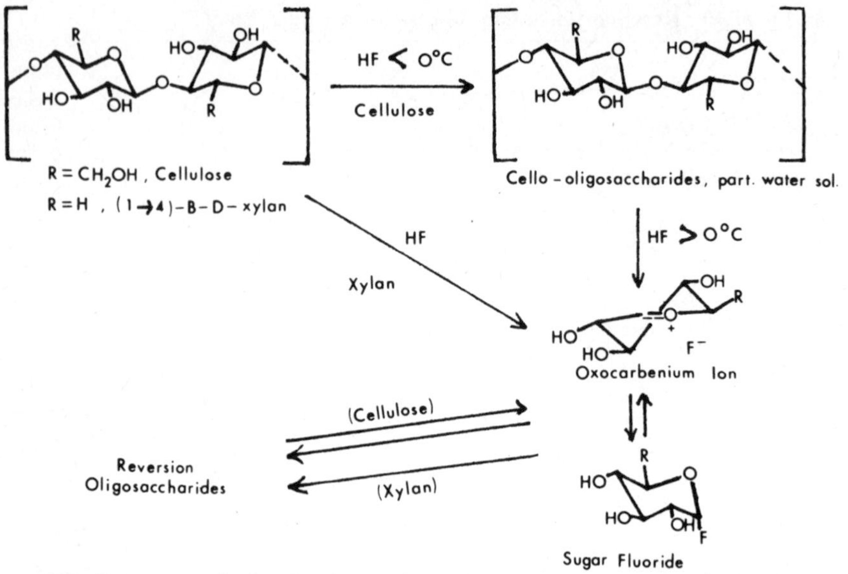

Figure 17.7 Proposed mechanism for the reaction of HF with polysaccharides, (Defaye et al, *J. Appl. Polym. Sci. Appl. Polym. Symp.*, **37**, 662 (1983)).

HF generated virtually identical ^{13}C NMR spectra after 45 min at 20°C. In addition, the effect of HF on amylose was studied to investigate the reaction with polysaccharides. As with D-glucose, the ^{13}C NMR spectra obtained for amylose after reaction for 45 min at 20°C was virtually identical to that of cellulose. Defaye et al. conclude that α-D-glucopyranosyl fluoride is the common intermediate, with the degree of fluoride and reversion oligomers being dependent on temperature and concentration conditions (12).

It is important to note that the reaction products contained no furan decomposition compounds, which are potent fermentation inhibitors. The formation of degradation furaldehyde products is believed to involve an open-chain enediol as an intermediate which is not formed in an HF system because of the stability of the proposed oxocarbonium ion (12).

GAS-PHASE HF SORPTION STUDIES

An economically viable HF saccharification process requires an efficient HF recovery/recycle system. Franz et al. at Hoechst have conducted adsorption/desorption studies of HF on lignocellulosic substrates (prehydrolyzed hardwoods, softwoods, and straw) aimed toward development of a simple HF recovery/recycle process (20). Figure 17.8 shows the equilibrium sorption isotherms of an HF/lignocellulose system. According to Franz et al. (20) the thick curved line represents discontinuous HF desorption. If a lignocellulose with an initial HF content of 60 parts HF to 100 parts lignocellulose at 40°C is heated

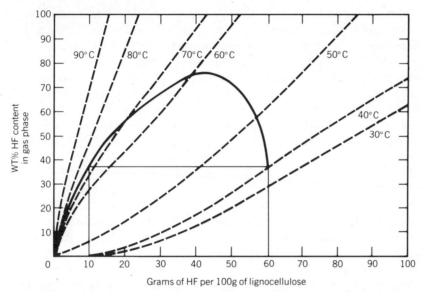

Figure 17.8 Equilibrium sorption isotherm of an HF/lignocellulose system. (R. Franz et al., *Energy from Biomass*, Appl. Sci. Pub., London, 1982, p. 877.) (Courtesy of John Wiley & Sons, publisher).

to 80°C, the initial desorption can only progress to a residual HF content of 10 parts HF to 100 parts lignocellulose at about 72°C. This must be the case for this gas mixture to be able to reestablish the required HF loading of 60 parts HF to 100 parts lignocellulose. A sequence of desorption stages is therefore required, producing in the later stages gas mixtures useful only for a presorption stage because of low concentration (20).

Franz et al. (20) have also studied the heats associated with these sorption steps. They observed differences between the isothermal and nonisothermal heats of HF sorption on lignocellulose explained by the unusually high heat capacity of gaseous HF between the normal boiling point of 19.5°C and 70°C. This high heat capacity will help maintain isothermal conditions during the exothermic saccharification reaction (30).

Heats of desorption are being studied at Hoechst (30) for recycle energy requirement determination. Two methods under investigation yield essentially the same results. The first involves use of the process in Figure 17.8 for the dependency of partial vapor pressure on temperature and calculation of the heat of desorption from the Clausius–Clapeyron equation. The second involves careful calorimetric studies (30).

SUMMARY AND CONCLUSIONS

HF saccharification was initially investigated in Germany in the 1930s. Modern investigations began in 1979 when the oil shortage and renewed interest in alternative energy technologies led several researchers independently to redis-cover HF saccharification as an attractive route to chemicals from biomass. The major process advantages stem from the very high (90% +) sugar yields ob-tainable and the relative ease of recycling an acid that boils at near-ambient temperatures (19.5°C). Compared to alternative hydrolysis technologies, this leads to an encouraging analysis of process economics. In addition, no appreciable degradation of either pentose or hexose sugars occurs in HF reactions, thus eliminating a major barrier to fermentation of the product sugars. Research on HF saccharification technology is currently under way at Michigan State Uni-versity and in Germany, Canada, France, and Denmark.

The reaction of HF with a lignocellulosic substrate is very complex. Sugar fluorides are formed via an oxocarbonium ion intermediate and can then further react with water to form free sugars or with other sugar molecules to form oligomeric reversion products, in either case liberating HF for continued reaction or recovery and recycle. In liquid-phase reactions the cellulose also dissolves in HF as hydrogen bonding with HF disrupts the internal hydrogen bonding between the polysaccharides. The lignin fraction is largely insoluble in HF. It undergoes a significant degree of condensation, either Friedel–Krafts type auto condensation or condensation between liquid and carbohydrate, but ester and ether linkages in the lignin are not cleaved.

Recovery and recycle of HF is an important determinant of process economics. Because HF boils at 19.5°C, operation of either a gas-phase or a liquid-phase reaction is possible with a high recycle efficiency. The major loss of HF is related to reaction of the ash components of the wood, forming fluorides. HF recycle efficiencies of 98% and better have already been reported.

The feasibility of production of high yields of sugar from lignocellulose uti-lizing an HF catalyst that is recovered and recycled has been demonstrated, and preliminary economic analyses are very encouraging. However, much remains to be done before this technology can be commercially applied.

REFERENCES

1. E. C. Sherrard and F. W. Kressman, *Ind. and Eng. Chem.*, **37**, 4 (1945).
2. G. Gore, *J. Chem. Soc.*, **22**, 396–406 (1869).
3. K. Fredenhagen and G. Cadenbach, *Agnew. Chem.*, **46**, 113 (1933).
4. Z. A. Rogovin and Y. L. Pogosov, *Gidroliz. i. Lesokhim. Prom.*, **11**(1), 4 (1958); *Chem. Abstr.*, **52**, 8453i (1958).
5. Z. A. Rogovin and Y. L. Pogosov, *Nauch Doklady Vysshei Shkoly, Khim. i. Khim. Tekhnol.*, **2**, 368 (1959); *Chem. Abstr.*, **53**, 22912h (1959).
6. B. Helferich and S. Bottger, *Justus Liebigs Annal. Chem.*, **476**, 150 (1929).
7. H. Luers, *Holz Roh Werkstoff*, **1**, 35 (1937).
8. H. Luers, *Holz Roh Werkstoff*, **1**, 342 (1938).
9. A. J. Mort and D. T. A. Lamport, *Anal. Biochem.*, **82**, 289 (1977).
10. J. F. Bartholic, J. Hanover, L. Tombaugh, K. Downey, M. Hawley, and H. Koenig, in *Wood and Agricultural Residues, Research on Use for Feed, Fuels, and Chemicals*, E. Soltes, ed., Academic, New York, 1983, pp. 529–565.

11. J. D. Wright, A. J. Power, and P. W. Bergeron, "Evaluation of Concentrated Halogen Acid Hydrolysis Processes for Alcohol Fuel Production," SERI Tech Report, TR-232-2386, Aug. 1984.

12. J. Defaye, A. Gadelle, J. Papadopoulos, and C. Pedersen, *J. Appl. Polym. Sci. Appl. Polym. Symp.*, **37**, 653 (1983).

13. S. M. Selke, Ph.D. Thesis, Michigan State University, East Lansing, 1983.

14. S. M. Selke, M. C. Hawley, H. Hardt, D. T. A. Lamport, G. Smith, and J. Smith, *I & EC Prod. Res. Dev.*, **21**, 11 (1982).

15. S. M. Selke, M. C. Hawley, and D. T. A. Lamport, *Wood and Agricultural Residues, Research on Use for Feed, Fuels, and Chemicals*, in E. Soltes, ed., Academic, New York, 1983, pp. 329–349.

16. K. W. Downey, S. M. Selke, M. C. Hawley, and D. T. A. Lamport, in Proceedings of the Seventh International FPRS Industrial Wood Energy Forum, Nashville, TN, Sept. 19–21, 1983.

17. B. L. Browning, *The Chemistry of Wood*, Wiley, New York, 1963.

18. M. C. Hawley, S. M. Selke, and D. T. A. Lamport, *Energy Agric.*, **2**, 219 (1983).

19. I. S. Goldstein, *Science*, **189**, 847 (1975).

20. R. Franz, R. Erckel, T. Riehm, R. Woerule, and H. Deger in *Energy from Biomass*, A. Strub, P. Chartier, and G. Schleser, eds., Applied Science Publishers, London, 1982, pp. 873–878.

21. C. M. Ostrovski, H. E. Duckworth, and J. C. Aitken, "Canertech's Ethanol-From-Cellulose Program," presented at the VI International Symposium on Alcohol Fuels Technology, Ottawa, Canada, May 21–25, 1984, in preparation.

22. T. Bentsen, *Afdelinger for Bioteknologi*, **3–4**, 40 (1982).

23. K. Kringstad, in *Future Sources of Organic Raw Materials: CHEMRAWN I.*, J. St. Pierre and G. R. Brown, eds., Pergamon Press, New York, 1980.

24. H. Janshekar and A. Fiechter, *Adv. Biochem. Eng. Biotechnol.*, **27**, 119 (1983).

25. E. Grulke, personal communication, May 1980.

26. C. Schuerch, *J. Am. Chem. Soc.*, **74**, 5061 (1952).

27. J. A. Smith, D. T. A. Lamport, M. C. Hawley, and S. M. Selke, *J. Appl. Polym. Sci. Appl. Polym. Symp.*, **37**, 641 (1983).

28. H. Hardt and D. T. A. Lamport, *Biotech. Bioeng.*, **24**, 903 (1982).

29. T. Hardy, D. Hoerger, and A. Jenks, personal communication, 1983.

30. H. Lohaus and L. Schlingmann, Hoechst Aktiengesellschaft, personal communication, Feb. 1984.

31. H. Hardt and D. T. A. Lamport, *Photochemistry*, **21**, 2301 (1982).

18

Continuous Pretreatment and Enzymatic Saccharification of Lignocellulosics

S. MORIYAMA and T. SAIDA
Research Center, Toyo Engineering Corporation, Mobara, Chiba, Japan

Japan depends chiefly on foreign countries for energy. The degree of external dependency for oil in Japan in 1981 was 99.8% and for energy from coal, cokes, oil, natural gas, and hydro- and nuclear electricity, 84.8% (1). Energy is a serious problem confronting Japan.

The increase in the oil price was dramatic in the 1970s. The official price of Arabian Light was about 2 $/bbl before the first oil crisis of 1973, then increased drastically to 10 $/bbl in 1974 and finally reached 30 $/bbl at the second oil crisis in 1979. Since then, much interest has been focused on the development of alternative energy resources to replace petroleum.

Lignocellulosic biomass is the most abundant organic substance on the earth, and furthermore renewable. One hundred billion tons of cellulose are produced annually (2). Recently much attention has been paid to utilizing lignocellulosic biomass for fuel ethanol (3, 4); unfortunately, there is yet no commercial process for this conversion. However, grains and molasses have been already utilized as a source of fuel ethanol. The program of petroleum alternatives for ethanol derived from cane sugar, cassava, and so on is in a fairly advanced stage in Brazil. In the United States, 10% ethanol derived from grains is employed to extend gasoline for automobiles without engine changes and with improved operating efficiency. However, a worldwide shortage of food has focused attention on utilizing lignocellulosics instead of grains.

The Ministry of International Trade and Industry (MITI) of Japan established the Research Association for Petroleum Alternatives Development (RAPAD) in 1980 on an 8-year program. RAPAD, organized by 23 private companies with a budget of approximately $95 million, has developed a process for producing fuel ethanol from biomass (5). A pilot plant with a production capacity of 150–200 L of ethanol per day has been constructed and operated at Yamaguchi Prefecture in Japan since 1984 using rice straw and cane sugar bagasse as raw materials. The construction of the plant will be accomplished in 1987.

Annual productivity of rice in Japan amounts to about 10 Mt. Since the weight ratio of the straw to the total rice is 57%, rice straw of 5.7 Mt is produced annually, which can be converted into 1.1 Mt of ethanol. This amount of ethanol corresponds to 8–9 times as much ethanol (excluding brewers) as was produced from cane sugar, starch, and ethylene in Japan in 1984. Contrary to our expectation, however, the cost of rice straw is not inexpensive taking into account collection, transportation, handling, and storage costs. According to an estimation by Heichel (6), sugar cane is the most efficient biomass from the viewpoint of cultural energy and total energy yield. He compared 11 modern cropping systems and estimated the ratio of total energy to cultural energy. For sugar cane, the ratio was about 22, and only 3–4 for sugar beets, peanuts, rice, and potatoes.

PRETREATMENTS

Lignocellulosic biomass is mainly composed of cellulose, hemicellulose, lignin, and ash. Biomass is resistant to enzymatic hydrolysis because of the presence of lignin (7–9) and the crystalline structure of cellulose (10–13). Some of the pretreatment methods already developed are efficient, such as ball milling (2, 14–21), steam explosion (22, 23), and alkali treatment (23–28), but they are quite costly. The pretreatment is classified into physical, chemical, and biological methods.

Physical Pretreatments

Mechanical grinding, a typical physical pretreatment, not only decreases the crystallinity of cellulose (10–13) but also increases the surface-to-volume ratio, thus making cellulose susceptible to hydrolysis. Matsumura et al. (10) found that the relationship between the crystallinity index and enzymatic hydrolysis rate of woods ground and sieved to be 40–60 mesh was almost linear. With an increase of crystallinity of cellulosic materials treated by compression milling, the initial rate of hydrolysis decreased linearly and the apparent Michaelis constant increased linearly (13). According to Caulfield and Moore (24), the overall increase in digestibility is apparently a result of decreased particle size and increased available surface rather than reduced crystallinity.

To reduce cellulosics to finer particles, various kinds of mills have been investigated—for example, a ball mill (2, 14–21), a vibratory ball mill (29, 30), a hammer mill (15, 17), a colloid mill (15, 16), and a roll mill (21, 31). Ball milling gave the most efficient results (15). The impact and abrasion forces of ball milling cause a reduction in crystallinity, a decrease in the mean degree of polymerization, an increase in bulk density, and a decrease in particle size. Mandels et al. (15) pointed out that the hydrolysis rate of newsprint increased with an increase in ball milling times. They reported a high conversion ratio of 75% for 7-day ball-milled newsprint compared with lower conversion ratio of 25% for newsprint without ball milling. However, as Fan et al. (12) mentioned, the pretreatment time and processing cost might make ball milling impractical on a large scale in spite of its predominant pretreatment effect.

Nonmechanical physical pretreatments such as steam explosion (22, 23) and high-energy irradiation (32) have also been extensively investigated. Interest has recently been focused on steam explosion. The raw material is first heated with high-pressure steam and then instantaneously released to atmospheric pressure. Steam explosion breaks alpha and beta allyl ether bonds and renders lignin soluble in methanol or ethanol solution (also see Chap. 15).

Chemical Pretreatments

Sodium hydroxide has been employed to enhance the digestibility of lignocellulosic materials for ruminants (25, 26) because the effect of NaOH is dominant compared with other alkaline reagents. Linko (33) implied that alkali treatment of bagasse would be more efficient than ball milling. The pretreatment effect of NaOH is due to some swelling (7, 8) and to disruption (27) of the crystalline structure of cellulose. The optimum amount of NaOH required to enhance the saccharification rate was shown by Dunlap et al. (34). The saccharification rate increased almost linearly up to 0.1 g NaOH per gram of biomass (wheat straw, oat straw, and maize cobs). Fan et al. (8) observed that the hydrolysis rate for wheat straw increased with an increase in the degree of delignification up to about 50% delignification and increased slightly at higher degrees of delignification. Delignification alone is not sufficient, but it is necesary for the increase in digestibility of lignocellulosic biomass (35).

Recently Tanaka et al. (36) demonstrated a pretreatment method using n-butylamine (n-BA). The method is characterized by efficient and easy recovery of the solvent, because the boiling point of n-BA is lower than that of water, and the latent heat of vaporization of n-BA is only one-fifth that of water. Holtzapple and Humphrey (35) pretreated poplar of a particle size of 20–40 mesh with ethanol and butanol in the presence of a catalyst, such as sulfuric acid, sodium hydroxide, and aqueous ammonia at 160–190°C. They concluded that pretreatment by NaOH dissolved in ethanol was the most effective. In the pulp-and-paper industry, additives such as sodium sulfide, anthraquinone, and hydrogen peroxide have also been employed with NaOH to increase the rate of delignification. Conventional pulping processes, such as kraft and sulfite pulping, are, however, too costly as bioconversion pretreatments.

Biological Pretreatments

Lignin performs a number of functions essential to the life of plants, such as an energy storage system, a permanent bonding agent between cells, and a water-proofing agent (37). The biological degradation of lignin is one of the most important parts of the biospheric carbon–oxygen cycle. Lignin degradation in nature is the result of the cooperative action between different fungi and bacteria in the soil. Microorganisms that can degrade lignocellulosics are classified into four groups: white-rot fungi, brown-rot fungi, soft-rot fungi, and bacteria (38). Above all, white-rot fungus is the most promising for pretreatment of lignocel-lulosics (38, 39). According to Platt et al. (40), a 65% reduction in lignin content in 3 weeks by white-rot fungi was observed in mushroom production on cotton straw. Although the degradation rate of lignin by microorganisms is very low (8, 39–41), the biological pretreatment could possibly offer a low-cost process to enhancing the digestibility of lignocellulosics.

SACCHARIFICATION

Acid hydrolysis of cellulosic materials has been studied for many years (30, 42–45). In the Soviet Union, a plant with a two-stage acid hydrolysis process has produced industrial alcohol and fodder yeast from wood (44). Acid hydrolysis, however, requires acid-resistant equipment and yields a poor grade of sugar.

Much attention has recently been paid to enzymatic hydrolysis, because the enzymatic reaction can be carried out at relatively mild conditions. Enzymatic hydrolysis of cellulosic materials has also been intensively investigated (13, 14, 44, 46–56). Wilke et al. (47) estimated the cost for cellulase production at more than 50% of the total cost for the production of glucose from lignocellulosics. Spano et al. (57) have indicated that cellulase production can be as much as 40% of the process cost.

Various sources of cellulolytic enzymes have been studied in detail (58–60). The fungus *Trichoderma viride* has proved to be the most effective source to date in spite of its poor β-glucosidase activity. To enhance the productivity of cellulase, various kinds of culture methods have been investigated, such as continuous (60, 61), feed-batch (61, 62), and cell-holding culture (63).

The kinetic model is helpful in designing the saccharification process. The exact mechanism involved in the enzymatic degradation of cellulosic materials is not fully understood because of the complexity of the physical structure of substrates and the multiplicity of the cellulase enzyme complex. Many empirical equations have been presented to elucidate the reaction mechanism (44, 52–55). Time-course analysis of saccharification of cellulosics shows that the hydrolysis rate decreases gradually with an increase in time. The reduction in rate has been attributed to product inhibition, adsorption of cellulase on cellulosics, heat and shear inactivation of cellulase, and decreased susceptibility of the residual cellulose. Moreover, since cellulose is an insoluble substrate, some of its kinetic characteristics are substantially different from those of the usual enzyme-catalyzed homogeneous reactions (64).

Inactivation of cellulase is one of the most important factors to merit consideration in the design of equipment for the enzymatic hydrolysis of cellulose. Many enzymes are protected against thermal inactivation by their substrates to form an enzyme–substrate complex (65–67). In the case of cellulase, however, as Howell et al. (53) and Okazaki et al. (68) have suggested, an enzyme–substrate complex might be denatured irreversibly. According to Reese and Mandels (49), shaking greatly reduced avicelase activity. Tanaka et al. (48) and Mukataka et al. (51) demonstrated that the hydrolysis rate of cellulose was reduced by agitation with a gas-liquid interface. Cellulase is also inactivated by shear. According to Reese and Ryu (50), the inactivation constant by shear is a function of the flow rate of the enzyme solution through a fine capillary tube. They suggested that the inactivation constant increased much more rapidly when the shear stress was greater than 15 dyne/cm^2 at 50°C, pH 5.0.

Since the cost of cellulase is ecomically prohibitive for technical use (47, 57), cellulase should be reused effectively. To reuse cellulase, attention has been paid to adsorption and ultrafiltration. Cellulose adsorbs cellulase efficiently at the optimal range for enzymatic activity. Adsorption depends on the concentration and particle size of cellulose and on the enzyme component. β-Glucosidase was not adsorbed onto bagasse up to 30 min at 5°C (54). Castanon and Wilke (69) found that C_x enzyme was adsorbed preferentially to C_1 enzyme in the early stage of hydrolysis but that afterward this situation was reversed.

Wilke et al. (70) proposed a scheme for the enzymatic hydrolysis of cellulose. They adsorbed cellulase onto fresh cellulosic materials to reuse the enzyme. Recovery of cellulase by adsorption by contact with the substrate is simple and attractive. Peitersen et al. (71) suggested that the adsorption of cellulase was largely independent of pH but that the maximum adsorbed enzyme decreased as the temperature increased. They used a Langmuir adsorption isotherm type of equation to relate the adsorbed enzyme concentration to free enzyme concentration. In practice, not more than 50% enzyme recovery could be expected from the adsorption systems (33).

Product inhibition is also an important limitation to the practical use of cellulases in biomass conversion processes. Inhibition of cellulases by soluble products is also poorly understood. Some researchers believe the inhibition to be competitive (53, 72), and others believe it to be noncompetitive (52, 55, 68). The products, cellobiose and glucose, are inhibitory even at low concentrations. The use of β-glucosidase has been proposed to convert highly inhibitory cellobiose to glucose, which is much less inhibitory (73).

To eliminate the inhibitory effect, it would be advantageous to remove the products continuously. Hahn-Hägerdal et al. (28, 56) hydrolyzed cellulosics in a UF membrane reactor and found that product inhibition could be eliminated by continuous removal of products through the UF membrane, thus retaining the macromolecular substrate and cellulase. Ghose and Kostick (46) developed a continuous hydrolysis system that removed glucose from the product while permitting enzyme and unused substrate to be recirculated to the reactor. However, Mandels et al. (74) subsequently found that at high substrate concentrations, nearly all of the cellulase existed in a form bound to the cellulose, with very little remaining free in the solution. Reese (75) suggested that this finding eliminated the need for a special membrane. Nevertheless, a UF membrane is quite suitable for the continuous removal of products, especially at relatively low substrate concentrations. High conversion, even complete hydrolysis, is observed at low substrate concentrations (76). Since most cellulase is free in the solution at low substrate concentrations, the enzyme has to be recovered with a UF membrane.

Recently several workers (18–20) have reported that enzymatic hydrolysis of cellulosic materials can be significantly improved by simultaneous attrition milling. Ryu and Lee (20) stated that simultaneous wet attrition milling and hydrolysis helped mass transfer of cellulase and product in the reactor. Although the power consumption of attrition milling is very high compared with that of regular stirring without milling media, the high power consumption can be compensated for by the increased extent of saccharification and the dramatically reduced operating time of the attrition bioreactor. Bose et al. (43) saccharified groundnut shell pulp with sulfuric acid using small glass beads to ensure thorough mixing and good agitation during shaking.

MATERIALS AND METHODS

The enzymes cellulase Onozuka 3S and R-10, derived from *Trichoderma viride*, were employed without further purification. The activity of R-10 is much higher than that of 3S. The raw material (rice straw and cane sugar bagasse) was first chopped with an ensilage cutter to 1–2 cm and then cut with a multiblade cutter to smaller than 0.4 cm. The total volume of the saccharification tank was 80 L. The first tank was 40 L, and the second and the third were 20 L. The cellulose content was determined by the Updegraff method (77), hemicellulose by the Chen-Anderson method (78), and lignin by the JIS (Japanese Industrial Standard) method (79).

Pretreatment effects were estimated by enzymatic saccharification using cellulase Onozuka R-10 or 3S at 40°C, pH 5.0. Conversion of cellulosics into glucose was measured by the glucose oxidase–peroxidase test. The conversion ratio was determined by the following equation.

$$\text{Conversion ratio (\%)} = \frac{\text{glucose produced}}{\text{cellulose}} \times \frac{162}{180} \times 100$$

CONTINUOUS PRETREATMENT AND SACCHARIFICATION

Pretreatment

Milling to finer particles requires higher energy consumption and higher costs. Nystrom (16) estimated the milling costs in the mesh region of 40–270 mesh. Reineke (80) also summarized energy requirements for comminuting dry white pine shavings to various screen sizes. Figure 18.1 demonstrates the relationship between the power consumption of rice straw with a length of 1–2 cm after cutting with an ensilage cutter and the mean particle size after grinding with various types of mills. As Figure 18.1 indicates, the power consumption to comminute rice straw was almost constant up to 0.4–0.6 mm in diameter; however, it increased drastically for smaller particles.

On the other hand, a major difference in the conversion ratio of rice straw was not observed in the power consumption region examined (Fig. 18.2). The conversion ratio of rice straw pretreated by colloid milling was about 65% after hydrolysis for 48 hr under the conditions described in the legend to Figure 18.2. Since the cellulose contained in 1 kg of rice straw yields 0.20 kg ethanol, rice straw with a conversion ratio of 65% yields 0.13 kg ethanol. Therefore, the heat capacity of 0.13 kg ethanol corresponds to 920 kcal, assuming that the combustion heat of ethanol is 7.1 kcal/g. The power consumption for mechanical pretreatment is 0.60 kWh/kg-straw, which corresponds to 1500 kcal, taking into consideration power generating efficiency of 2500 kcal/kWh. Obviously, the finding that the input energy due to mechanical pretreatment is higher than the output energy of 920 kcal suggests it impractical that lignocellulosic biomass be pretreated by colloid milling (81).

Combination of mechanical grinding with alkali treatment might enhance the pretreatment effect. Alkali treatment changes the composition of lignocellulosics. The change of composition of rice straw after pretreatment in a 0.25-(sc) NaOH

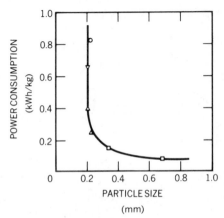

Figure 18.1 Relationship between power consumption of milling and particle size after milling. Rice straw with a length of 1–2 cm was comminuted with several kinds of mills: colloid mill (dry) (○); vibratory mill (▽); hammer mill (△); multiblade cutter (□).

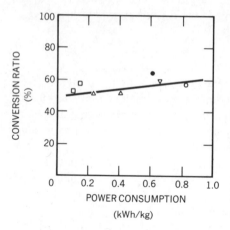

Figure 18.2 Enzymatic estimation of the effect of pretreatment by milling. Rice straw was pretreated by milling with several kinds of mills: colloid mill (dry) (○); colloid mill (wet) (●); vibratory mill (▽); hammer mill (△); multiblade cutter (□). The pretreatment effect was estimated by enzymatic hydrolysis using cellulase Onozuka 3S (enzyme concentration 0.018%, substrate concentration 1.0% at 40°C, pH 5.0, in a 0.1 M acetate buffer solution, reaction time 48 hr).

solution at 80°C for 1 hr is summarized in Table 18.1. Almost no difference in cellulose content was observed between before and after treatment; however, more than half of the contents of hemicellulose, lignin, and ash were extracted with a NaOH solution. The finding that the change in composition results in weight loss led to the supposition that the process of alkali treatment prior to fine grinding might reduce the total energy consumption. When biomass is comminuted after alkali treatment, a wet grinder should be utilized, such as a colloid mill, ball mill or attrition mill. As Figure 18.2 shows, wet colloid milling gave a higher pretreatment effect than dry colloid milling in spite of the lower power consumption. The relationship of the conversion ratio after hydrolysis for 48 hr versus the power consumption for several pretreatment processes are illustrated in Figures 18.2 and 18.3. The combination pretreatment, colloid milling after NaOH treatment, shows a higher conversion ratio and much lower energy consumption than colloid milling alone. The energy consumption of colloid milling after NaOH treatment decreased to one-tenth that of colloid milling.

Disadvantages of alkali treatment are the need of wash water and corrosion-resistant equipment and the disposal of spent alkali solution. The cost of re-

TABLE 18.1 Composition of Rice Straw before and after NaOH Treatment at 80°C for 1 hr

Composition	Before (%)	After (%)
Cellulose	40	38
Hemicellulose	25	11
Lignin	20	6
Ash	15	3
Total	100	58

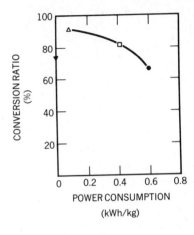

Figure 18.3 Comparison of enzymatic digestibility and power consumption among several pretreatment processes. Pretreatment effects were estimated by enzymatic hydrolysis and power consumption. Alkali treatment (∇); colloid mill (wet) (●); colloid mill (wet) → alkali treatment (□); alkali treatment → colloid mill (wet) (△).

covering NaOH is a key factor to saving pretreatment energy. To reduce the amount of alkaline waste effluent, biomass was treated with a countercurrent extractor. Biomass was treated with a NaOH solution at 80°C in the first half section and washed with hot water in the second half section. Alkali treatment using a countercurrent extractor had distinct advantages over a stirred tank reactor for the high solid-to-liquid ratio and for ease of continuous operation.

The effect of alkali treatment is dependent on temperature, pressure, alkali concentration, size of raw material, the catalysts (e.g., sodium sulfide), and retention time. Taking into consideration the installation and operation costs and the sterilization effect, the extractor was operated at 80°C with a retention time for alkali treatment of 1–3 hr. The ratio of biomass to NaOH was 0.6–1.5:10. Not much consideration has been given to utilizing the hemicellulose contained in the biomass. Since the biomass was treated with NaOH at 80°C for more than 1 hr, almost no contamination with microorganisms was observed in the slurry before fine grinding. The observations that milling tends to increase the ease of swelling (14) and that the energy consumption of cutting with a multiblade cutter is low (Fig. 18.1) led to the introduction of cutting before alkali treatment.

Consequently, the pretreatment system is as follows (Fig. 18.4). Rice straw or cane sugar bagasse is chopped into small pieces of 1–2 cm and then to smaller than 0.4 cm in length by two-stage cutting; the particles are fed to the countercurrent extractor for treatment with a NaOH solution and then ground with a wet attrition mill to give about a 5% slurry. The slurry concentration of 5% was optimum for handling, such as pumping.

Saccharification

In general, the plug flow reactor (PFR) is preferable to the continuous stirred tank reactor (CSTR) to hydrolyze cellulosic materials for the following reasons: (a) The hydrolysis reaction is inhibited by the product; (b) the retention time

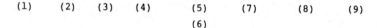

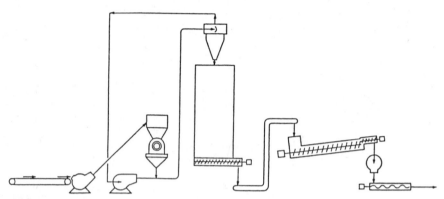

Figure 18.4 Apparatus for continuous pretreatment. (1) Conveyor; (2) first cutter; (3) blower; (4) second cutter; (5) hopper (silo); (6) screw feeder; (7) conveyor; (8) extractor; (9) attrition mill.

must be limited, because the amount of cellulase lost by adsorption onto unsaccharified residue at the separation section by centrifugation has to be reduced; and (c) PFR is economically more favorable than CSTR at a low substrate concentration, considering the vessel volume, treatment quantity, and enzyme concentration.

On the other hand, CSTR is advantageous over PFR for the following reasons: (a) The reaction conditions (e.g., stirring and pH) can be easily controlled; (b) noncellulosic materials can be easily removed; and (c) the substrate is insoluble.

In consideration of the merits of each process, a system composed of CSTRs in series was designed by a graphical procedure (Fig. 18.5, 18.6). The solid line in Figure 18.6 is the extrapolated curve from experimental data, and the broken line is the simulated curve based on glucose inhibition assuming that the inhibition is competitive, the Michaelis constant is 0.40%, the inhibitor constant is 0.35%, and the glucose concentration is 0.50%. Inhibition by cellobiose was neglected because the enzyme had high activity of β-glucosidase and only a small amount of cellobiose was observed by analysis with high-performance liquid chromatography (HPLC). The characteristics of the saccharification system are: (a) continuous reaction, (b) high conversion ratio to glucose, (c) continuous removal of noncellulosic materials such as lignin and ash, (d) immobilized enzyme with a UF membrane, and (e) continuous removal of glucose and cellobiose using a UF membrane.

The continuous saccharification system shown in Figure 18.5 was controlled by a process computer and was operated under aseptic conditions as follows. After grinding by colloid milling, rice straw or cane sugar bagasse (5% slurry) was sterilized at 120°C for 3 sec using a plate heat exchanger and then fed to the first saccharification tank. The raw material was hydrolyzed by cellulolytic

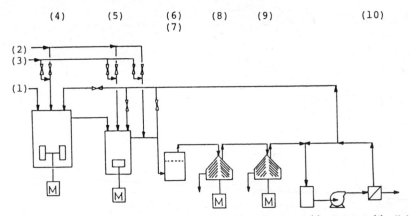

Figure 18.5 Apparatus for continuous saccharification. (1) Raw material; (2) cellulase 1; (3) cellulase 2; (4) first saccharification tank; (5) second saccharification tank; (6) third saccharification tank; (7) vibratory screen; (8) first centrifuge; (9) second centrifuge; (10) UF membrane.

enzymes in the saccharification tanks at 45°C, pH 5.0. The mean retention time in the tanks was 4–15 hr. The unsolubilized straw was returned to the first tank after sieving with the vibratory screen. The colloidal solution passed through the vibratory screen was separated with a two-stage centrifugation. In the first stage, ash-rich solid was removed by relatively mild centrifugation, and then the residue was centrifuged at higher speed. Since the substrate concentration was low, the loss of the enzyme by adsorption onto the unsaccharified residue was minimized. The supernatant was then filtered with a UF membrane equipped

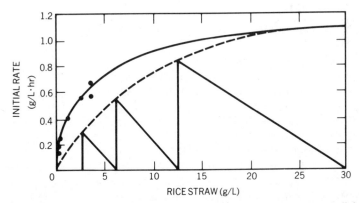

Figure 18.6 Graphical procedure for designing reactor(s). The enzyme used was cellulase Onozuka 3S, and the concentration was 0.018%. The concentration of the substrate, rice straw, was 1.0%. The enzymatic hydrolysis was carried out at 45°C, pH 5.0. The initial rate was followed using a glucose oxidase-peroxidase test kit. The solid line represents the extrapolated curve from experimental data; and the broken line, the simulated curve based on glucose inhibition as described in the text. The circles signify the experimental data.

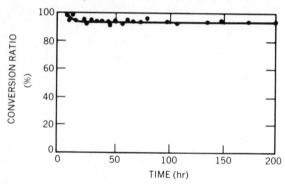

Figure 18.7 Conversion ratio of rice straw in continuous saccharification. Rice straw in a 5% slurry was fed continuously. Hydrolysis was carried out at 45°C, pH 5.0.

with automatic cleaning, and the enzyme was recirculated into the first saccharification tank. The enzyme was complemented continuously on the basis of HPLC analyses and the remaining activities of the enzyme components.

Figure 18.7 shows the conversion ratio to glucose at the outlet of the third saccharification tank versus operating time. More than 90% of the rice straw was converted into glucose. The retention time will be further shortened, and the vessel volume will be decreased when higher-activity cellulase is used. We are designing the continuous saccharification system in a pilot scale using a pipe-type reactor (PFR) and a high-activity cellulase selected and improved by RAPAD (5).

REFERENCES

1. *Japan and the World in Statistics 1984*, Tokyo Chamber of Commerce and Industry, Tokyo, Japan, 1984.

2. L. A. Spano, J. Medeiros, and M. Mandels, *Resour. Recovery Conserv.*, **1**, 279 (1976).

3. D. L. Miller, *Biotechnol. Bioeng. Symp.*, **5**, 345 (1975).

4. N. Kosaric, Z. Duvnjak, and G. G. Stewart, in *Advances in Biochemical Engineering*, A. Fiechter, ed., Vol. 20, Springer-Verlag, New York, 1981, p. 119.

5. *Research and Development on Synfuels 1984*, Research Association for Petroleum Alternatives Development, Tokyo, Japan, 1984.

6. G. H. Heichel, *Biotechnol. Bioeng. Symp.*, **5**, 43 (1975).

7. M. A. Millett, A. J. Baker, and L. D. Satter, *Biotechnol. Bioeng. Symp.*, **5**, 193 (1975).

8. L. T. Fan, Y.-H. Lee, and M. M. Gharpuray, in *Advances in Biochemical Engineering*, A. Fiechter, ed., Vol. 23, Springer-Verlag, New York, 1982, p. 157.

9. M. J. Neilson, F. Shafizadeh, S. Aziz, and K. V. Sarkanen, *Biotechnol. Bioeng.*, **25**, 609 (1983).

10. Y. Matsumura, K. Sudo, and K. Shimizu, *Mokuzai Gakkaishi*, **23**, 562 (1977).

11. M. Tanaka, M. Taniguchi, T. Morita, R. Matsuno, and T. Kamikubo, *J. Ferment. Technol.*, **57**, 186 (1979).

12. L. T. Fan, Y.-H. Lee, and D. H. Beardmore, in *Advances in Biochemical Engineering*, A. Fiechter, ed., Vol. 14, Springer-Verlag, New York, 1980, p. 101.

13. D. D. Y. Ryu, S. B. Lee, D. H. Tassinari, and C. Macy, *Biotechnol. Bioeng.*, **24**, 1047 (1982).

14. J. A. Howsmon and R. H. Marchessault, *J. Appl. Polym. Sci.*, **1**, 313 (1959).

15. M. Mandels, L. Hontz, and J. Nystrom, *Biotechnol. Bioeng.*, **16**, 1471 (1974).

16. J. Nystrom, *Biotechnol. Bioeng. Symp.*, **5**, 221 (1975).

17. R. K. Andren, R. J. Erickson, and J. E. Medeiros, *Biotechnol. Bioeng. Symp.*, **6**, 177 (1976).

18. R. G. Kelsey and F. Shafizadeh, *Biotechnol. Bioeng.*, **22**, 1025 (1980).

19. M. J. Neilson, R. G. Kelsey, and F. Shafizadeh, *Biotechnol. Bioeng.*, **24**, 293 (1982).

20. S. K. Ryu and J. M. Lee, *Biotechnol. Bioeng.*, **25**, 53 (1983).

21. M. M. Gharpuray, Y.-H. Lee, and L. T. Fan, *Biotechnol. Bioeng.*, **25**, 157 (1983).

22. R. H. Marchessault, S. Coulombe, H. Morikawa, and D. Robert, *Can. J. Chem.*, **60**, 2372 (1982).

23. M. J. Playne, *Biotechnol. Bioeng.*, **26**, 426 (1984).

24. D. F. Caulfield and W. E. Moore, *Wood Sci.*, **6**, 375 (1974).

25. F. Rexen and K. V. Thompson, *Anim. Feed Sci. Technol.*, **1**, 73 (1976).

26. Y. W. Han, P. L. Yu, and S. K. Smith, *Biotechnol. Bioeng.*, **20**, 1015 (1978).

27. M. Taniguchi, M. Tanaka, R. Matsuno, and T. Kamikubo, *Eur. J. Appl. Microbiol. Biotechnol.*, **14**, 35 (1982).

28. I. Ohlson, G. Träggårdh, and B. Hahn-Hägerdal, *Biotechnol. Bioeng.*, **26**, 647 (1984).

29. B. A. Dehority and R. R. Johnson, *J. Dairy Sci.*, **44**, 2261 (1961).

30. M. A. Millett, A. J. Baker, and L. D. Satter, *Biotechnol. Bioeng. Symp.*, **6**, 125 (1976).

31. T. Tassinari and C. Macy, *Biotechnol. Bioeng.*, **19**, 1321 (1977).

32. M. Kamakura and I. Kaetsu, *Biotechnol. Bioeng.*, **24**, 991 (1982).

33. M. Linko, in *Advances in Biochemical Engineering*, T. K. Ghose, A. Fiechter, and N. Blakebrough, eds., Vol. 5, Springer-Verlag, New York, 1977, p. 25.

34. C. E. Dunlap, J. Thompson, and L. C. Chiang, *AIChE Symp. Ser.*, No. 158, **72**, 58 (1976).

35. M. T. Holtzapple and A. E. Humphrey, *Biotechnol. Bioeng.*, **26**, 670 (1984).

36. M. Tanaka, G.-J. Song, R. Matsuno, and T. Kamikubo, *Appl. Microbiol. Biotechnol.*, **22**, 19 (1985).

37. H. Janshekar and A. Fiechter, in *Advances in Biochemical Engineering Biotechnology*, A. Fiechter, ed., Vol. 27, Springer-Verlag, New York, 1983, p. 119.

38. P. Ander and K. E. Eriksson, in *Progress in Industrial Microbiology*, M. J. Bull, ed., Vol. 14, Elsevier, New York, 1978, p. 1.

39. H. H. Yang, M. J. Effland, and T. K. Kirk, *Biotechnol. Bioeng.*, **22**, 65 (1980).

40. M. W. Platt, I. Chet, and Y. Henis, *Eur. J. Appl. Microbiol. Biotechnol.*, **13**, 194 (1981).

41. T. K. Kirk, *Biotechnol. Bioeng. Symp.*, **5**, 139 (1975).

42. J. Saeman, *Ind. Eng. Chem.* **37**, 43 (1945).

43. B. Bose, T. R. Ingle, and J. L. Bose, *Ind. J. Technol.*, **11**, 391 (1973).

44. A. E. Humphrey, *Adv. Chem. Ser.*, **181**, 25 (1979).

45. R. S. Roberts, D. K. Sondhi, M. K. Bery, A. R. Colcord, and D. J. O'Neil, *Biotechnol. Bioeng. Symp.*, **10**, 125 (1980).

46. T. K. Ghose and J. A. Kostick, *Biotechnol. Bioeng.*, **12**, 921 (1970).

47. C. R. Wilke, R. D. Yang, A. E. Sciamanna, and R. P. Freitas, *Biotechnol. Bioeng.*, **23**, 163 (1981).

48. M. Tanaka, S. Takenawa, R. Matsuno, and T. Kamikubo, *J. Ferment. Technol.*, **56**, 108 (1978).

49. E. T. Reese and M. Mandels, *Biotechnol. Bioeng.*, **22**, 323 (1980).

50. E. T. Reese and D. Y. Ryu, *Enzyme Microb. Technol.*, **2**, 239 (1980).

51. S. Mukataka, M. Tada, and J. Takahashi, *J. Ferment. Technol.*, **61**, 615 (1983).

52. J. A. Howell and J. D. Stuck, *Biotechnol. Bioeng.*, **17**, 873 (1975).

53. J. A. Howell and M. Mangat, *Biotechnol. Bioeng.*, **20**, 847 (1978).

54. T. K. Ghose and V. S. Bisaria, *Biotechnol. Bioeng.*, **21**, 131 (1979).

55. M. T. Holtzapple, H. S. Caram, and A. E. Humphrey, *Biotechnol. Bioeng.*, **26**, 753 (1984).

56. B. Hahn-Hägerdal, E. Andersson, M. López-Leiva, and B. Mattiasson, *Biotechnol. Bioeng. Symp.*, **11**, 651 (1981).

57. L. Spano, T. Tassinari, D. Ryu, A. Allen, and M. Mandels, in *Proceedings Biogas and Alcohol Fuels Production Seminar*, G. Goldstein, ed., J. G. Press, Emmaus, PA, 1980, p. 62.

58. E. T. Reese, *Biotechnol. Bioeng. Symp.*, **6**, 91 (1976).

59. B. Montenecourt and D. Eveleigh, *Appl. Env. Microbiol.*, **34**, 777 (1977).

60. A. L. Allen and R. E. Andereotti, *Biotechnol. Bioeng. Symp.*, **12**, 451 (1982).

61. G. Mitra and C. R. Wilke, *Biotechnol. Bioeng.*, **17**, 1 (1975).

62. N. Hendy, C. R. Wilke, and S. Blanch, *Biotechnol. Lett.*, **4**, 785 (1982).

63. T. K. Ghose and V. Sahai, *Biotechnol. Bioeng.*, **21**, 283 (1979).

64. Y.-H. Lee and L. T. Fan, *Biotechnol. Bioeng.*, **24**, 2383 (1982).

65. N. Citri, *Adv. Enzymol.*, **37**, 397 (1973).

66. R. D. Schmidt, in *Advances in Biochemical Engineering*, A. Fiechter, ed., Vol. 12, Springer-Verlag, New York, 1979, p. 41.

67. S. Moriyama, S. Kataoka, K. Nakanishi, R. Matsuno, and T. Kamikubo, *Agric. Biol. Chem.*, **44**, 2737 (1980).

68. M. Okazaki and M. Moo-Young, *Biotechnol. Bioeng.*, **20**, 637 (1978).

69. M. Castanon and C. R. Wilke, *Biotechnol. Bioeng.*, **22**, 1037 (1980).

70. C. R. Wilke, R. D. Yang, and U. V. Stockar, *Biotechnol. Bioeng. Symp.*, **6**, 155 (1976).

71. N. Peitersen, J. Medeiros, and M. Mandels, *Biotechnol. Bioeng.*, **19**, 1091 (1977).

72. C. P. Dwivedi and T. K. Ghose, *J. Ferment. Technol.*, **57**, 15 (1979).

73. M. S. Pemberton, R. D. Brown, and G. H. Emert, *Can. J. Chem. Eng.*, **58**, 723 (1980).

74. M. Mandels, J. Kostick, and R. Parizek, *J. Polym. Sci.* (Part C), **36**, 445 (1971).

75. E. T. Reese, *Biotechnol. Bioeng. Symp.*, **6**, 9 (1976).

76. M. Mandels and E. T. Reese, *Dev. Ind. Microbiol.*, **5**, 5 (1964).

77. D. M. Updegraff, *Anal. Biochem.*, **32**, 420 (1969).

78. W. P. Chen and A. W. Anderson, *Biotechnol. Bioeng.*, **22**, 519 (1980).

79. JIS (Japanese Industrial Standard) P8008-1961.

80. L. H. Reineke, U.S. Forest Service Res. Note, FPL-0113, 1966.

81. T. Saida, S. Moriyama, H. Ishibashi, and K. Matsumura, *Abstr. Pan-Pacific Synfuels Conf.*, Tokyo, Japan, 1982, p. 440.

19

Enzymatic Breakdown of Cellulose Crystals

B. HENRISSAT and H. CHANZY
Centre de Recherches sur les Macromolécules Végétales, CNRS,‡ Saint-Martin-D'Hères, France

Despite the large number of reports on cellulase production, purification, possible utilization, molecular cloning, and so on, published during the past 30 years, the exact mechanism of the biodegradation of crystalline cellulose is still poorly understood. *Endo*-1,4-β-D-glucan glucanohydrolases, (E.C. 3.2.1.4), also called endoglucanases, have been found to degrade soluble carboxymethyl cellulose (CMC) and amorphous acid-swollen cellulose (1). As they were also found to have little or no action on cotton or Avicel cellulose, their role has been attributed to the *endo*-hydrolysis of amorphous domains in cellulose (1). On the other hand, 1,4-β-D-glucan cellobiohydrolases (E.C. 3.2.1.91, cellobiohydrolases, CBH) have been shown to produce almost exclusively cellobiose from insoluble celluloses while being rather ineffective in the hydrolysis of CMC (2–4). On this basis they were classified as *exo*-enzymes, and their role was attributed to the recurrent removal of cellobiosyl residues from the nonreducing cellulose chain ends precisely created by endoglucanases (2–4). These two hypotheses provided a logical explanation of the observed synergism between these two types of enzymes when degrading cotton or Avicel cellulose (1).

However, one of the two CBH from *Trichoderma reesei*, namely CBHI, has been shown to lack specificity for the penultimate glycosidic bond of oligocellodextrins (5). Several authors have also shown that cellobiohydrolases can degrade highly crystalline and ordered cellulose without the help of any *endo* activity (6–9). Other observations on purified 1,4-β-D-cellobiohydrolases give additional insight: Fägerstam and Pettersson (8) have found that the two cel-

‡Affiliated with the Scientific and Medical University of Grenoble.

lobiohydrolases, CBHI and CBHII, from *Trichoderma reesei* exhibit a remarkable synergistic action difficult to interpret in terms of chain ends attack or *exo–exo* cooperation. Rabinovitch et al. (10) classify the cellulolytic enzymes in their ability to bind to the crystalline regions of cellulose. They have shown that the ability of some enzymes to bind tightly to cellulose is related to their ability to hydrolyze crystalline domains. The other enzymes, which do not bind tightly to cellulose, restrict their action to the less crystalline zones (10). These apparently conflicting results point out the need for further investigations on the mechanism of enzymatic degradation of crystalline cellulose.

Native cellulose occurs as extremely long microfibrils of various widths, depending on the species considered (11, 12; also see Chaps. 1, 6). These microfibrils are crystalline, and they are believed to consist of extended-chain crystals (13–15) whose lateral dimensions are comparable to those of the microfibrils. In the case of *Valonia* cellulose, which is considered to be one of the most perfect native cellulose materials (16), it was shown that each microfibril was an individual crystal of cellulose (17–19). The crystalline perfection of *Valonia* microcrystals even allowed the recording of lattice images of cellulose by electron microscopy (20, 21). On this basis, electron microscopy and diffraction experiments on highly crystalline and well-defined substrates may represent a useful tool in obtaining new information on the enzymatic breakdown of cellulose. Such studies will be the object of this chapter.

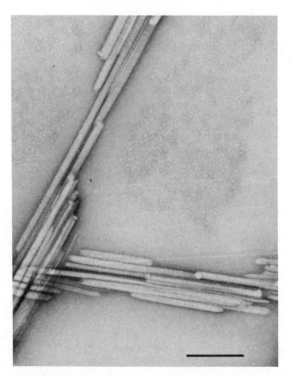

Figure 19.1   Cellulose microcrystals from *Valonia macrophysa*. Bar = 200 nm.

ELECTRON MICROSCOPY AND ENZYMATIC HYDROLYSIS OF CELLULOSE

With the exception of the early work of Wardrop and Jutte (22) on the degradation of *Valonia ventricosa* cellulose by *Helix pomatia* cellulase, high-resolution microscopy observations of the enzymatic hydrolysis of cellulose have been reported only recently. White and Brown (23) have observed the behavior of bacterial cellulose ribbons in the presence of purified cellulases from *Trichoderma reesei*. They noticed a splaying of the ribbon when incubated with endoglucanase, whereas cellobiohydrolase (CBHI) adsorbed all along the microfibrils. Chanzy and Henrissat (24) have shown that cellulase complexes from *Schizophyllum commune* and *Trichoderma reesei* can degrade the highly crystalline cellulose from the algae *Valonia macrophysa*. The enzymatic hydrolysis of the 20-nm-broad algal microfibrils leads to their fibrillation into subelements having a lateral dimension reduced to a few nanometers. Electron diffraction of the digested samples provided typical patterns displaying a dramatic intensity decrease for the equatorial 020 reflection whereas the meridional 004 reflection was unaffected (indices used throughout this chapter refer to the two-chain unit cell described by Gardner and Blackwell (25)). Such morphological and crystallographic alteration has been also found to occur when purified 1,4-β-D-

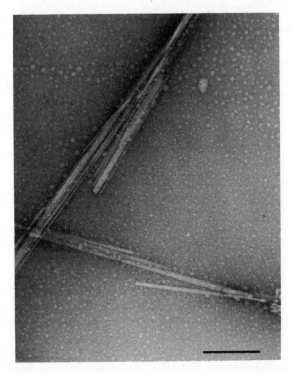

Figure 19.2 Cellulose microcrystals from *Valonia macrophysa* after 24 hr of incubation with CBHI. Bar = 200 nm.

glucan cellobiohydrolase I from *Trichoderma reesei* acts on *Valonia* cellulose microcrystals (9). Recently, colloidal gold labeling of CBHI from *Trichoderma reesei* with 4- to 6-nm Au particles has been successfully performed and provided evidence for an *endo* adsorption pattern of the enzyme onto the cellulose surface (26). Moreover, a certain "crystalline" specificity of the CBHI–Au complex has been recorded since it displayed a preferential adsorption on the 110 face of the microcrystals (26).

MATERIALS AND METHODS

The enzymes 1,4-β-D-glucan cellobiohydrolase I (CBHI), purified from Celluclast (27) (a commercial preparation of cellulase from *Trichoderma reesei*), and the crude cellulase complex from *Humicola insolens* (28) were generous gifts from Dr. M. Schülein. For substrates, *Valonia macrophysa* cellulose microcrystals were prepared as described by Chanzy and Henrissat (24) and bacterial cellulose was obtained from a culture of *Acetobacter xylinum* (29). The latter was converted to microcrystalline cellulose by an acidomechanical treatment (24) and then

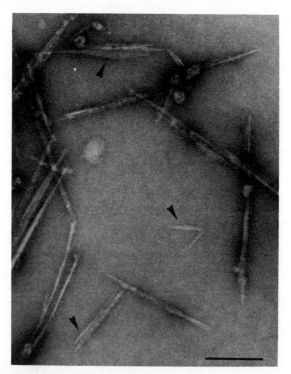

Figure 19.3 Cellulose microcrystals from *Valonia macrophysa* after 16 hr of incubation with *Humicola insolens* cellulase complex. Bar = 200 nm.

purified from noncellulosic contaminants by alternately refluxing in 1% aqueous sodium hydroxide and washing in dilute hydrochloric acid (25).

All the degradation experiments were carried out at a substrate concentration of 1 mg/mL in 50 mM sodium acetate buffer (pH 4.8) and an enzyme concentration of 100 μg/mL. Temperature of incubation was 45°C. Cellulose samples were centrifuged (3000 rpm), and the sediment was treated with 0.1% aqueous sodium hydroxide to remove the adsorbed enzymes as preparation for electron microscopy (9, 30). The cellulose was then extensively washed with distilled water. Drops of the aqueous suspension of cellulose were deposited on carbon-coated electron microscope grids, and excess of liquid was drawn off with filter paper. The specimens were used without further treatment for electron diffraction or negatively stained with 1.5% uranyl acetate for imaging.

A Philips EM 400T electron microscope was used at an accelerating voltage of 80 kV for imaging and 120 kV for electron diffraction. "Powder" electron diffraction diagrams were recorded on Ilfoset (Ilford) films while moving the specimen under the beam. This technique produces a uniform intensity distribution around the diffraction rings and minimizes the dose per surface area by constantly irradiating fresh material.

The sediment of the washed specimens was allowed to dry in a small test tube under rigorously dust-free conditions to avoid possible contamination by

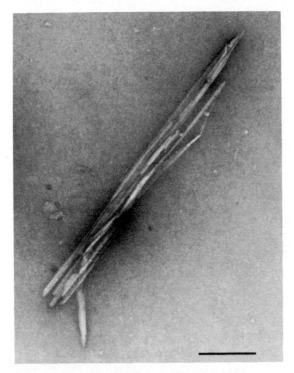

Figure 19.4   Microcrystalline cellulose from *Acetobacter xylinum*. Bar = 200 nm.

extraneous crystalline material. Thin pellets of the dried specimen were then mounted onto 0.20-mm-diameter collimators and exposed *in vacuo* to Cu K_α ($\lambda = 1.5418$ Å) X rays in a Warhus camera. Diagrams were recorded on Kodak No-Screen x-ray film, and the microdensitometer tracings of the radial intensity distribution were collected with a Joyce-Loebl microdensitometer.

OBSERVATIONS OF ENZYME DEGRADATION

After incubation with CBHI, most of the initially smooth and well-defined *Valonia* microcrystals (Fig. 19.1) display their typical fibrillated morphology (Fig. 19.2) (24). This transformation, which leads progressively to microcrystals of much narrower diameters than the initial crystals, is already observable after only a few hours of digestion. However, as the hydrolysis is very heterogeneous, a percentage of intact microcrystals can still be found, even after extended reaction times (48 hr). In contrast to the narrowing of the crystals, the attack of CBHI does not seem to affect the length of the digested crystals, even to the limit of total disappearance of the crystals. Their initial length is maintained, and short elements are scarcely observed.

The samples incubated with the crude cellulase complex from *Humicola insolens* follow a somewhat different digestion pattern. When hydrolyzed, the cellulose crystals become both narrower and shorter, and they diplay a characteristic spindlelike appearance (see arrows in Fig. 19.3), denoting both a longitudinal and lateral attack of the crystals. Such an effect, which was initially observed with *Valonia* microcrystals, was also recorded with other substrates such as bacterial cellulose microcrystals (Fig. 19.4, initial sample, and Fig. 19.5, sample digested for 16 hr with the *Humicola insolens* cellulase complex).

DIFFRACTION DATA ON DIGESTED CELLULOSE CRYSTALS

When the digested crystals are studied by electron and x-ray diffraction experiments, the cellulose pattern is dramatically affected by the hydrolysis process. Unlike the above ultrastructural observations, however, the samples hydrolyzed with either CBHI or *Humicola insolens* appeared to give identically modified diffraction diagrams. In Figure 19.6, corresponding to an electron diffraction experiment, it is seen that with a well-digested *Valonia* microcrystalline sample, all the equatorial diffraction lines have vanished, whereas the meridional and the off-meridional reflections are still present and apparently unaffected. This is confirmed by an x-ray experiment performed on an 80% degraded sample of bacterial cellulose (Fig. 19.7). In that case, the x-ray diffraction pattern of the digested sample indicates a strong decrease of the intensity of the $1\bar{1}0$, 110, 102, and especially 020 reflections, whereas the 004 reflection appears unaffected. Another feature observed in the comparison of the two tracings in Figure 19.7 is that the diminished intensity of the maxima in Figure 19.7*B* occurs without noticeable line broadening.

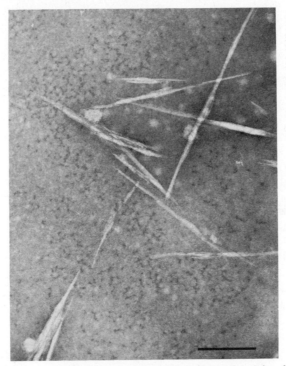

Figure 19.5 Microcrystalline cellulose from *Acetobacter xylinum* after 8 hr of incubation with *Humicola insolens* cellulase complex. Bar = 200 nm.

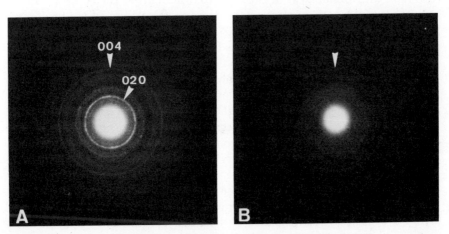

Figure 19.6 (*A*) Electron diffraction diagram of cellulose microcrystals from *Valonia macrophysa*. (*B*) Same as (*A*) but after 16 hr of incubation with *Humicola insolens* cellulase complex. **343**

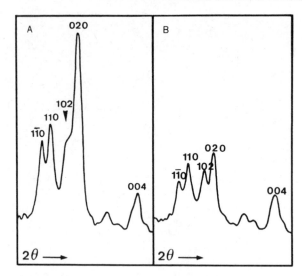

Figure 19.7 (*A*) Microdensitometer tracings of the radial intensity distribution of an x-ray diffractogram of bacterial microcrystalline cellulose. (*B*) Same as (*A*) but after 8 hr of incubation with *Humicola insolens* cellulase complex.

Despite enzyme-dependent differences in their morphology, the cellulose crystals derived from either CBHI or *Humicola insolens* cellulase action exhibit the same unusual alteration in their diffraction patterns. Close observation of the above diffraction data, together with numerous other patterns, brings the following conclusion. As a consequence of enzymatic degradation, the diffraction maxima of the digested specimen undergo a decrease of intensity inversely related to their azimuthal position in the fiber diffraction diagram. This is illustrated in Figure 19.8, which summarizes our results: The equatorial and off-equatorial reflections with azimuthal angle close to zero (white circles) are the most reduced. On the other hand, those close to the meridian (black circles) are unaffected. Intermediate spots such as 102 and 204, with an azimuthal angle close to 45°, are only slightly affected by hydrolysis (black-and-white circles).

These diffraction data, together with the absence of significant line-broadening effects, are consistent with a model involving a bimodal distribution of two types of crystals: the unaffected ones and those degraded to the limit. As the degradation proceeds, there is a decrease in the percentage of the undegraded crystals and a corresponding increase of the degraded ones. A concomitant appearance of partially degraded crystals which would contribute to diffraction line broadening is not observed. The degraded crystals, which constitute the overall solid sample when degradation reaches 95%, are remarkable in that they no longer have any detectable equatorial diffraction lines. With such crystals, the diffraction interferences are concentrated in a narrow azimuthal zone close to their c^* axis. As a consequence, these degraded crystals must be extremely small in cross section yet relatively large and well defined along their axis. They can be described as nearly perfect unidimensional cellulose crystals. The persistence of such crys-

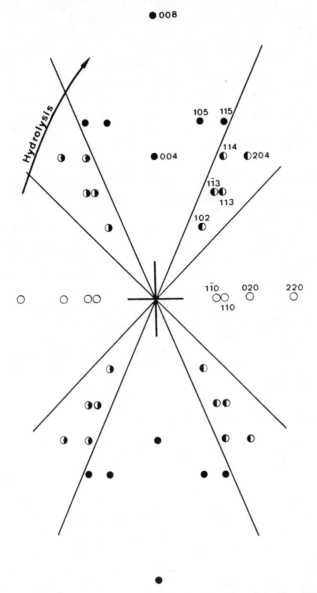

Figure 19.8 Schematic illustration of the progressive alteration of the fiber diffraction diagram of cellulose during enzymatic hydrolysis.

tals has certainly to deal with the kinetics of hydrolysis. The absence of partially degraded crystals favors a model involving a direct fibrillation of the starting metastable cellulose crystal into unidimensional subelements as first consequence of the surface attack catalyzed by the enzymes. This may reflect a selective hydrolysis of certain edges of planes of the native crystal which would be disrupted by the release of internal forces.

The morphological observations presented in Figures 19.1–19.5 are in good agreement with the diffraction data, although the diffraction method provides better averaging than the display of selected areas in electron micrographs. The degraded linear crystals, whose existence is deduced from electron diffraction data, are obviously the fibrillated or eroded crystals seen in the various micrographs and would correspond to the description of unidimensional cellulose crystals. Their physical shape and their relationship to the initial crystal remain somewhat unclear at present, as ultimately one would like to visualize them in three dimensions. This could be achieved in the future through micrographs of their cross sections.

A final interesting observation resulting from this work is that a complete cellulase complex such as that from *Humicola insolens* leads to much shorter crystals than that of either the purified CBHI from *Trichoderma reesei* or even the crude enzyme complex of this fungus (24). This difference is certainly related to differences in the enzyme composition of the complexes secreted by these different microorganisms. From our results, the action of CBHI, the most abundant enzyme of the *Trichoderma reesei* complex, as a true *exo*-enzyme appears doubtful. One would expect such an enzyme to reduce the length of the cellulose crystals by digesting them from their end. This should lead to shorter "spindlelike" crystals, as in the case of the crude complex from *Humicola insolens*. The absence of such shortening together with an accumulation of observations (5, 9, 22) indicate that the action of CBHI on cellulose is very peculiar and to date incompletely understood. It remains to be seen whether other enzymes classified as *exo*-acting, such as for instance CBHII from *Trichoderma reesei*, will also behave as CBHI or follow the expected mode of attack shortening the cellulose crystals by a digestion from their nonreducing end.

REFERENCES

1. T. M. Wood and S. I. McCrae, *Adv. Chem. Ser.*, **181**, 181 (1979).
2. T. M. Wood and S. I. McCrae, *Biochem. J.*, **128**, 1183 (1972).
3. L. E. R. Berghem and L. G. Pettersson, *Eur. J. Biochem.*, **37**, 21 (1973).
4. K. E. Eriksson and B. Pettersson, *Eur. J. Biochem.*, **51**, 213 (1975).
5. H. van Tilbeurgh, M. Claeyssens, and C. K. de Bruyne, *FEBS Lett.*, **149**, 152 (1982).
6. G. Halliwell and M. Griffin, *Biochem. J.*, **135**, 587 (1973).
7. T. Sasaki, T. Tanaka, N. Nanbu, Y. Sato, and K. Kainuma, *Biotechnol. Bioeng.*, **21**, 1031 (1979).
8. L. G. Fägerstam and L. G. Pettersson, *FEBS Lett.*, **119**, 97 (1980).
9. H. Chanzy, B. Henrissat, R. Vuong, and M. Schülein, *FEBS Lett.*, **153**, 113 (1983).
10. M. L. Rabinovitch, N. V. Viet, and A. A. Klesov, *Biokhimiya*, **47**, 465 (1982).
11. J. R. Colvin, *CRC Crit. Rev. Macromol. Sci.*, **1**, 47 (1972).
12. R. H. Marchessault and P. R. Sundararajan, in *The Polysaccharides*, G. O. Aspinall, ed., Vol. 2, Academic Press, New York, 1983, p. 11.
13. V. E. Stockmann, *Biopolymers*, **11**, 251 (1972).
14. A. Sarko, *Tappi*, **61**, 59 (1978).

15. J. Blackwell, F. J. Kolpak, and K. H. Gardner, *Tappi,* **61**, 71 (1978).

16. A. K. Kulshreshtha and N. E. Dweltz, *J. Polym. Sci. Polym. Phys. Ed.,* **11**, 487 (1973).

17. A. Bourret, H. Chanzy, and R. Lazaro, *Biopolymers,* **11**, 893 (1972).

18. H. Chanzy, in *Structure of Fibrous Polymers, Colston Papers No. 26,* E. D. T. Atkins and A. Keller, eds., Butterworths, London, 1975, p. 417.

19. J. F. Revol, *Carbohydr. Polym.,* **2**, 123 (1982).

20. J. Sugiyama, H. Harada, Y. Fujiyoshi, and N. Uyeda, *Mokuzai Gakkaishi,* **30**, 98 (1984).

21. H. Harada, *Mokuzai Gakkaishi,* **30**, 513 (1984).

22. A. B. Wardrop and J. M. Jutte, *Wood Sci. Technol.,* **2**, 105 (1968).

23. A. R. White and M. R. Brown, Jr., *Proc. Natl. Acad. Sci. USA,* **78**, 1047 (1981).

24. H. Chanzy and B. Henrissat, *Carbohydr. Polym.,* **3**, 161 (1983).

25. K. H. Gardner and J. Blackwell, *Biopolymers,* **13**, 1975 (1974).

26. H. Chanzy, B. Henrissat, and R. Vuong, *FEBS Lett.,* **172**, 193 (1984).

27. M. Schülein, H. E. Schiff, P. Schneider, and C. Dambmann, in *Bioconversion and Biochemical Engineering Symposium 2,* T. K. Ghose, ed., Vol. 1, Thomson Press, Faridabad, Haryana, India, 1981, p. 97.

28. N. W. Lützen, M. H. Nielsen, K. M. Oxenboell, M. Schülein, and B. Stentebjerg-Olesen, *Philos. Trans. R. Soc. Lond. B,* **300**, 283 (1983).

29. S. Hestrin, *Methods Carbohydr. Chem.,* **3**, 4 (1963).

30. E. T. Reese, *Proc. Biochem.,* **17**, 2 (1982).

20

Thermal Decomposition of Wood Pulps at Different Degrees of Delignification

ERDOĞAN KIRAN

Department of Chemical Engineering, University of Maine, Orono, Maine

Thermal decomposition and characterization of lignocellulosic materials, in particular wood and its constituents, have been the subject of a large number of studies. Several review articles describe the general phenomena (1–7). The interest in the area stems primarily from the fact that lignocellulosic materials are renewable resources of fuel, chemicals (e.g., methanol, acetic acid, charcoal), and industrial materials (e.g., pulp and paper, or building materials for houses). An understanding of the decomposition phenomena is needed to either enhance or regulate the decompositions to improve the pyrolytic product distributions for better fuel efficiency and chemicals production, or to retard decompositions to achieve longer thermal stability, permanence, and fire retardancy. Previous studies on the thermal degradation behavior of pulps for example have been primarily concerned with the assessment of paper permanence, stability, and aging (8–10).

The chemical composition and the physical characteristics of the cellulosic materials are the most important parameters that influence their thermal behavior. From this perspective, wood pulps form a useful model system, since, depending on the nature of the raw material and the pulping conditions, the level of delignification can be regulated and a variety of chemical composition and physical characteristics can be obtained.

Detailed discussions of the principles of delignification and related processes are available in the literature (11). From a chemical perspective, wood consists of primarily cellulose (40–45%), lignin (20–30%), and hemicelluloses (20–30%), which are all polymeric in nature, and low-molecular-weight extractives (2–4%) and inorganic matter (1% or less). The actual amounts of each constituent depend on the nature of the wood species. From a physical perspective, it is composed of elongated cells (fibers) that are held together by the middle lamella. In pulping, these fibrous components are separated into a form suitable for papermaking which is achieved by either mechanical or chemical means, or by a combination of both. In chemical pulping operations, the lignin that is mainly located in the middle lamella and the secondary wall of the fiber is removed by reaction and conversion to a soluble derivative. In the process, extractives are dissolved, and some degradation and dissolution of cellulose and hemicellulose fractions also occur. The extent of the dissolution of the various constituents of wood depends, however, on the operational conditions (such as temperature, concentration of active chemicals in the cooking liquor, duration of pulping) and the type of chemical action (such as kraft or sulfite pulping conditions). Thus, depending on the nature of the raw material and the pulping conditions, wood pulps of different compositions can be readily prepared. Constituents that remain in the pulps are better representations of their natural forms as they exist in the wood matrix than their isolated forms, which are often structurally different. As such, pulps can be used as logical model systems to study the *in situ* behavior of the constituents in the wood matrix in a controlled way.

This paper presents the results of a systematic study of the thermal decomposition of red spruce (softwood) and maple (hardwood) and their pulps obtained by kraft and sulfite procedures at various levels of delignification. Thermal decomposition of each sample has been characterized using thermogravimetry and pyrolysis–molecular weight chromatography. Thermogravimetric analyses and the pyrolytic product distributions both display characteristic features that are closely associated with the composition (i.e., level of delignification) of the pulp samples and the pulping process by which they are obtained.

The differences in the decomposition phenomena of the pulps are significant and open up new possibilities for pulp characterization and/or development of alternative sensors for controlling the digesters for desired delignification (or yield) levels in pulping operations. Furthermore, the differences in the decomposition behavior of wood and its pulps illustrate the relative significance and contributions of the various constituents of wood to the overall thermal process. The interactive influence of the constituents on the thermal behavior of each other as they exist in the lignocellulosic matrix and the consequences on the distribution of volatile decomposition products are also illustrated.

MATERIALS AND METHODS

Red spruce and maple were selected as representatives of softwood and hardwood species, respectively. Pulps were prepared in a 10-L laboratory digester under either kraft or sulfite cooking conditions at a liquor-to-wood ratio of 10:1. Kraft cooks were conducted at 175°C using a liquor of 30% sulfidity [Na_2S/

(NaOH + Na$_2$S)] and 22% active alkali [NaOH + Na$_2$S] content. Pulps of different lignin levels were obtained by varying the cooking time at 175°C. In the preparation of sulfite pulps, a magnesium base liquor [Mg (HSO$_3$)$_2$] containing 30 g/L total SO$_2$ was used. Cooking was conducted by first heating the digester to 110°C and holding for 1 hr after which the temperature was raised to 170°C. Pulps with different lignin levels were obtained by varying the cooking time at 170°C. After each cooking, pulps were thoroughly washed with water and then dried in air. Kappa and/or permanganate numbers were determined according to Tappi standard procedures (12). Powdered samples were prepared using a Thomas–Wiley laboratory mill with a 40-mesh screen.

Thermogravimetric analyses were conducted using a DuPont 951 unit which has been interfaced with a digital PDP 11/60 minicomputer for real-time data collection and analysis in our laboratories. All the analyses were carried out under an inert flowing nitrogen atmosphere (at 50 mL/min) using relatively small (∼ 10 mg) powdered samples. To minimize the variations arising from initial moisture levels, sample weights were normally adjusted to 100% after 105°C.

A special instrumental system consisting of a programmable pyrolyzer coupled in series with a thermal conductivity cell, a trap, and a mass chromatograph was used for selective trapping and analysis of the decomposition products. The details of this system have been fully described in the literature (13, 14). The thermal conductivity detector monitors the evolution of volatile products (to permit selective trapping) and provides data complimentary to derivative thermogravimetry. The trap is a column packed with Porapak Q from which the trapped fractions are thermally desorbed and introduced to the mass chromatograph. This instrument is a dual column gas chromatograph that directly provides mass numbers of the resolved components of a mixture by means of a pair of gas density detectors.

In the present study small powdered samples (∼ 10 mg) were pyrolyzed in helium atmosphere at a heating rate of 20°C/min from room temperature to 500°C. Only the fraction of volatiles that were trapped without subambient cooling were analyzed in the mass chromatograph. The matched columns were 15 ft long and packed with Dexsil-300. Column carriers were SF$_6$ and CO$_2$.

THERMOGRAVIMETRIC ANALYSES

Behavior of Wood Species and Their Isolated Components

Thermogravimetric (TG) and derivative TG behavior of the wood species red spruce and maple are shown in Figure 20.1. The behavior of a sample of α-cellulose and loblolly pine holocellulose and milled wood lignin samples are also included. A number of general features are immediately noted from Figure 20.1A where residual weight is plotted as a function of temperature. The temperature at which the initial weight loss is observed (i.e., the onset of decomposition)

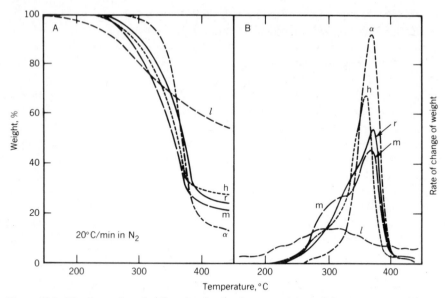

Figure 20.1 Thermogravimetric (*A*) and derivative thermogravimetric (*B*) behavior lignocellulosic materials. *l*, Loblloly pine milled wood lignin; h, loblolly pine holocellulose; r, red spruce; m, maple; α, α-cellulose. In the derivative plots (*B*), peak heights are in relative units.

increases in going from lignin to cellulose. With each sample, there is a residual weight that remains even after reaching high temperatures, and this residual weight decreases in going from lignin to cellulose. The overall behavior of the woods is intermediate but closer to that of α-cellulose and is more similar to that of holocellulose. This is likely due to the fact that cellulose and hemicellulose make up a larger fraction of wood than does lignin. There are observable differences between red spruce and maple, such as the lower residual weight in maple which appears to be consistent with the lower lignin levels of hardwoods.

In the derivative TG curves shown in Figure 20.1*B*, where rate of change in weight is plotted as a function of temperature, the onset of decompositions is reflected by the temperatures at which the initial deviations from the base lines are observed. The intensity of the peaks represent the rate of decompositions, and the temperatures corresponding to the peak maxima correspond to the temperatures where the rates reach their highest values. Even though the onset of decomposition shifts to higher temperatures, the rate of decomposition becomes greater in going from lignin to cellulose as displayed by the increase in the derivative peak intensities. Decomposition of lignin is complex and occurs at many stages, as evidenced by the multiple maxima in the derivative plot.

The derivative plots show major differences in the behavior of maple and red spruce. Both species, in contrast with α-cellulose, show more than one maximum in the derivative plots, and the lower temperature peak is more pronounced in maple. These lower temperature decompositions are associated with the more readily degradable noncrystalline hemicellulose fractions of the species. Typically, hardwoods contain larger amounts of hemicelluloses than softwoods and may account for the greater intensity of the lower temperature decomposition peak in maple.

Behavior of Wood Pulps

Kraft Pulps

Figure 20.2 shows the TG and derivative TG behavior of red spruce kraft pulps. The pulps cover a wide range of kappa numbers (from 145 to 30). The degree of delignification increases in going from pulp a to f. As can be seen in Figure 20.2A, the onset of decomposition shifts to higher temperatures (i.e., pulps become more stable) with increasing lignin removal. Once initiated, decomposition proceeds first rapidly resulting in a significant weight loss after which an abrupt change occurs where weight loss proceeds at a much smaller rate. The characteristic residual weight (char) level (which can be identified by drawing tangents to the weight loss curve in the rapid and slow decomposition regimes) decreases with increasing degree of delignification and approaches that of α-cellulose. It is interesting to note that the behavior of wood is intermediate to its low and high lignin content pulps.

The derivative plot shows that with increasing delignification, decompositions start at higher temperatures but proceed at much faster rates (which is reflected

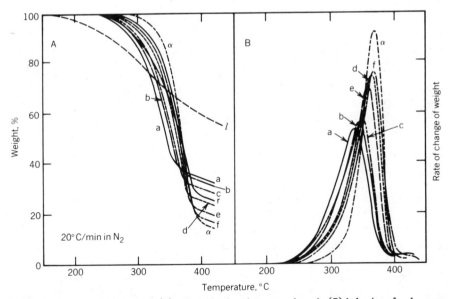

Figure 20.2 Thermogravimetric (A) and derivative thermogravimetric (B) behavior of red spruce kraft pulp. Kappa numbers of pulps are $a = 143$, $b = 132$, $c = 110$, $d = 57$, $e = 43$, $f = 30$; ℓ, r, and α represent milled wood lignin, red spruce, and α-cellulose. In the derivative plot (B), peak heights are in relative units. (% lignin in pulps $\simeq 0.15 \times$ Kappa number).

by the peak intensities). With respect to the temperatures corresponding to maximum rate of decompositions and the relative peak intensities, high and low kappa number pulps (ie., pulps *a, b, c* versus *d, e, f*) appear to behave as two distinct groups.

As was observed for red spruce and maple in Figure 1*B*, the presence of hemicelluloses, which are amorphous polysaccharides with low thermal stability, significantly affects the thermal behavior of wood. The presence of amorphous regions are known to affect the behavior of cellulose as well. More specifically, literature on cotton cellulose indicates that with increasing amorphous content, intensity of the derivative TG curve decreases and shifts to lower temperatures and char formation becomes greater (15, 16). If the lower decomposition temperatures, lower peak intensities, and higher char levels that are also observed in Figure 20.2 for the kraft pulps can be attributed to the amorphous fractions, then the observation of two distinct groups suggests that below a critical kappa number, not only lignin but especially the hemicellulose removal becomes very significant, and the thermal behavior of the pulps shifts to a different regime that is influenced more by the crystalline domain of the matrix.

The trends in Figure 20.2 indicate that thermogravimetric data can differentiate kraft pulps with respect to their compositions. This has been further substantiated by the results obtained for red spruce and maple pulps prepared under identical pulping conditions (i.e., cooked simultaneously in the same digester using two screen baskets, one containing red spruce and the other maple wood chips) (17). At short cooking times maple gives pulps with lower lignin content than red spruce, and this is reflected in the TG behavior with lower residual weight and higher derivative peak intensities for the maple pulp (17). However, the differences become less pronounced at longer cooking times, corresponding to comparable lignin (or cellulose) levels, suggesting that the levels of crystallinity of the cellulose in maple or red spruce kraft pulps of comparable lignin levels are probably very similar and that the level of delignification is the major factor that influences the overall TG behavior of the kraft pulps. Similarities in levels of crystallinity have in fact been verified by x-ray measurements (18).

Sulfite Pulps

Figure 20.3 shows the comparative TG and derivative TG behavior of red spruce sulfite pulps. The general trends are similar to those observed with kraft pulps. With increased level of delignification, residual weight decreases, and the onset of decomposition shifts to higher temperatures. However, at comparable kappa numbers (i.e., similar levels of delignification), sulfite pulps display consistently lower residual weights, the onset of decomposition occurs at higher temperatures, and the rates of decomposition (i.e., derivative peak intensities) are greater. These are illustrated in Figure 20.3*C*, which combines TG and derivative TG plots for sulfite pulp S1 and the kraft pulp of comparable kappa numbers. (A different scale factor was chosen in the derivative plots for clarity.).

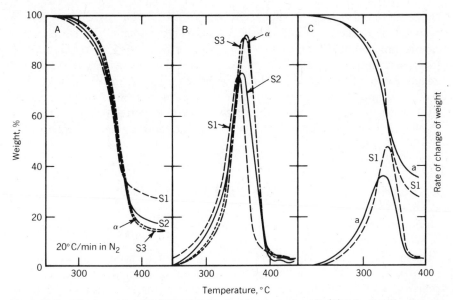

Figure 20.3 Thermogravimetric (A) and derivative thermogravimetric (B) behavior of red spruce sulfite pulps. Kappa numbers of pulps are S1 = 143, S2 = 95, S3 = 41. In (C), a represents kraft pulp with kappa number 143. The scale factor for the derivative plot in (C) is different from that in (B).

Similar trends are observed for a variety of other pulp pairs (17). Since cellulose contents are comparable, these differences can be attributed to structural differences such as the level of crystallinity and or the molecular weight of cellulose. There is indication in the literature that sulfite pulps are more crystalline than kraft pulps (19). X-ray measurements conducted in our laboratories show without ambiguity that at comparable lignin levels sulfite pulps are consistently more crystalline than kraft pulps (18). The different TG behavior of the sulfite pulps is indeed consistent with high level of crystallinity, which is further substantiated by the similarity of the behavior of pulp S3 (kappa number = 41) to that of α-cellulose. This is to be contrasted with the behavior of kraft pulp e in Figure 20.2 for which the kappa number is 43 but the behavior is far removed than that of α-cellulose. In the literature the differences in TG behavior of a number of commercial-paper grade pulps have also been attributed to variations in their cellulose content and crystallinity (8, 20).

Summary—Thermogravimetric Analyses

The dynamic TG data for the various pulps presented in the foregoing sections show that the derivative peak intensities and residual char levels are sensitive indicators of the lignin level and/or the physical state (i.e., amorphous vs. crystalline content) of the pulps. As such, thermogravimetry can be used to characterize pulps and/or pulping processes. Quantitative correlations that incorporate information on full chemical analysis (actual lignin content and characterization, cellulose content) of the pulps and their x-ray crystallinities need to be developed. Such correlations can then be used to determine the level of delignification, quality, and stability of pulps and the yields in pulping operations. In the full quantitative treatment, kinetic parameters such as the apparent activation energies for decomposition may be another correlating parameter. As the lignin levels decrease and the pulps become more stable, the apparent activation energies become larger.

ANALYSIS OF PYROLYSIS PRODUCTS

Decomposition Products from Wood Species

The responses from one channel (using CO_2 as the carrier gas) of the mass chromatograph for the decomposition products of maple, red spruce, and α-cellulose are compared in Figure 20.4. These pyrograms correspond to the fraction of decomposition products formed in the temperature range 125–500°C that could be collected in the Porapak trap without subambient cooling. The peaks with same retention times have been given the same symbols. A number of characteristic features are noted. Peak a, which is a major component in maple pyrogram, is not observed in that of red spruce or α-cellulose. The product distributions in the range e–f are different. In contrast to peak e in maple, peak f is more significant in red spruce. The relative intensities of peaks g and h in maple are different from those in red spruce. Peak h appears only as a shoulder in maple. A similar change in intensities is observed with respect to peaks k and l. Formation of peak l is not significant in red spruce. In contrast to maple, peaks in the range n–o are not formed to the same extent in red spruce. The intensities of peaks l, m, and p decrease in maple, but in red spruce the intensity of peak p is greater than those of l and m. Peaks k–r are not observed in the pyrogram for α-cellulose.

These differences in the pyrolysis products of red spruce (softwood) and maple (hardwood) are understandable since softwoods and hardwoods differ in their chemical compositions (11). More specifically, softwood lignins contain exclusively guaiacyl propyl units (which contain one methoxyl group), whereas hardwood lignins contain not only guaiacyl but also syringyl propyl moieties (with two methoxyl groups). Hemicellulose contents are also different. Whereas xylans are the main hemicellulose of hardwoods, mannans are the principal hemicelluloses of softwoods. Hardwoods contain larger amounts of hemicelluloses but are lower in their lignin content.

The literature on the products of decomposition of wood and its primary constituents, namely cellulose, hemicellulose, and lignin, is extensive (3, 5, 21–31). Excellent reviews of the literature are available (3, 29, 30). Cellulose pyrolysis leads to the formation of char, tar, and gaseous products, and the tar fraction

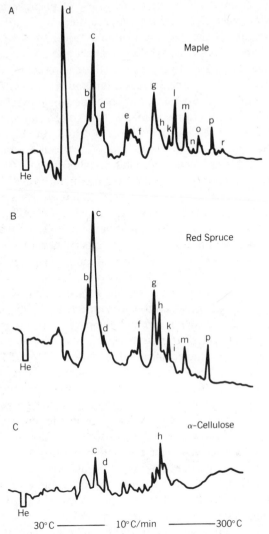

Figure 20.4 Products of pyrolysis from maple (*A*), red spruce (*B*), and α-cellulose (*C*).

contains levoglucosan and other anhydroglucose compounds as the major components. Lignin produces a tarry mixture of aromatic phenolic compounds that are homologues to phenol, guaiacol or 2,6-dimethoxyphenol. Substituent groups are usually para to the phenolic hydroxyl and are typically 3-carbons or less such as methyl, ethyl, propyl, and allyl. Hemicelluloses, in comparison with cellulose, produce less tar and no levoglucosan. Acetic acid is a major decomposition product of 5-carbon sugars associated with hemicelluloses, especially in hardwoods. The products in wood tars cover a wide spectrum of phenolic and nonphenolic compounds including a variety of phenols, acids, aldehydes, ketones, alcohols, esters, hydrocarbons, and furans.

Molecular weights of the major peaks in Figure 20.4 were estimated using the ratio of peak intensities from the CO_2 channel and SF_6 channel (not shown) of the mass chromatogram. Even though these molecular weights are only estimates (because of nonideal peak separations and associated limitations in the mass chromatographic calculations), they provide valuable information.

Peak a, which primarily forms in maple, has a molecular weight of about 61, which compares with acetic acid (MW = 60) and can be attributed to the hemicellulose fraction in maple. Peaks c and d, which are observed in wood and cellulose pyrograms, have molecular weights about 106 and 96, which compares with the molecular weights for 5-methylfurfural (MW = 110) and furfural (MW = 96), respectively. Their formation is clearly associated with the cellulosic fraction of the wood species. Peak g has a molecular weight of about 137. This peak is formed in largest amount in red spruce, yet it is not significant in α-cellulose and therefore appears to be associated with the guaiacol lignin moieties of softwoods. 4-Methyl guaiacol has a molecular weight of 138, which compares with that of peak g. Since softwood lignins contain only guaiacol moieties, the formation of products related to such groups is expected to be greater in red spruce. Peak h, which becomes especially significant in α-cellulose, has a molecular weight of about 146 and compares with the molecular weight of 1,4-3,6-dianhydro-D-glucopyranose (MW = 144), known to form from cellulose. Peaks k and l have molecular weights of about 150 and 159, respectively. They are absent in a α-cellulose pyrogram and must therefore be associated with the lignin moieties. Peak k is observed in larger amounts in red spruce, which suggests that it involves guaiacol groups. 5-Ethyl guaiacol and vanillin each have a molecular weight of 152. Peak l in maple must have its origin in the syringyl moieties of the hardwood lignin structure. It is absent in red spruce (i.e., peak l is not significant). This compound is likely to be 2,6-dimethoxyphenol, which has a molecular weight of 154 and has been identified as the major product from the hardwood milled wood lignins (28). Peaks m and p must be arising primarily from the guaiacol groups, since they are present in both wood species. Their calculated weights are 165 and 182, respectively. Propyl guaiacol has a molecular weight of 166, and coniferyl alcohol has a molecular weight of 180. However, it is possible that 4-methyl-2,6-dimethoxyphenol (MW = 168) and syringylaldehyde (MW = 182), which arise from syringyl moieties, may also be masked under these peaks in the maple pyrogram. It is clear that peaks n to o in maple must arise from the syringyl moieties, since they are not significant in red spruce pyrograms.

Decomposition Products from Kraft Pulps

Pyrolysis products (generated in the temperature range 125–500°C) of maple kraft pulps of different lignin levels are compared with those of initial wood and α-cellulose in Figure 20.5. The following characteristic features are observed. The formation of the decomposition product a is not observed in the pulps,

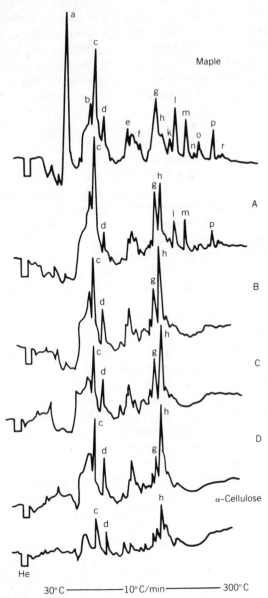

Figure 20.5 Products of pyrolysis from maple and its kraft pulps with Kappa numbers 44 (*A*), 20 (*B*), 13 (*C*), and 11 (*D*).

indicating that the portion of the wood matrix (i.e., hemicelluloses) leading to this product is either dissolved or decomposed in the early stages of cooking. The products corresponding to peaks *k* or *r* decrease in intensity as cooking is continued and disappears for well-cooked pulps, indicating that these products indeed originate from the lignin moieties and thus their generation decreases with increased delignification.

In the pyrogram of original wood, peak g is a major decomposition product and peak h appears as a shoulder. With cooking, however, the intensities of these peaks undergo a reversal, and peak h gradually dominates this pair with increased delignification, substantiating the observation that its origin is the cellulose moieties of the matrix. Some changes in the distribution of products corresponding to peaks e to f are also noted. A final observation is the decrease in the intensity of peak c with pulping and the general similarity of the pyrolysis pattern of the well-cooked maple pulp with that of α-cellulose. The peaks c, d, and h are characteristic products arising from the cellulosic portion of the wood matrix and remain with α-cellulose.

The behavior of red spruce and its pulps is shown in Figure 20.6. The general trends are similar to those observed for maple pulps. Intensity of peak c decreases with delignification in a manner similar to maple pyrolysis. The intensities of peaks g and h show the most significant differences upon delignification in red spruce pulps also. As in maple pulps, with increased pulping, peak h becomes the dominant product. The intensities of peaks k to p decrease with delignification, and the behavior of well-cooked pulps becomes similar to that of α-cellulose. The intensity of peak c also decreases with delignification.

Summary—Pyrolytic Techniques

The foregoing discussions illustrate the capability of pyrolytic techniques in differentiating different wood species and pulps. Even though no attempt was made to analyze all the decomposition products that are formed, the fraction that is collected at room temperature and then thermally desorbed (at 230°C) from the trap provides sufficient characteristic differences between wood species and their pulps of different lignin levels. The absence of very low and very high molecular weight compounds simplifies the pyrograms and highlights the unique features.

The systematic variations that take place in characteristic product distributions, especially with respect to peaks g and h with delignification, are most remarkable. Peak resolutions can be improved by using capillary columns, and the ratio of intensities of peaks g and h can be precisely determined and correlated with the chemical composition (i.e., lignin content) of pulps. Fast procedures for pulp characterization and/or digester control by pyrolytic methods can thus be developed. The ability of the pyrolytic technique to detect guaiacol- and syringyl-type lignin products has been reported by other workers also (26, 28, 31) and should be of further analytical value. Work is now being planned in our laboratories for pyrolytic characterization of the cooking liquors corresponding to the different levels of delignification. These results will not only complement the behavior of the corresponding pulps but will be of special value since cooking liquors are easier to sample from digesters during cooking, and any characteristic information on pyrolysis of dissolved fragments can be more readily adaptable for digester control operations. It should be further noted that pyrolytic behavior of concentrated cooking liquors is currently of industrial im-

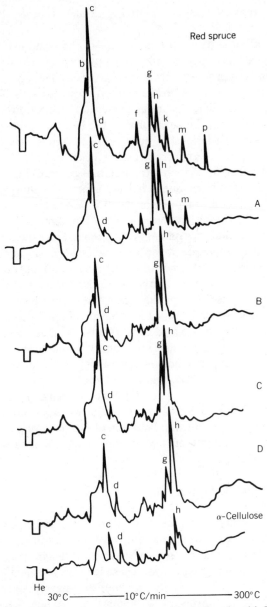

Figure 20.6 Products of pyrolysis from red spruce and its kraft pulp with Kappa numbers 127 (*A*), 75 (*B*), 49 (*C*), and 29 (*D*).

portance in the recovery cycles in pulping operations, especially in the kraft process.

Even though cellulose makes up about 50% of the wood matrix, the relative simplicity of the cellulose pyrogram is not observed in wood (see Fig. 20.4). Furthermore, pyrolysis product distribution of various lignin preparations shows significant variations depending on the method of preparation. The behavior of mixtures of cellulose and lignin is not a simple sum of the individual pyrograms of cellulose and lignin (32). These emphasize the fact that pyrolytic behavior of lignocellulosic materials is influenced by various interactive effects, and the usual assumption that is often inferred from TG analyses that the pyrolysis of wood is the sum of the pyrolysis of its major (isolated) components must be treated with caution. An example of interactive influence of lignin moieties on the decomposition of cellulose fractions of wood is suggested by the behavior of peak *h* in maple and red spruce pyrograms (Fig. 20.4) which, as already discussed, is associated with the cellulosic content of the species. Since the cellulose contents of softwoods and hardwoods are about the same, the behavior of peak *h* suggests either a strong suppressive influence of the syringyl–propane moieties in maple or an enhancing influence of the guaiacyl–propane moieties in red spruce on decomposition of cellulose. The prediction of the overall behavior of the composite material (such as wood) from the behavior of its isolated constituents is further complicated since the isolated constituents, particularly lignins, are often structurally different from their natural form in wood. From this perspective, pulps at different levels of delignification form a logical model system and provide a way of focusing on the *in situ* behavior of the constituents in the complex network. These effects can and will be further investigated by studying the pyrolytic behavior of pulps of known lignin, cellulose, and hemicellulose content and comparing the results with the behavior of mixtures of cellulose, hemicellulose, and milled wood lignin of same overall composition as the pulp.

CONCLUDING REMARKS

Thermogravimetric analyses, especially derivative TG and analysis of product distributions by pyrolysis–chromatography of wood and pulps, show a number of characteristic features that depend on the level of delignification, the type of wood species, and the pulping process. These features are significant and open up new characterization schemes for the pulp-and-paper industry. From the perspective of pyrolytic conversion of biomass to fuel and chemicals, use of pulps of different lignin levels provides a logical approach to the study of the interactive effects of various components of the lignocellulosics as they exist in the complex network. Because of the strong dependence of char level on lignin-to-cellulose ratio and/or the ratio of amorphous to crystalline domains, pulps of different levels of delignification and their chars can also be used as model systems to elucidate the relative significance of lignin/cellulose content on the resultant char morphology and properties.

REFERENCES

1. F. C. Beall and H. W. Eickner, USDA Forest Service Research Paper FPL 130, May 1970.

2. T. Nguyen, E. Zavarin, and E. M. Barall II, *J. Macromol. Sci. Rev. Macromol. Chem.*, C 20(1), 1 (1981).

3. E. J. Soltes and T. J. Elder, in *Organic Chemicals from Biomass*, I. S. Goldstein, ed., CRC Press, Boca Raton, FL, 1981, p. 63.

4. F. J. Kilzer, in *Cellulose and Cellulose Derivatives*, N. M. Bikales and L. Segal, eds., Part V, Wiley, New York, 1971, p. 1015.

5. F. Shafizadeh, *Appl. Polym. Symp.* **28**, 153 (1975).

6. G. G. Allan and T. Mattila, in *Lignins: Occurrence, Formation, Structure and Reactions*, K. V. Sarkanen and C. H. Ludwig, eds., Wiley, New York, 1971, p. 575.

7. S. L. LeVan, *Adv. Chem. Ser.*, **207**, 531 (1984).

8. R. D. Cardwell and P. Luner, *Adv. Chem. Ser.*, **164**, 382 (1977).

9. A. A. Duswalt, *Adv. Chem. Ser.*, **164**, 352 (1977).

10. S. M. Saad and A. E. El-Kholy, *Holzforshung,* **34**(4), 147 (1980).

11. E. Sjöström, *Wood Chemistry. Fundamentals and Applications*, Academic, New York, 1981.

12. Tappi Standards T 236 os-76 and T 214 m-50.

13. E. Kiran and J. K. Gillham, *J. Appl. Poly. Sci.*, **20**, 931 (1976).

14. E. Kiran and J. K. Gillham, in *Developments in Polymer Degradation—2*, N. Grassie, ed., Applied Science Publishers, London, 1979, p. 1.

15. K. E. Cabradilla and S. H. Zeronian, in *Thermal Uses and Properties of Carbohydrates and Lignins*, F. Shafizadeh, K. V. Sarkanen, and D. A. Tillman, eds., Academic, New York, 1976, p. 73.

16. M. Lewin, A. Basch, and C. Roderig, in *Proc. Int. Symp. on Macromolecules*, Rio de Janeiro, July 26–31, 1974, E. B. Marro, ed., Elsevier, Amsterdam, 1975, p. 225.

17. E. Kiran, C. C. Marsano and L. Tuttle, to be published.

18. E. Kiran and J. V. Chernosky, to be published.

19. D. H. Page, *J. Pulp Paper Sci.* (Canada), pp. TR 15–12, March 1983.

20. R. D. Cardwell and P. Luner, *Adv. Chem. Ser.*, **164**, 362 (1977).

21. F. Shafizadeh, K. V. Sarkanen, and D. A. Tillman (eds.), *Thermal Uses and Properties of Carbohydrates and Lignins*, Academic Press, New York, 1976.

22. R. F. Schwenker, Jr., and L. R. Beck, *J. Polym. Sci. Part C,* **2**, 331 (1963).

23. T. A. Milne, R. J. Evans, and M. N. Soltys, "Biomass Energy Systems. Fundamental Pyrolysis Studies," Quarterly Report 1 March 1983–30 May 1983, SERI, Golden, CO, July 1983.

24. F. Martin, C. Saiz-Jimenez, and F. J. Gonzalez-Vila, *Holzforshung,* **33**, 210 (1979).

25. D. J. Gardner, T. P. Shultz, and G. D. McGinnis, *J. Wood Chem. Tech.*, **5**(1), 85 (1985).

26. Von O. Faix and W. Schweers, *Holzforshung,* **29**(6), 224 (1975).

27. F. P. Petrocelli and M. T. Klein, *Macromolecules,* **17**, 161 (1984).

28. J. R. Obst, *J. Wood Chem. Tech.*, **3**(4), 377 (1983).

29. W. J. Irwin, *Analytical Pyrolysis. A Comprehensive Guide*. Chromatographic Science Series, Vol. 22, Marcel Dekker, New York, 1982, p. 333.

30. F. Shafizadeh, in *Handbook of Physical and Mechanical Testing of Paper and Paperboard*, R. E. Mark, ed., Dekker, New York, 1984, p. 259.

31. T. J. Fullerton and R. A. Franich, *Holzforshung* **37**, 267 (1983).

32. E. Kiran, to be published.

APPENDIX I

Listing of BASIC Program Used to Simulate the Hydrolysis of Cellulose with Dilute Sulfuric Acid

```
0100 REM ********************************************
0110 REM BASIC VERSION OF CELLULOSE HYDROLYSIS MODEL
0120 REM ********************************************
0130 REM
0140 REM LIST OF VARIABLES
0150 REM
0160 REM TA - AMBIENT TEMPERATURE (DEG C)
0170 REM TB - BATH TEMPERATURE (DEG C)
0180 REM CA - CONCENTRATION OF APPLIED SULFURIC ACID
         SOLUTION (%)
0190 REM AM - CONCENTRATION OF SULFURIC ACID
         (MOLALITY)
0200 REM HM - H-ION CONCENTRATION (MOLALITY)
0210 REM CM - CONCENTRATION OF CATIONS (MOLALITY)
0220 REM LS - LIQUID TO SOLID RATIO
0230 REM CEC - CELLULOSE CONTENT OF LIGNOCELLULOSE (%)
0240 REM COP - PROPORTION OF CELLULOSE WHICH IS
         RESISTANT (%)
0250 REM EQ - NEUTRALIZING CAPACITY OF LIGNOCELLULOSE
         (EQUIVALENTS/KG)
0260 REM X - TIME (MINUTES)
0270 REM Y(1,-) - PROPORTION OF EASILY HYDROLYZABLE
         CELLULOSE REMAINING (%)
0280 REM Y(2,-) - PROPORTION OF RESISTANT CELLULOSE
         REMAINING (%)
0290 REM Y(3,-) - STIOCHIOMETRIC YIELD OF MONOMERIC
         GLUCOSE (%)
0300 REM Y(4,-) - STIOCHIOMETRIC YIELD OF DIMER (%)
0310 REM Y(5,-) - STIOCHIOMETRIC YIELD OF GLUCOSIDE
         (%)
0320 REM Y(6,-) - STIOCHIOMETRIC YIELD OF LEVOGLUCOSAN
         (%)
```

```
0330 REM G - STIOCHIOMETRIC YIELD OF TOTAL
     ANHYDROGLUCOSE UNITS IN SOLUTION (%)
0340 REM C - STIOCHIOMETRIC YIELD OF GLUCOSE IN
     COMBINED FORM (%), i.e. AS, REVERSION PRODUCTS
0350 REM PC - PROPORTION OF TOTAL GLUCOSE IN COMBINED
     FORM (%)
0360 REM CON - CONCENTRATION OF TOTAL GLUCOSE (AS
     GLUCOSE) IN SOLUTION (%)
0370 REM CR - PROPORTION OF ORIGINAL CELLULOSE
     REMAINING (%)
0380 REM R - STIOCHIOMETRIC YIELD OF REDUCING SUGARS
     (%)
0390 REM CER - CELLULOSE CONTENT OF REMAINING
     LIGNOCELLULOSE (%)
0400 REM ALPHA - EQUIPMENT DEPENDENT HEAT TRANSFER
     CONSTANT (1/MINUTES)
0410 REM NI - NUMBER OF INTEGRATION INTERVALS
0420 REM NPRT - PRINT INDEX
0430 REM KC - RATE CONSTANT FOR HYDROLYSIS OF THE
     RESISTANT CELLULOSE
0440 REM KG - RATE CONSTANT FOR THE DEHYDRATION OF
     GLUCOSE
0450 REM KD - EQUILIBRIUM CONSTANT FOR THE FORMATION
     OF DIMER FROM GLUCOSE
0460 REM KL - EQUILIBRIUM CONSTANT FOR THE FORMATION
     OF "LEVOGLUCOSAN"
0470 REM
0480 REM INPUT STARTING CONDITIONS
0490 REM
0500 PRINT "***** CELLULOSE HYDROLYSIS MODEL *****"
0510 PRINT :PRINT "    BATH TEMPERATURE = ";:INPUT TB
0520 PRINT :PRINT "   ACID CONCENTRATION = ";:INPUT CA
0530 PRINT :PRINT "LIQUID TO SOLID RATIO = ";:INPUT LS
0540 PRINT :PRINT "CELLULOSE CONTENT OF
                     STARTING MATERIAL= ";:INPUT CEC
0550 PRINT :PRINT "  RESISTANT CELLULOSE = ";:INPUT
     COP
0560 PRINT :PRINT "   ASH NEUTRALIZING
                      CAPACITY= ";:INPUT EQ
0570 DIM X(2),Y(6,2),DY(6,4),V(6,4),A$(1),TEMP$(7),
     PRNT$(80)
0580 DERIV=1640
0590 ALPHA=13.8:NI=500:TA=25:YO=100
0600 CONST=1620*LS/CEC
0610 REM
0620 REM CALCULATION OF INTEGRATION INCREMENT
0630 REM
0640 CM=EQ/(LS*(1-CA/100)):REM MOLAL CONCENTRATION OF
     CATIONS
0650 AM=CA/(9.8*(1-CA/100)):REM MOLAL CONCENTRATION OF
     SULFURIC ACID
0660 HM=AM-CM:REM MOLAL CONCENTRATION OF [H]
0670 KC=2.87E+20*HM^1.218
     *EXP(-42900/(1.987*(TB+273.1)))
```

```
0680 KG=5.58E+15*HM*EXP(-33800/(1.987*(TB+273.1)))
0690 TMAX=LOG(KC/KG)/(KC-KG)
0700 DX=2.5*TMAX/NI:REM INTEGRATION INCREMENT
0710 XFINAL=2.5*TMAX-LOG(0.01)/ALPHA:REM INTEGRATION
     RANGE
0720 NIT=INT(XFINAL/DX):REM NUMBER OF INTEGRATION
     INTERVALS
0730 NPRT=INT(NIT/41):REM PRINT INDEX
0740 REM
0750 REM PREPARING PRINTOUT PAGE
0760 REM
0770 OPEN #3,8,0,"P:"
0780 PRINT #3:PRINT #3:PRINT #3
0790 PRINT #3;"                              CELLULOSE
     HYDROLYSIS"
0800 PRINT #3;"                       YIELDS
     CALCULATED FROM MODEL"
0810 PRINT #3:PRINT #3;"------------------------------
     -------------------------------------------------
     ---"
0820 PRINT #3;"      ***   GLUCOSE YIELDS   ***
        GLUCOSE  CELLULOSE  REDUCING CELLULOSE"
0830 PRINT #3;" TIME   TOTAL   MONOMER   COMBINED
        CONC    REMAINING  YIELD IN RESIDUE"
0840 PRINT #3;"------------------------------------
     -----------------------------------"
0850 PRINT #3;"  MIN    %        %          %
           %        %          %         %"
0860 PRINT #3
0865 REM
0870 REM RUNGE-KUTTA NUMERICAL INTEGRATION
0875 REM
0880 X(1)=0
0890 Y(1,1)=YO*(1-COP/100)/162
0900 Y(2,1)=YO*COP/16200
0910 Y(3,1)=0
0920 Y(4,1)=0
0930 Y(5,1)=0
0940 Y(6,1)=0
0950 MAXTEST=0
0960 FOR K=0 TO NIT
0970 A$=" "
0980 FOR J=1 TO 6
0990 IF Y(J,1)<1.0E-15 THEN V(J,1)=0:GOTO 1010
1000 V(J,1)=Y(J,1)
1010 NEXT J
1020 X=X(1):N=1:GOSUB DERIV
1030 X2=X(1)+DX/2
1040 FOR J=1 TO 6
1050 IF Y(J,1)<1.0E-15 THEN V(J,2)=0:GOTO 1070
1060 V(J,2)=Y(J,1)+DY(J,1)*DX/2
1070 NEXT J
1080 X=X2:N=2:GOSUB DERIV
1090 FOR J=1 TO 6
```

```
1100 IF Y(J,1)<1.0E-15 THEN V(J,3)=0:GOTO 1120
1110 V(J,3)=Y(J,1)+DY(J,2)*DX/2
1120 NEXT J
1130 X=X2:N=3:GOSUB DERIV
1140 X3=X(1)+DX
1150 FOR J=1 TO 6
1160 IF Y(J,1)<1.0E-15 THEN V(J,4)=0:GOTO 1180
1170 V(J,4)=Y(J,1)+DY(J,3)*DX
1180 NEXT J
1190 X=X3:N=4:GOSUB DERIV
1200 FOR J=1 TO 6
1210 Y(J,2)=Y(J,1)+(DY(J,1)+2*(DY(J,2)+DY(J,3))
     +DY(J,4))*DX/6
1220 NEXT J
1230 X(2)=X3
1240 IF MAXTEST THEN 1260
1250 IF Y(3,2)<Y(3,1) THEN A$="*":MAXTEST=1:GOTO
     1270:REM TEST FOR MAXIMUM
1260 IF INT(K/NPRT)<>K/NPRT THEN 1550
1270 Y(1,1)=16200*Y(1,1)/(YO*(1-COP/100)):REM EASILY
     HYDROLYZED CELLULOSE REMAINING (%)
1280 Y(2,1)=1620000*Y(2,1)/(YO*COP):REM RESISTANT
     CELLULOSE REMAINING (%)
1290 Y(3,1)=Y(3,1)*CONST:REM MONOMERIC GLUCOSE YIELD
     (%)
1300 Y(4,1)=Y(4,1)*2*CONST:REM DIMER YIELD (%)
1310 Y(5,1)=Y(5,1)*CONST:REM GLUCOSIDE YIELD (%)
1320 Y(6,1)=Y(6,1)*CONST:REM LEVOGLUCOSAN YIELD (%)
1330 REM
1340 REM FORMAT OUTPUT FOR PRINTING
1350 REM
1360 PRNT$(1,1)=" ":PRNT$(80,80)=" ":PRNT$(2)=PRNT$:
     TEMP$="       "
1370 IF X(1)<0.01 THEN PRNT$(2,5)="0.00":GOTO 1390
1380 TEMP$=STR$(INT((X(1)+5.0E-03)*100)/100+1.0E-03):
     PRNT$(7-LEN(TEMP$),6)=TEMP$(1,LEN(TEMP$)-1)
1390 IF G<0.01 THEN PRNT$(9,12)="0.00":GOTO 1410
1400 TEMP$=STR$(INT((G+5.0E-03)*100)/100+1.0E-03):
     PRNT$(14-LEN(TEMP$),13)=TEMP$(1,LEN(TEMP$)-1)
1410 IF Y(3,1)<0.01 THEN PRNT$(17,20)="0.00":GOTO 1430
1420 TEMP$=STR$(INT((Y(3,1)+5.0E-03)*100)/100
     +1.0E-03):PRNT$(22-LEN(TEMP$),21)
     =TEMP$(1,LEN(TEMP$)-1)
1430 IF C<0.01 THEN PRNT$(28,31)="0.00":GOTO 1450
1440 TEMP$=STR$(INT((C+5.0E-03)*100)/100+1.0E-03):
     PRNT$(33-LEN(TEMP$),32)=TEMP$(1,LEN(TEMP$)-1)
1450 IF CON<0.01 THEN PRNT$(39,42)="0.00":GOTO 1470
1460 TEMP$=STR$(INT((CON+5.0E-03)*100)/100+1.0E-03):
     PRNT$(44-LEN(TEMP$),43)=TEMP$(1,LEN(TEMP$)-1)
1470 IF CR<0.01 THEN PRNT$(48,51)="0.00":GOTO 1490
1480 TEMP$=STR$(INT((CR+5.0E-03)*100)/100+1.0E-03):
     PRNT$(53-LEN(TEMP$),52)=TEMP$(1,LEN(TEMP$)-1)
1490 IF R<0.01 THEN PRNT$(59,62)="0.00":GOTO 1510
1500 TEMP$=STR$(INT((R+5.0E-03)*100)/100+1.0E-03):
     PRNT$(64-LEN(TEMP$),63)=TEMP$(1,LEN(TEMP$)-1)
```

```
1510 IF CER<0.01 THEN PRNT$(70,73)="0.00":GOTO 1530
1520 TEMP$=STR$(INT((CER+5.0E-03)*100)/100+1.0E-03):
     PRNT$(75-LEN(TEMP$),74)=TEMP$(1,LEN(TEMP$)-1)
1530 PRNT$(75)=A$
1540 PRINT #3;PRNT$
1550 X(1)=X(2):FOR J=1 TO 6:Y(J,1)=Y(J,2):NEXT J
1560 NEXT K
1570 PRINT #3;"-------------------------------------
     -------------------------------------------"
1580 PRINT #3;"Temperature = ";TB;" C ** L/S = ";LS;
     " ** Added Acid = ";CA;" %"
1590 PRINT #3;"Neutralizing capacity of ash = ";EQ;
     " meq/gm"
1600 PRINT #3;"Cellulose content of starting material
     = ";CEC;" % * Resistant cellulose = ";COP;" %"
1610 PRINT #3;"All yields are anhydroglucose based as
     a % of the original"
1620 PRINT #3;"  cellulose.  Concentration reported as
     % glucose"
1630 CLOSE #3:GOTO 1920
1635 REM
1640 REM SUBROUTINE DERIV
1645 REM
1650 REM WHEN N=1 SUBROUTINE CALCULATES DEPENDENT
     VARIABLES FOR OUTPUT
1660 T=(TB-TA)*(1-EXP(-X*ALPHA))+TA:REM TEMPERATURE AT
     TIME X
1670 KC=2.87E+20*HM^1.218
     *EXP(-42900/(1.987*(T+273.1)))
1680 KG=5.58E+15*HM*EXP(0.183*V(3,N))
     *EXP(-33800/(1.987*(T+273.1)))
1690 TEMP=T
1700 IF T<180 THEN T=180
1710 IF T>230 THEN T=230
1720 KD=7.902E-07*EXP(5444/(T+273.1)):REM EQUILIBRIUM
     CONSTANT FOR DIMER
1730 KL=991*EXP(-4521/(T+273.1)):REM EQUILIBRIUM
     CONSTANT FOR LEVOGLUCOSAN
1740 T=TEMP
1750 DY(1,N)=-50*KC*V(1,N)
1760 DY(2,N)=-KC*V(2,N)
1770 DY(3,N)=-(DY(1,N)+DY(2,N))*CEC/(LS*10)
     -KG*V(3,N)-35*KL*KG*V(3,N)
     -70*KD*KG*V(3,N)^2+70*KG*V(4,N)
1780 DY(3,N)=DY(3,N)+35*KG*(V(5,N)+V(6,N))
1790 DY(4,N)=-35*KG*V(4,N)-KG*V(4,N)
     +35*KG*KD*V(3,N)^2
1800 DY(5,N)=-35*KG*V(5,N)+KG*V(4,N)
1810 DY(6,N)=-35*KG*V(6,N)+35*KL*KG*V(3,N)
1820 IF N>1 THEN RETURN
1830 C=CONST*(2*V(4,1)+V(5,1)+V(6,1)):REM COMBINED
     GLUCOSE YIELD (%)
1840 R=CONST*(V(3,1)+V(4,1)):REM REDUCING SUGAR YIELD (%)
1850 CR=16200*(V(1,1)+V(2,1))/YO:REM CELLULOSE
     REMAINING (%)
```

```
1860 TS=V(3,1)+2*V(4,1)+V(5,1)+V(6,1):G=CONST*TS:REM
     TOTAL ANHYDROGLUCOSE YIELD (%)
1870 CON=18000*TS/(TS*180+1000):REM CONCENTRATION AS %
     GLUCOSE
1880 CER=16200*CEC*(V(1,1)+V(2,1))/(162*CEC
     *(V(1,1)+V(2,1))+YO*(100-CEC)):REM
     CELLULOSE CONTENT OF HYDROLYSIS RESIDUE
1890 IF (C+G)<=0 THEN RETURN
1900 PC=100*C/(C+G):REM TOTAL GLUCOSE COMBINED (%)
1910 RETURN
1920 END
```

Index